Application of Fermentation Technology in Animal Nutrition

Guest Editors
Wansup Kwak
Konkuk University
Chun-Ju
Korea, South

Siran Wang
Jiangsu Academy of
Agricultural Sciences
Nanjing
China

Editorial Office
MDPI AG
Grosspeteranlage 5
4052 Basel, Switzerland

This is a reprint of the Special Issue, published open access by the journal *Fermentation* (ISSN 2311-5637), freely accessible at: https://www.mdpi.com/journal/fermentation/special_issues/C0WN18XMV6.

For citation purposes, cite each article independently as indicated on the article page online and as indicated below:

Lastname, A.A.; Lastname, B.B. Article Title. *Journal Name* **Year**, *Volume Number*, Page Range.

ISBN 978-3-7258-2769-5 (Hbk)
ISBN 978-3-7258-2770-1 (PDF)
https://doi.org/10.3390/books978-3-7258-2770-1

Contents

fermentation

MDPI

Editorial

Application of Fermentation Technology in Animal Nutrition

Siran Wang [1,2]

[1] Institute of Animal Science, Jiangsu Academy of Agricultural Sciences, Nanjing 210014, China; wangsiran@jaas.ac.cn

[2] Key Laboratory for Crop and Animal Integrated Farming, Ministry of Agriculture and Rural Affairs, Nanjing 210014, China

Fermentation technology has been utilized in animal nutrition worldwide for an extended period, with particular focus on animal feed. The principal objective of such methods is to preserve feedstuffs over an extended period. Harvested forages, putrefactive agro-industrial or food processing by-products, and total mixed rations are frequently ensilaged to ensure that the stability, palatability, and safety standards of animal feed are maintained over time [1]. Enhancing the nutritive value or safety of underutilized feed resources through fermentation technology to eliminate the risk of toxicity or low digestibility is also of considerable interest. To achieve those objectives, a range of microbiological, chemical and physical techniques have been developed. This Special Issue on "Application of Fermentation Technology in Animal Nutrition" comprises 11 original research articles and 1 communication. Collectively, these contributions illustrate the promising future of fermentation technology in animal nutrition research, one that is committed to both tradition and innovation, with the objective of advancing the frontiers of fermentation science.

In light of the growing popularity of Greek yogurt, significant quantities of acid whey are being generated on a global scale. However, the utilization of yogurt acid whey in animal nutrition remains constrained. Palamidi et al. [2] investigated the impact of incorporating yogurt acid whey powder (YAWP) into forage maize prior to ensilage on the microbial, nutritional, and fermentation quality of the resulting maize silage and found that adding 5 and 10% YAWP during silage preparation improved the fermentative and nutritional quality.

Improper disposal of vegetable waste can cause significant environmental contamination. However, due to their high moisture and organic matter content, these materials are not suitable for disposal via landfill or incineration. However, vegetable waste contains a substantial quantity of nutrients and, when combined with rice straw, exhibits certain complementary effects in terms of its moisture levels, nutrients, and physical structure. Lu et al. [3] assessed the safety profile of the mixed silage by quantifying the mycotoxin, vitamin, heavy metal, and pesticide residue content, as well as conducting a feeding trial with Hu sheep. The feed safety index content of the mixed silage complied with the Chinese feed safety and hygiene standards and had no adverse effect on the growth of the Hu sheep. Furthermore, the silage enhanced the digestive tract and immune performance of the sheep, facilitating their healthy development.

Dhakal et al. [4] examined the influence of in vitro rumen fermentation on the microbial community compositions and structure of rumen fluid, comparing samples taken before and after fermentation assays. The in vitro procedure did not impact the bacterial community structure in comparison to the original rumen fluid inoculum. It is important to note that evaluating the microbiome at a single endpoint (i.e., 48 h) was insufficient, as it did not provide a comprehensive understanding of the microbiome profile dynamics. Nevertheless, this study established a basis for future microbiome-based in vitro fermentation experiments and confirmed that this technique permits a high degree of species diversity that approximates the function of the rumen in vivo.

Citation: Wang, S. Application of Fermentation Technology in Animal Nutrition. *Fermentation* **2024**, *10*, 596. https://doi.org/10.3390/fermentation10120596

Received: 13 November 2024
Accepted: 20 November 2024
Published: 22 November 2024

Sugarcane (*Saccharum officinarum*) bagasse (SCB) is one of the most extensively produced lignocellulosic biomasses and has significant potential for sustainable food production if recycled as ruminant animal feed. Nevertheless, due to the presence of severe lignification, namely lignin–(hemi)–cellulose complexes, ruminants are only able to utilize a small portion of the polysaccharides in such recalcitrant lignocellulosic biomasses. Khan et al. [5] systematically evaluated the improvement in nutritive value, the in vitro dry matter digestibility, and the extent and rate of in vitro total gas and methane production during the 72-hour in vitro ruminal fermentation of SCB, in which it underwent bioprocessing with *Pleurotus djamor*, *Agaricus bisporus*, *Pleurotus ostreatus* and *Calocybe indica* under solid-state fermentation for 0, 21 and 56 days. Treating SCB with lignin-degrading white-rot fungi enhanced its nutritive value and digestibility. In particular, *C. indica* shows excellent potential for selective, extensive and rapid degradation of lignin, as well as for enhancing SCB's nutritive value and digestibility for ruminant nutrition.

Silage is a fundamental element of a ruminant diet, and it plays a pivotal role in ruminant production in terms of both productivity and animal health. Nevertheless, there is a paucity of data concerning the impact of mixed silage as an unconventional roughage on the fecal microbiota and metabolites in Hu sheep. Li et al. [6] examined the impact of mixed silage comprising rice straw and Chinese cabbage waste (mixed silage) on the fecal microbiota and metabolites in Hu sheep, employing metabolomic and Illumina sequencing analysis, and concluded that adding mixed silage to the diet of Hu sheep could change the microbial community structure of hindgut and regulate the metabolism of nucleotides and amino acids, which influences the animal performance.

Pineapple residue (PR), which contains substantial amounts of water-soluble carbohydrates, protein, and fiber, is another potential food source for ruminants. However, its high moisture levels make storage challenging. Corn straw (CS) is a common roughage source for ruminants and is frequently utilized for ensilage; thus, co-ensiling of CS and PR might prove an effective solution to the storage issue. Li et al. [7] evaluated the fermentation quality, chemical and microbial compositions of CS silage mixed with PR; said mixture exhibited a reduced pH value and elevated acetic and lactic acid concentrations. Furthermore, the addition of PR enhanced the relative abundance of *Lactobacillus* in the mixed silage. It was concluded that PR can enhance the quality of CS ensilage. The optimal proportion of PR was determined to be 15% based on fresh weight.

Lactobacillus delbrueckii has attracted considerable interest due to its contributions to diarrheal treatment and prevention, gut microecological balance, growth promotion, immune modulation, and meat quality enhancement in livestock. Nevertheless, there is a paucity of documented evidence concerning its effect on the liver metabolism and gut microbiota in weaned piglets. Wang et al. [8] studied the impact of *Lactobacillus delbrueckii* on liver metabolism changes in weaned piglets, investigated the mechanisms by which it influences their productive performance and supplied key theoretical support and reference data for applying *Lactobacillus delbrueckii* formulations in antibiotic-free diets and actual production for weaned piglets. In comparison to the control, *Lactobacillus delbrueckii* augmented liver glycogen content, mitigated weaning stress-induced liver damage and metabolic disorders, and reinforced liver antioxidant function by optimizing the metabolism of carbohydrates and lipids, improving liver function.

Essential oils have been found to effectively control fungal growth in silage. Nevertheless, the efficacy of essential oils may be contingent upon the chemical constituents and concentration of each oil. Utilizing essential oil combinations could result in additive, antagonistic, or synergistic effects. Aniceto et al. [9] assessed the influence of those essential oils and their combinations on the fermentation characteristics, microbial numbers, losses by effluents and gases, aerobic stability, and nutritive value of sorghum silage. The researchers concluded that the essential oils and their blends had disparate impacts on sorghum silage. Among the essential oils, rosemary (Ros) and blends containing Ros demonstrated superior efficacy. The addition of rosemary (Ros) was found to enhance the aerobic stability and nutritive value of sorghum silage.

To supplement existing feed resources, Li et al. [10] studied the influences of additives, moisture level, and their interactions on the fermentative profile, aerobic stability, and in vitro digestibility of mixed silage composed of cornmeal and amaranth. The optimal fermentation quality, aerobic stability and in vitro digestibility of cellulase+ *Lactobacillus plantarum*-treated cornmeal and amaranth mixed silage were achieved at a moisture level of 60%. This study corroborates the potential of amaranth as a silage feedstock and its potential application for enhancing feed quality and animal performance. These findings could provide farmers with a greater range of feed options and opportunities for resource utilization, thereby promoting the sustainable development of the animal husbandry.

Plant extracts are a promising alternative to synthetic drugs due to the secondary compounds they contain, including tannins, saponins, essential oils, and flavonoids, which exhibit a range of biological activities. Freitas et al. [11] assessed the impact of hydroalcoholic (HE) and ethanolic (EE) extracts of *Urochloa brizantha* at varying concentrations on rumen fermentation through an in vitro gas production technique. Both the HE and EE extracts of *U. brizantha* influenced rumen fermentation kinetics, with the HE extract demonstrating a greater level of protodioscin. More research is required to optimize extraction methodologies, comprehensively profile secondary compounds and evaluate the efficacy of different doses to determine the viability of *U. brizantha* extract as an additive.

Co-ensiling has been extensively applied to improve fermentation quality. Nevertheless, there is still a lack of clarity regarding the potential for producing quality silages by co-ensiling Napier grass (NG) with sugarcane tops (STs). Xie et al. [12] assessed the fermentative profile and the chemical and microbial compositions of silage produced from mixtures of ST and NG in different proportions. This study suggests that quality silage could be produced with NG:ST ratios of 40:60 to 20:80, and that these silages could optimize the nutrient supply for ruminants.

The soybean (*Glycine max* (L.) Merr.) is a significant oilseed crop which is renowned for its abundant nutritive value and quality protein. To enlarge the feed protein resources and enhance the utilization of soybeans as a raw material, Meng et al. [13] studied the feasibility of employing whole-plant soybean as silage and identified the optimal fermentation duration. WPS silage is a potential feed source; however, under natural fermentation conditions, it is susceptible to undesirable fermentation.

Together, this Special Issue presents several original studies in the field of animal nutrition, including research on co-ensiling technology, silage additives, rumen fermentation, and related topics. These findings could help to improve feed resources, optimize silage quality and improve animal performance. As this Special Issue draws to a close, it is obvious that the field of silage research is not merely evolving but undergoing a revolution. The challenges that lie ahead are considerable, but so are the opportunities. By addressing these challenges and pursuing the research directions outlined above, it is possible to ensure that fermentation technology will remain a critical aspect of animal nutrition, meeting the evolving needs of animal husbandry.

Funding: This study was financially supported by National Natural Science Foundation of China (32301500), the earmarked fund for Jiangsu Agricultural Industry Technology System (JATS[2023]398), the open project of Key Laboratory for Crop and Animal Integrated Farming, Ministry of Agriculture and Rural Affairs (202303).

Acknowledgments: We appreciate the support of *Fermentation* in hosting this Special Issue "Application of Fermentation Technology in Animal Nutrition" and thank Wansup Kwakall, as well as all of the editors and authors whose valuable work was published in this issue and thus contributed to its success.

Conflicts of Interest: The authors declare no conflicts of interest.

References

1. Wang, S.; Liu, H.; Zhao, J.; Dong, Z.; Li, J.; Shao, T. Influences of Organic Acid Salts and Bacterial Additives on Fermentation Profile, Aerobic Stability, and In Vitro Digestibility of Total Mixed Ration Silage Prepared with Wet Hulless Barley Distillers' Grains. *Agronomy* **2023**, *13*, 672. [CrossRef]
2. Palamidi, I.; Paraskeuas, V.V.; Kotsampasi, B.; Hadjigeorgiou, I.; Politis, I.; Mountzouris, K.C. Effect of Yogurt Acid Whey on the Quality of Maize Silage. *Fermentation* **2023**, *9*, 994. [CrossRef]
3. Lu, Z.; Li, C.; Hou, L.; Zhang, Z.; Lu, Y.; Xue, C.; Qi, R.; Zafar, M.H.; Xu, J.; Wang, M. Safety of Chinese Cabbage Waste and Rice Straw Mixed Silage and Its Effect on Growth and Health Performance of Hu Sheep. *Fermentation* **2024**, *10*, 47. [CrossRef]
4. Dhakal, R.; Neves, A.L.A.; Sapkota, R.; Khanal, P.; Hansen, H.H. Prokaryote Composition and Structure of Rumen Fluid before and after In Vitro Rumen Fermentation. *Fermentation* **2024**, *10*, 108. [CrossRef]
5. Khan, N.A.; Khan, M.; Sufyan, A.; Saeed, A.; Sun, L.; Wang, S.; Nazar, M.; Tan, Z.; Liu, Y.; Tang, S. Biotechnological Processing of Sugarcane Bagasse through Solid-State Fermentation with White Rot Fungi into Nutritionally Rich and Digestible Ruminant Feed. *Fermentation* **2024**, *10*, 181. [CrossRef]
6. Li, C.; Lu, Z.; Qi, R.; Zhang, Z.; Lu, Y.; Zafar, M.H.; Yang, K.; Wang, M. Illumina Sequencing and Metabolomic Analysis Explored the Effects of the Mixed Silage of Rice Straw and Chinese Cabbage Waste on Fecal Microorganisms and Metabolites in Hu Sheep. *Fermentation* **2024**, *10*, 233. [CrossRef]
7. Li, D.; Xie, H.; Zeng, F.; Luo, X.; Peng, L.; Sun, X.; Wang, X.; Yang, C. An Assessment on the Fermentation Quality and Bacterial Community of Corn Straw Silage with Pineapple Residue. *Fermentation* **2024**, *10*, 242. [CrossRef]
8. Wang, X.; Ma, L.; Liu, Z.; Huang, X. The Impact of *Lactobacillus delbrueckii* Hepatic Metabolism in Post-Weaning Piglets. *Fermentation* **2024**, *10*, 286. [CrossRef]
9. Aniceto, E.S.; Oliveira, T.S.; Meirelles, J.R., Jr.; Silva, I.N.; Mozelli Filho, E.J.L.; Gomes, R.S.; Arévalo, J.P.; Moraes, P.R. Evaluation of Essential Oils and Their Blends on the Fermentative Profile, Microbial Count, and Aerobic Stability of Sorghum Silage. *Fermentation* **2024**, *10*, 335. [CrossRef]
10. Li, X.; Jin, Y.; Li, F.; Yu, M.; Du, J.; Yi, Q.; Zhao, T.; Yuan, B.; Wang, P. Effects of Lactobacillus plantarum and Cellulase on Mixed Silages of *Amaranthus hypochondriacus* and Cornmeal: Fermentation Characteristics, Nutritional Value, and Aerobic Stability. *Fermentation* **2024**, *10*, 378. [CrossRef]
11. Freitas, R.S.X.d.; da Silva, J.S.; Furtado, A.J.; Perna Junior, F.; de Oliveira, A.L.; Bueno, I.C.d.S. The Production of Marandu Grass (*Urochloa brizantha*) Extracts as a Natural Modifier of Rumen Fermentation Kinetics Using an In Vitro Technique. *Fermentation* **2024**, *10*, 447. [CrossRef]
12. Xie, H.; Zeng, F.; Luo, X.; Li, Z.; Pan, Y.; Guo, Y.; Peng, L.; Liang, L.; Li, J.; Liang, Y.; et al. Silage Making of Napier Grass and Sugarcane Top at Different Proportions: Evolution of Natural Fermentation Characteristics, Chemical Composition, and Microbiological Profile. *Fermentation* **2024**, *10*, 525. [CrossRef]
13. Meng, H.; Jiang, Y.; Wang, L.; Li, Y.; Wang, S.; Tong, X.; Wang, S. Dynamic Analysis of Fermentation Quality, Microbial Community, and Metabolome in the Whole Plant Soybean Silage. *Fermentation* **2024**, *10*, 535. [CrossRef]

fermentation

MDPI

Article

Biotechnological Processing of Sugarcane Bagasse through Solid-State Fermentation with White Rot Fungi into Nutritionally Rich and Digestible Ruminant Feed

Nazir Ahmad Khan [1,2,*], Mussayyab Khan [2], Abubakar Sufyan [3], Ashmal Saeed [2], Lin Sun [4], Siran Wang [5], Mudasir Nazar [5], Zhiliang Tan [1], Yong Liu [1,*] and Shaoxun Tang [1,6,*]

1 Key Laboratory for Agro-Ecological Processes in Subtropical Region, Institute of Subtropical Agriculture, Chinese Academy of Sciences, Changsha 410125, China; zltan@isa.ac.cn

2 Department of Animal Nutrition, The University of Agriculture, Peshawar 25130, Pakistan; khanmussayyab2016@gmail.com (M.K.); ashmalsaeed123@gmail.com (A.S.)

3 Department of Livestock and Poultry Production, Bahauddin Zakariya University, Multan 60800, Pakistan; linksufyan@bzu.edu.pk

4 Inner Mongolia Engineering Research Center of Development and Utilization of Microbial Research in Silage, Inner Mongolia Academy of Agricultural and Animal Husbandry Sciences, Hohhot 010031, China; sunlin2013@126.com

5 Institute of Ensiling and Processing of Grass, College of Agro-Grassland Science, Nanjing Agricultural University, Nanjing 210095, China; wangsiran@njau.edu.cn (S.W.); mudasirnazar50@yahoo.com (M.N.)

6 College of Advanced Agricultural Sciences, University of Chinese Academy of Sciences, Beijing, 100049, China

* Correspondence: nazir.khan@aup.edu.pk (N.A.K.); y.liu@isa.ac.cn (Y.L.); sxtang@isa.ac.cn (S.T.)

Abstract: Sugarcane (*Saccharum officinarum*) bagasse (SCB) is one of the most widely produced lignocellulosic biomasses and has great potential to be recycled for sustainable food production as ruminant animal feed. However, due to severe lignification, i.e., lignin-(hemi)-cellulose complexes, ruminants can only ferment a minor fraction of the polysaccharides trapped in such recalcitrant lignocellulosic biomasses. This study was therefore designed to systematically evaluate the improvement in nutritional value, the in vitro dry matter digestibility (IVDMD), and the rate and extent of in vitro total gas (IVGP) and methane (CH_4) production during the 72 h in vitro ruminal fermentation of SCB, bioprocessed with *Agaricus bisporus*, *Pleurotus djamor*, *Calocybe indica* and *Pleurotus ostreatus* under solid-state fermentation (SSF) for 0, 21 and 56 days. The contents of neutral detergent fiber, lignin, hemicellulose and CH_4 production (% of IVGP) decreased ($p < 0.05$), whereas crude protein (CP), IVDMD and total IVGP increased ($p < 0.05$) after the treatment of SCB for 21 and 56 days with all white-rot fungi (WRF) species. The greatest ($p < 0.05$) improvement in CP (104.1%), IVDMD (38.8%) and IVGP (49.24%) and the greatest ($p < 0.05$) reduction in lignin (49.3%) and CH_4 (23.2%) fractions in total IVGP were recorded for SCB treated with *C. indica* for 56 days. Notably, *C. indica* degraded more than ($p < 0.05$) lignin and caused greater ($p < 0.05$) improvement in IVDMD than those recorded for other WRF species after 56 days. The increase in IVGP was strongly associated with lignin degradation ($R^2 = 0.72$) and a decrease in the lignin-to-cellulose ratio ($R^2 = 0.95$) during the bioprocessing of SCB. Our results demonstrated that treatment of SCB with (selective) lignin-degrading WRF can improve the nutritional value and digestibility of SCB, and *C. indica* presents excellent prospects for the rapid, selective and more extensive degradation of lignin and, as such, for the improvement in nutritional value and digestibility of SCB for ruminant nutrition.

Keywords: sugarcane bagasse; lignocellulosic biomass; fungal treatment; solid-state fermentation; lignin degradation; delignification; digestibility

Citation: Khan, N.A.; Khan, M.; Sufyan, A.; Saeed, A.; Sun, L.; Wang, S.; Nazar, M.; Tan, Z.; Liu, Y.; Tang, S. Biotechnological Processing of Sugarcane Bagasse through Solid-State Fermentation with White Rot Fungi into Nutritionally Rich and Digestible Ruminant Feed. *Fermentation* **2024**, *10*, 181. https://doi.org/10.3390/fermentation10040181

Academic Editor: Konstantinos G. Kalogiannis

Received: 7 February 2024
Revised: 28 February 2024
Accepted: 11 March 2024
Published: 26 March 2024

1. Introduction

The efficient utilization of agriculture crop residues and the fiber-rich co-products of agro-based industries as an economical and sustainable feed or biofuel stock has become a major research challenge in recent decades [1,2]. In ruminant nutrition, the optimal utilization of lignocellulosic biomass (LCB) could mitigate the growing food–feed competition, reduce feed costs and ensure the more complete utilization and recycling of nutrients in agriculture production systems [3,4]. Sugarcane (*Saccharum officinarum*) is among the most widely produced crops with annual productions of 2026 million tons [5], representing 21% of world's crop production. Likewise, sugarcane bagasse (SCB) is one of the most generated agro-industrial residues with a global production of 700 million tons [6], and has the potential to be recycled for sustainable food production via feeding to ruminant animals [7]. However, the utilization of SCB as ruminant feed is limited due to its low crude protein (CP) content (<3%) and large contents of highly recalcitrant lignin (>10%) and structural polysaccharides (>85%), cellulose and hemicellulose [8,9]. Ruminant animals have the ability to extract energy from cellulose and hemicellulose through symbiosis with rumen microbes. However, the lignin molecule is highly recalcitrant to microbial degradation, particularly in the low-oxygen environment of rumen. Moreover, lignin forms complexes with (hemi)cellulose, which are tightly bound together via direct (covalent) or indirect (ester or ether) linkages [2,10]. These physical and chemical complexes with lignin make the whole structure extremely dense, recalcitrant and mostly inaccessible for microbial degradation in the rumen [2,11]. Therefore, despite a very specialized digestive system, ruminants can ferment only a minor fraction of the polysaccharides embedded in the very dense recalcitrant structures of LCB such as SCB.

In recent decades, several chemical, physical, physicochemical and biotechnological pretreatment methods have been developed to degrade lignin and/or breakdown lignin–polysaccharide complexes to loosen the rigid structures of LCB and increase the accessibility of cellulose and hemicellulose for microbial fermentation in the rumen [12]. The chemical, physical and physicochemical methods are, however, expensive, require high energy input and specialized equipment and are often implicated in terms of safety issues and the production of toxic waste. In recent years, the biotechnological treatment of LCB with a wood-decaying basidiomycete white-rot fungi (WRF) in their natural growing and solid-state fermentation (SSF) conditions has emerged as a preferable treatment method [3,12]. The ecological nature, low energy requirements and greater potential for delignification compared to the other treatment methods are key features of the enormous scope of this biotechnology. This background provides a strong impetus to fully exploit this biotechnology for the optimum utilization of SCB in ruminant nutrition.

The physical structure and chemical composition of the biomass, WRF species and the SSF period are the most important factors influencing the efficiency of fungal treatment [11,13,14]. For instance, *Lentinula edodes* degraded lignin only by 4.8% in corn stover while the same fungus caused 65.4% and 65.0% delignification in rice straw and oil palm fronds, respectively [15]. This further demonstrates that the WRF species and substate combination is important for improving the nutritional value of the biomass through fungal treatment. During the biotechnological processing of biomass under SSF, the WRF not only degrade lignin but also consume cell wall polysaccharides as an energy source. Fungal species that degrade maximum lignin and minimum cellulose during the mycelial colonization stage result in maximum improvement in the nutritional value of the biomass for ruminant nutrition [16–18]. It is also evident from recent studies that the degradation of lignin during fungal treatment is strongly correlated with the rate and extent of the ruminal fermentation of the treated biomass [11,15,19]. Moreover, the extent of lignin degradation and improvement in nutritional value and fermentability of the fungal-treated biomass is positively associated with the lignin content of the untreated biomass [17]. Owing to the high lignin content (>10%) of SCB [8,9,15] compared to the commonly treated LCB, such as wheat straw (6.1–9.1%) [14], corn stover (4.95%) and rice straw (3.93%) [15], it is hypothesized that SCB has greater potential for delignification and improvement in its nutritional worth, if bioprocessed with potential WRF species. Four WRF species, i.e., *Agaricus bisporus*, *Pleurotus djamor*, *Calocybe indica* and *Pleurotus ostreatus*, were selected for this study. *Pleurotus* species were selected due to their better colonization and delignification potential [14,20]. *C. indica* was selected as the fungus requires a low spawning rate [21] and substrate moisture level [22] and was thus expected to have greater rate and extent of substrate colonization and delignification. *A. bisporus* was selected due to its higher Mn peroxidases production potential [23], which plays a key role in lignin degradation and the early formation of mycelia [24]. Overall, the species selected for this study have been tested for treatment of cereal straws [11,25,26]. However, to our knowledge, the potential of these WRF species and treatment periods for delignification and the improvement in nutritional value and ruminal fermentation of SCB has not been evaluated.

This study was therefore designed to systematically evaluate the nutrient losses and changes in the chemical composition, the in vitro dry matter digestibility (IVDMD), and the rate and extent of in vitro total gas (IVGP) and CH_4 production during the 72 h in vitro ruminal fermentation of SCB, bioprocessed with *A. bisporus*, *P. djamor*, *C. indica* and *P. ostreatus* under SSF for 0, 21 and 56 days. The overall aim was to identify the most promising WRF species and SSF period in terms of improvement in the availability of nutrients in treated SCB for ruminant nutrition.

2. Materials and Methods

The biotechnological treatment of SCB under SSF with four WRF species was carried out at the University of Agriculture, Peshawar, Pakistan, and chemical analysis and in vitro experiments were carried out at the rumen laboratory, Department of Animal Science, Southern Illinois University, Illinois, USA. The animals used in this study were handled and cared for according to the guidelines of the ethical committee of the Southern Illinois University and the National Research Council. A brief description of the methods is given below.

2.1. White-Rot Fungi Species and Grain Spawn Preparation

The spores of Basidiomycete WRF, *A. bisporus*, *P. djamor*, *C. indica* and *P. ostreatus* were procured from the Department of Plant Pathology, University of Agriculture, (Peshawar, Pakistan). The spores of each fungal species were aseptically cultured on a pre-sterilized (121 °C for 20 min in autoclave) malt agar extract (pH = 5.6) plate. The agar culture contained 20.0 g/L malt extract, 0.5 g/L KH_2PO_4, 0.5 g/L $MgSO_4{\cdot}7H_2O$ and 0.5 g/L Ca $(NO_3)_2{\cdot}4H_2O$. After inoculation, the culture of each fungal species was incubated at 24 °C, until the mycelia had fully colonized the agar plate. The fully mycelial colonized agar plates were preserved in a refrigerator at 4 °C for grain spawn preparation.

For grain spawn production, wheat grains were first rinsed with water and then boiled in water for 20 min. The water was drained using a mesh strainer, and the moist grains were loaded into spawn mycobags (20 (w) × 60 (L) cm) until three-quarters full and subsequently autoclaved (121 °C for 60 min). The sterilized grains in mycobags were allowed to cool to room temperature (24 °C), then aseptically inoculated with 15 pieces of the mycelium-coated agar (1 cm^2) and sealed. The grains in mycobags were thoroughly mixed to homogenously distribute mycelium in the grains. For each fungal species, 3 separate spawn bags were prepared. All bags were then kept in an incubator (IMC18, Thermo Fisher Scientific, Waltham, MA, USA) at 24 °C until the mycelium completely colonized the grains. The prepared spawns were kept in a refrigerator at 4 °C until used for the inoculation of SCB.

2.2. Substrate Preparation, Pasteurization and Mycobag Processing

Sugarcane bagasse was collected from three consecutive batches (100 kg from each batch) from Khazan Sugar Mill Peshawar (Haryana Payan, Khazana, Pakistan). The bagasse was thoroughly mixed, and 50 kg representative biomass was collected and chopped to 2 cm particle size. The chopped biomass was submerged (using netted bags) in cold water for 24 h, to allow the maximum penetration of water into the biomass. The excess water was drained, and the moist bagasse was pasteurized using steam (90 °C for 2 h). After steaming, the bagasse was taken out and allowed to cool to room temperature and dry out to a moisture content of 75%. The pasteurized bagasse (500 ± 5 g) was aseptically transferred into 64 mycobags (20 (w) × 60 (L) cm), which were further divided into four subgroups, and each subgroup (n = 16 bags per fungal species) was aseptically inoculated with grain spawn (3% on DM basis) of *A. bisporus*, *P. djamor*, *C. indica* or *P. ostreatus*. Four day-0 (control) bags, inoculated but not incubated, of each fungal species were immediately transferred to the laboratory. Subsequently, samples were collected for the analysis of dry matter (DM) content, and the remaining bagasse of each replicate day-0 bag was air dried in hot air oven at 70 °C to stop the growth of mycelium. The other inoculated bags (n = 8) of each fungal species were incubated for 21 and 56 days under the SSF conditions in a climate-controlled chamber at 24 °C with a relative humidity of 75–85%. Before the placement of the bags, 15 holes (about 0.5 cm diameter) were aseptically made in each bag for aeration and plugged with sterile cotton. At the end of each incubation period, the bags were removed and subsamples were collected for fresh DM content. The remaining bagasse of each replicate bag was air-dried in a hot air oven and processed for chemical analysis and in vitro digestibility studies.

2.3. Chemical Analysis

The DM content of the fresh day-0 and fungal-treated bagasse was determined by oven drying the samples at 70 °C overnight and then at 103 °C until they reached a constant weight (International Organization for Standardization (ISO), method 6496). For chemical analysis and in vitro studies, samples were air dried at 70 °C for 72 h and ground in a Thomas-Wiley Laboratory Mill (Model 4, Thomas Co., Philadelphia, PA, USA) using a 1 mm sieve. The air-dried samples were analyzed for DM content by oven drying at 103 °C until reaching a constant weight (ISO, method 6496). The CP (N × 6.25) content was analyzed with a Kjeltec 2400 autoanalyzer (Foss Analytical A/S, Hillerød, Denmark), using the Kjeldahl method (ISO, method 5983). The content of ash was determined after the complete incineration of samples in a muffle furnace at 550 °C (ISO method 5984). The ANKOM 200 Fiber Analyzer (ANKOM Technology Corps., Macedon, NY, USA) was used to analyze the contents of neutral detergent fiber (NDF) and acid detergent fiber (ADF) according to Van Soest et al. [27] methods. The lignin content was determined by the 3 h extraction of the ADF residues with 72% sulfuric acid. The NDF, ADF and lignin contents were sequentially analyzed on an ash-free basis, and the values of cellulose (ADF-lignin) and hemicellulose (NDF-ADF) were computed. The aflatoxin B1 (AFB1) rapid test kit (RingBio, Beijing, China) was used to screen the bioprocessed bagasse for AFB1 using the method described by Zhang et al. [28] for validation. The limit of detection of the kit was 5 μg/kg. The nutrient loss after 21- and 56-day treatment period with respect to the respective control was computed using the following equation:

$$Nutrient\ loss(\%) = 100 - \frac{[(100 - DM_L) \times PNP]}{PNC}$$

where DM_L is the percent DM loss, PNP is the percentage of nutrient after the treatment period and PNC is the percentage of nutrient in control.

2.4. In Vitro Dry Matter Digestibility

One day prior to the IVDMD trial, a fermentation buffer solution was prepared in demineralized water containing 3.72 g L^{-1} $NaHPO_4$, 9.82 g L^{-1} $NaHCO_3$, 0.47 g L^{-1} NaCl, 0.57 g L^{-1} KCl, 0.15 g L^{-1} CH_4N_2O, 0.12 g L^{-1} $MgSO_4{\cdot}7H_2O$ and 1 mL L^{-1} $CaCl_2$. The reducing solution was prepared by dissolving 320 mg NaOH and 1200 mg $Na_2S{\cdot}9H_2O$ in 200 mL of demineralized water under carbon dioxide (CO_2) flux. The fermentation buffer and reducing solutions were stored overnight at room temperature (24 °C) in Woulff bottles. One hour prior to the collection of rumen fluid, the buffer solution was placed in a Daisy jar assembly (SKU: D2) at 39 °C. For each run of in vitro incubation, about 500 mL of rumen fluid was obtained from each rumen-cannulated Holstein heifer (n = 4), before morning feeding in hot (39 °C) thermos bottles, pre-flushed with CO_2. A roughage-based total mixed ratio containing 35% SCB, 25% alfalfa hay, 10% wheat straw and 30% concentrate mixture was fed to the animals with overall chemical composition as 7.9% ash, 12.2% CP, 1.25 Mcal/kg metabolizable energy and 52.5% neutral detergent fiber (NDF) on a DM basis. The rumen fluid was transported to the rumen laboratory, composited for four donor heifers, homogenized in a laboratory blender (LB20ES, Shanghai Prime Science Co., Ltd., Shanghai, China) for 30 s, filtered through double layers of muslin cloth and mixed (1:2 *v*/*v*) with the pre-warmed buffer solution in Daisy jars. The air-dried ground day-0 and pre-treated bagasse samples (1 ± 0.03 g) were placed into Dacron bags (5 cm × 10 cm) with an average pore size of 50 µm in prewarmed buffered rumen fluid in Daisy jars. The jars were flushed immediately with CO_2 to ensure an anaerobic environment. The reducing agent was added to the buffered rumen fluid to determine the establishment of anerobic conditions. The change in color from blue to pink and finally to almost colorless indicated the absence of oxygen. Bagasse samples from each mycobag were incubated with buffered rumen fluid in duplicate bottles in two replicate runs in the Daisy incubator (Model: D200, ANKOM Technology Corps., Macedon, NY, USA). At the end of incubation period, the Dacron bags were removed from the incubator and rinsed with deionized water until the water was transparent. After washing, all bags were dried (at 65 °C) in oven until they reached a constant weight. The IVDMD was calculated by differences in dry weight of bagasse before and after incubation. The values were corrected for blanks.

2.5. *In Vitro Gas Production*

A fully automatic wireless system of ANKOM gas production modules (Model: RFS, ANKOM Technology Corps., Macedon, NY, USA) was used to measure IVGP, CH_4 and CO_2 production, according to the procedure described by Günal et al. [29]. Briefly, samples (ca. 3 g) were weighed and placed in ANKOM bottles (SKU: 7056, 250 mL). Then, 60 mL of fresh rumen fluid and 120 mL of buffer solution (prewarmed at 39 °C) was added to each bottle and subsequently flushed with CO_2. Each sample was incubated in duplicate bottles with the buffered rumen fluid in two replicate runs. All bottles, including four blanks containing only rumen fluid and buffer solution, were fitted in the ANKOM modules (SKU: RF1) equipped with a pressure/gas sensor, radio frequency sender and a microchip. The bottles fitted in the modules were incubated at 39 °C in a water bath, and manually shaken for 2 min after every 2 h. The IVGP was automatically recorded by a computer with a radio frequency receiver attached, for up to 72 h of fermentation. For CH_4 and CO_2 gas production, a small aliquot of gas (10 µL) was sampled from the top of each bottle using a gas-tight syringe (Hamilton 1701 N, point style 5 needles, 51 mm; Hamilton, Bonaduz, Switzerland). The CH_4 and CO_2 contents in the gas samples were analyzed using Shimadzu GC-2030 gas chromatograph system (Shimadzu Scientific Instruments AOC-20i Plus, Columbia, MD, USA), equipped with a 100 m Rt-2560 column (0.32 mm × 0.20 µm column; Restek, Bellefonte, PA, USA). The results of IVGP were corrected for blanks in which only buffer solution and rumen liquor were incubated without feed samples.

2.6. *Statistical Analysis*

Data on the effects of fungal species, treatment period and interactions with the changes in chemical composition, DM_L, nutrient losses, IVDMD, IVGP, CH_4, CO_2 and AFB1 were analyzed using the PROC MIXED procedure at SAS (SAS Inst., Inc., Cary, NC, USA). The fungal species, treatment period and their interactions were fixed effects, and replications were considered random effects.

$$Y_{ijk} = \mu + FS_i + PP_j + FS_i \times PP_j + \omega_{ijk}$$

where Y_{ijk} is the observation *j* in treatment *i*; μ is the overall mean; FS_i is the fixed effect of fungal species *i* (*A. bisporus*, *P. djamor*, *C. indica* and *P. ostreatus*); PP_j is the fixed effect of treatment period *j* (0, 21 and 56 days); $FS_i \times PP_j$ is the fixed effect of interaction between fungal species *i* and treatment period *j*; and ω_{ijk} is the random error. For parameters which showed an overall significant effect ($p < 0.05$), post hoc analyses were carried out on the least squares means adjusted for multiple comparisons using the Tukey–Kramer test to determine significant differences between the means.

3. Results

3.1. Changes in Chemical Composition of Sugarcane Bagasse Bioprocessed with WRF Species for Different Treatment Periods

Table 1 summarizes the data on the effect of WRF species and treatment periods on the changes in the chemical composition of SCB. Except for DM, ADF and cellulose, a significant ($p < 0.01$) interaction effect of fungal species and treatment period was observed for the contents of all measured chemical components. The contents of ash and CP increased ($p < 0.05$), while the contents of NDF, lignin and hemicellulose decreased ($p < 0.05$) with the SCB treatment for 21 and 56 days for all WRF species. The further comparison of the interaction data revealed that the greatest increase ($p < 0.05$) in CP content and the greatest decrease ($p < 0.05$) in NDF and lignin contents were observed for the treatment of SCB with *C. indica* for 56 days. Notably, SCB treated with *C. indica* for 21 days had lower ($p < 0.05$) lignin content (8.40%) than the value (9.30%) recorded for *A. bisporus* after 56 days, whereas the lignin contents SCB treated with *P. djamor* (8.33%) and *P. ostreatus* (8.27%) for 56 days were comparable ($p > 0.05$) to those of *C. indica* after 21 days. On the other hand, the CP content in SCB after 21 days of treatment with *C. indica* was comparable ($p > 0.05$) to the CP content recorded after 56 days of treatment with SCB for *A. bisporus*, *P. djamor* and *P. ostreatus*.

Table 1. Changes in chemical composition of sugarcane bagasse bioprocessed with four white-rot fungi (WRF) species under solid-state fermentation for 0, 21 and 56 days.

Treatment Period (Days)	WRF Species	DM (% FM)	Concentration (% DM)						
			Ash	CP	NDF	ADF	Lignin	HC	CEL
0	*A. bisporus*	24.9	5.09 ^{a}	1.99 ^{a}	86.3 ^{g}	58.5	12.9 ^{e}	27.8 ^{c}	46.5
	P. djamor	25.4	5.04 ^{a}	2.02 ^{a}	85.1 ^{g}	57.3	12.0 ^{e}	27.8 ^{c}	45.3
	C. indica	25.8	5.08 ^{a}	2.08 ^{a}	85.6 ^{g}	57.8	12.1 ^{e}	27.8 ^{c}	45.8
	P. ostreatus	25.2	5.09 ^{a}	1.98 ^{a}	86.5 ^{g}	58.4	12.0 ^{e}	28.0 ^{c}	46.4
21	*A. bisporus*	23.2	6.23 ^{b}	2.86 ^{b}	79.4 ^{f}	53.8	10.5 ^{d}	25.6 ^{bc}	43.3
	P. djamor	24.6	6.27 ^{b}	3.60 ^{c}	73.7 ^{d}	52.2	10.0 ^{cd}	21.5 ^{a}	42.2
	C. indica	22.9	6.34 ^{b}	4.17 ^{d}	72.0 ^{c}	50.7	8.40 ^{b}	21.6 ^{a}	42.3
	P. ostreatus	22.9	6.73 ^{bc}	3.77 ^{c}	73.9 ^{d}	53.0	10.1 ^{cd}	20.9 ^{a}	42.8
56	*A. bisporus*	22.8	7.11 ^{cd}	3.98 ^{d}	75.6 ^{e}	50.7	9.30 ^{c}	24.9 ^{b}	41.4
	P. djamor	24.1	7.75 ^{e}	4.24 ^{d}	69.1 ^{b}	49.0	8.33 ^{b}	20.1 ^{a}	40.7
	C. indica	21.8	7.45 ^{de}	5.02 ^{e}	67.2 ^{a}	47.5	7.23 ^{a}	19.7 ^{a}	40.3
	P. ostreatus	21.4	7.39 ^{de}	4.30 ^{d}	69.6 ^{b}	50.4	8.27 ^{b}	19.2 ^{a}	42.1
SEM		0.48	0.19	0.077	0.85	0.53	0.17	0.51	0.50
Overall mean fungal species									
A. bisporus		23.6 ^{a}	6.14	2.94 ^{a}	80.4 ^{a}	54.3 ^{c}	10.6 ^{c}	26.1 ^{b}	43.7 ^{b}
P. djamor		24.7 ^{b}	6.35	3.28 ^{b}	76.0 ^{c}	52.8 ^{ab}	10.1 ^{b}	23.1 ^{a}	42.7 ^{a}
C. indica		23.5 ^{a}	6.29	3.76 ^{c}	75.1 ^{d}	52.0 ^{a}	9.20 ^{a}	23.0 ^{a}	42.8 ^{a}
P. ostreatus		23.2 ^{a}	6.40	3.35 ^{b}	76.9 ^{b}	53.9 ^{bc}	10.1 ^{b}	22.7 ^{a}	43.8 ^{b}
SEM		0.38	0.16	0.044	0.41	0.31	0.09	0.30	0.29
Overall mean treatment period									
0 days		25.3 ^{c}	5.07 ^{a}	2.02 ^{a}	85.9 ^{c}	58.0 ^{c}	12.0 ^{c}	27.9 ^{c}	46.0 ^{c}
21 days		23.4 ^{b}	6.39 ^{b}	3.60 ^{b}	74.8 ^{b}	52.4 ^{b}	9.77 ^{b}	22.4 ^{b}	42.7 ^{b}
56 days		22.5 ^{a}	7.43 ^{c}	4.39 ^{c}	70.4 ^{a}	49.4 ^{a}	8.28 ^{a}	21.0 ^{a}	41.1 ^{a}
SEM		0.38	0.052	0.038	0.13	0.26	0.083	0.26	0.25
Significance									
WRF species		***	NS	***	***	***	***	***	*
Treatment periods		**	***	***	***	***	***	***	***
WRF species × incubation time		NS	**	***	***	NS	***	***	NS

Values with different superscripts in columns for treatment periods × fungal species, overall means of fungal species or overall means of treatment periods are significantly ($p < 0.05$) different. DM, dry matter; FM, fresh matter; CP, crude protein; NDF, neutral detergent fiber; ADF, acid detergent fiber; HC, hemicellulose; CEL, cellulose; *A. bisporus*, *Agaricus bisporus*; *P. ostreatus*, *Pleurotus ostreatus*; *C. indica*, *Calocybe indica*; *P. djamor*, *Pleurotus djamor*; SEM, standard error of the mean; NS, non-significant ($p > 0.05$); *, $p < 0.05$; **, $p < 0.01$; ***, $p < 0.001$.

Among the fungal species, on average, the highest ($p < 0.05$) content of CP and lowest ($p < 0.05$) contents of NDF and lignin were recorded for *C. indica*, whereas the lowest ($p < 0.05$) content of CP and highest ($p < 0.05$) contents of NDF, ADF, lignin and hemicellulose were recorded for *A. bisporus*. The mean content of ash and CP consistently increased ($p < 0.05$) with increases in treatment period (0 to 56 days), whereas the mean contents of NDF, ADF, lignin, cellulose and hemicellulose consistently decreased ($p < 0.05$) with increases in the treatment period.

3.2. Dry Matter and Nutrient Losses during Bioprocessing of Sugarcane Bagasse with Four WRF Species

Data on DM and nutrient losses after 21 and 56 days of the bioprocessing of SCB under SSF with the four WRF species, in comparison to the respective controls, are summarized in Table 2. Significant ($p < 0.01$) interaction effects of fungal species and treatment periods were observed for the losses of DM and all measured chemical components. The losses (degradation) of lignin increased ($p < 0.05$) with increases in treatment period from 21 to 56 days, irrespective of the fungal species, whereas the ash and CP content improved ($p < 0.05$) with increases in treatment period. The negative values of CP and ash indicated a relative increase in their contents (Table 2). The maximum ($p < 0.05$) improvement in CP (104.1%) and maximum losses of lignin (49.3%), NDF (33.6%) and cellulose (25.6%) were recorded for SCB treated with *C. indica* for 56 days. Further comparisons show that *C. indica* caused more ($p < 0.05$) losses of lignin (38.2%) during 21 days of treatment than those caused by *A. bisporus* (28.8%) and *P. djamor* (34.2%) after 56 days.

Table 2. Dry matter (DM) and nutrient losses during bioprocessing of sugarcane bagasse with four white-rot fungi (WRF) species under solid-state fermentation conditions for 21 and 45 days, in comparison to day-0 (inoculated but not incubated) bagasse.

Treatment Period (Days)	WRF Species	Percentage Losses of Nutrients in Comparison to Day 0							
		DM	Ash	CP	NDF	ADF	Lignin	HC	CEL
21	*A. bisporus*	6.69 ^{abc}	−14.3 ^{cd}	−33.0 ^{d}	14.2 ^{a}	14.1 ^{ab}	18.1 ^{a}	14.2 ^{a}	13.1 ^{ab}
	P. djamor	3.27 ^{a}	−20.5 ^{bcd}	−72.4 ^{c}	16.2 ^{b}	11.9 ^{a}	19.0 ^{a}	25.0 ^{bc}	10.0 ^{a}
	C. indica	11.2 ^{d}	−10.9 ^{d}	−77.9 ^{bc}	25.0 ^{e}	22.2 ^{cd}	38.2 ^{e}	30.8 ^{c}	17.9 ^{bc}
	P. ostreatus	9.11 ^{cd}	−20.2 ^{bcd}	−72.9 ^{bc}	22.3 ^{d}	17.6 ^{bc}	22.9 ^{b}	32.1 ^{cd}	16.1 ^{ab}
56	*A. bisporus*	8.4 ^{bcd}	−27.9 ^{b}	−83.6 ^{bc}	19.7 ^{c}	20.5 ^{c}	28.8 ^{c}	18.1 ^{ab}	18.4 ^{bc}
	P. djamor	5.4 ^{ab}	−45.7 ^{a}	−98.9 ^{bc}	23.2 ^{de}	19.1 ^{bc}	34.2 ^{d}	31.7 ^{cd}	15.1 ^{ab}
	C. indica	15.5 ^{e}	−24.0 ^{bc}	−104.1 ^{a}	33.6 ^{g}	30.5 ^{e}	49.3 ^{f}	40.0 ^{de}	25.6 ^{d}
	P. ostreatus	15.1 ^{e}	−23.3 ^{bcd}	−84.3 ^{b}	31.7 ^{f}	26.5 ^{d}	41.5 ^{e}	41.8 ^{e}	23.0 ^{cd}
SEM		0.74	2.58	2.41	0.39	1.06	0.71	1.67	1.03
Overall mean fungal species									
A. bisporus		7.56 ^{b}	−21.1 ^{b}	−58.2 ^{c}	17.0 ^{a}	17.3 ^{a}	23.5 ^{a}	16.1 ^{a}	15.8 ^{a}
P. djamor		4.31 ^{a}	−33.1 ^{a}	−85.6 ^{a}	19.8 ^{b}	15.5 ^{a}	26.6 ^{b}	28.3 ^{b}	12.5 ^{a}
C. indica		13.4 ^{c}	−17.4 ^{c}	−90.9 ^{a}	29.3 ^{d}	26.3 ^{c}	43.8 ^{d}	35.4 ^{c}	21.7 ^{b}
P. ostreatus		12.1 ^{c}	−21.7 ^{b}	−78.6 ^{b}	27.0 ^{c}	22.2 ^{b}	32.2 ^{c}	36.9 ^{c}	19.8 ^{b}
SEM		0.52	1.83	1.70	0.28	0.78	0.50	1.23	0.89
Overall mean treatment period									
21 days		7.58 ^{a}	−16.5 ^{b}	−64.3 ^{b}	19.4 ^{a}	16.4 ^{a}	24.6 ^{a}	25.5 ^{a}	21.0 ^{a}
56 days		11.1 ^{b}	−30.2 ^{a}	−92.7 ^{a}	27.1 ^{b}	24.2 ^{b}	38.5 ^{b}	32.9 ^{b}	49.1 ^{b}
SEM		0.37	1.29	1.20	0.21	0.75	0.70	0.87	0.63
Significance									
WRF species		***	***	***	***	***	***	***	***
Treatment periods		***	***	***	***	***	***	***	***
WRF species × treatment period		*	**	***	***	**	***	**	**

Values with different superscripts in columns for treatment periods × fungal species, overall means of fungal species or overall mean of treatment periods are significantly ($p < 0.05$) different. DM, dry matter; CP, crude protein; NDF, neutral detergent fiber; ADF, acid detergent fiber; HC, hemicellulose; CEL, cellulose; *A. bisporus*, *Agaricus bisporus*; *P. ostreatus*, *Pleurotus ostreatus*; *C. indica*, *Calocybe indica*; *P. djamor*, *Pleurotus djamor*; SEM, standard error of the mean; *, $p < 0.05$; **, $p < 0.01$; ***, $p < 0.001$.

The overall mean of fungal species revealed maximum ($p < 0.05$) improvement in CP content (90.9%) and maximum losses ($p < 0.05$) of lignin (43.8%), NDF (29.3%) and ADF (26.3%) for *C. indica*. The overall losses of DM, NDF, ADF, lignin, cellulose and hemicellulose increased ($p < 0.05$) with the increase in treatment period from 21 to 56 days. In contrast, ash and CP contents improved ($p < 0.05$) with increases in treatment period.

3.3. Effect of Bioprocessing of Sugarcane Bagasse with Different WRF Species on In Vitro Dry Matter Digestibility, In Vitro Total Gas and CH_4 Production

Data on the changes in IVDMD, IVGP and CH_4 production of SCB bioprocessed with the four WRF species under SSF for 0, 21 and 56 days are summarized in Table 3. Except for CH_4 (mL/g organic matter (OM)), significant ($p < 0.05$) interaction effects of fungal species and treatment period were observed for all measured in vitro fermentation parameters (Table 3). The IVDMD and total IVGP increased ($p < 0.05$) with the treatment of SCB for 21 and 56 days with all WRF species.

Table 3. Changes in in vitro dry matter degradation (IVDMD), in vitro gas production (IVGP), methane (CH_4) emission and aflatoxin B1 (AFB1) content sugarcane bagasse bioprocessed with different WRF species for different treatment periods.

Treatment Period (Days)	Fungal Species	IVDMD (g/100 g)	IVGP (mL/g OM)			CH_4 (% IVGP)	AFB1, μg kg^{-1}
			Total	CO_2	CH_4		
0	*A. bisporus*	46.0 ^{a}	156.3 ^{a}	130.5 ^{a}	22.8 ^{ab}	14.6 ^{ef}	<5.0
	P. djamor	46.8 ^{a}	159.0 ^{a}	133.7 ^{a}	22.4 ^{a}	14.1 ^{ef}	<5.0
	C. indica	46.9 ^{a}	159.0 ^{a}	132.2 ^{a}	22.6 ^{a}	14.2 ^{ef}	<5.0
	P. ostreatus	46.1 ^{a}	155.3 ^{a}	130.2 ^{a}	22.9 ^{ab}	14.7 ^{f}	<5.0
21	*A. bisporus*	53.8 ^{c}	176.0 ^{e}	148.2 ^{e}	24.4 ^{bcdef}	13.4 ^{bcd}	<5.0
	P. djamor	53.4 ^{c}	190.0 ^{d}	165.1 ^{d}	23.8 ^{cdef}	12.6 ^{de}	<5.0
	C. indica	64.6 ^{a}	203.7 ^{c}	177.2 ^{c}	23.7 ^{def}	11.6 ^{ef}	<5.0
	P. ostreatus	55.7 ^{bc}	193.3 ^{d}	165.9 ^{d}	24.6 ^{bcde}	12.7 ^{cde}	<5.0
56	*A. bisporus*	59.2 ^{c}	188.3 ^{c}	160.2 ^{c}	25.2 ^{cdef}	13.1 ^{cd}	<5.0
	P. djamor	57.9 ^{c}	208.0 ^{de}	180.5 ^{e}	25.7 ^{def}	12.4 ^{bc}	<5.0
	C. indica	65.1 ^{d}	237.3 ^{f}	205.9 ^{f}	25.9 ^{ef}	10.9 ^{a}	<5.0
	P. ostreatus	60.5 ^{c}	215.7 ^{e}	185.8 ^{e}	26.8 ^{f}	12.4 ^{bc}	<5.0
SEM		1.98	1.82	1.90	0.40	0.25	
Overall mean of WRF species							
A. bisporus		53.2 ^{ab}	173.6 ^{a}	148.2 ^{a}	24.1 ^{b}	13.9 ^{c}	<5.0
P. djamor		51.3 ^{a}	185.7 ^{b}	158.3 ^{b}	24.0 ^{b}	13.0 ^{b}	<5.0
C. indica		58.3 ^{c}	200.0 ^{c}	174.2 ^{d}	24.0 ^{b}	12.2 ^{a}	<5.0
P. ostreatus		54.2 ^{ab}	188.1 ^{b}	161.1 ^{c}	24.8 ^{a}	13.2 ^{b}	<5.0
SEM		1.05	0.23	0.35	0.14	0.21	
Overall mean treatment period							
0 days		45.0 ^{a}	157.4 ^{a}	132.1 ^{a}	22.7 ^{a}	14.4 ^{c}	<5.0
21 days		56.6 ^{b}	190.8 ^{b}	164.3 ^{b}	24.1 ^{b}	12.7 ^{b}	<5.0
56 days		61.0 ^{c}	212.3 ^{c}	186.4 ^{c}	25.9 ^{c}	12.3 ^{a}	<5.0
SEM		1.29	0.91	0.88	0.20	0.12	
Significance							
WRF species		*	***	***	NS	***	
Treatment period		***	***	***	***	***	
WRF species × treatment period		*	**	**	NS	*	

Values with different superscripts in columns for treatment periods × fungal species, overall means of fungal species or overall mean of treatment periods are significantly ($p < 0.05$) different. OM, organic matter; CH_4, methane; CO_2, carbon dioxide; AFB1, aflatoxin B1. NS, non-significant ($p > 0.05$); *, $p < 0.05$; **, $p < 0.01$; ***, $p < 0.001$.

Whereas CH_4 production (% of IVGP) decreased ($p < 0.05$). Further comparison of the interaction data revealed that the greatest ($p < 0.05$) increase in IVDMD and IVGP and greatest ($p < 0.05$) decrease in CH_4 (% of IVGP) were achieved through the bioprocessing of SCB for 56 days with *C. indica*. The lowest increase in IVGP and IVDMD and the lowest decrease in CH_4 (% of IVGP) after 21 and 56 days of treatment were observed for *A. bisporus*. Notably, SCB pretreated with *C. indica* for 21 days had greater ($p < 0.05$) IVGP (203.7 vs. 188.3 mL/g OM) and IVDMD (64.6 vs. 59.2%) values than those recorded for *A. bisporus* after 56 days. On the other hand, the decrease in CH_4 (% of IVGP) after 21 days of SCB treatment with *C. indica* was lower than the value recorded for *A. bisporus* after 56 days.

The overall mean of fungal species revealed the greatest ($p < 0.05$) IVDMD (58.3%) and IVGP (200 mL/g OM) and lowest ($p < 0.05$) CH_4 production (12.2% of IVGP) for SCB treated with *C. indica*. With the increase in treatment period from 0 to 56 days, the mean values of IVDMD (45.0 to 61.0%), IVGP (157.4 to 212.3 mL/g OM) and CH_4 production (22.7 to 25.9 mL/g OM) increased, whereas the proportion of CH_4 in total IVGP decreased (14.4 to 12.3%; $p < 0.05$) with the increase in the treatment period from 0 to 56 days (Table 3). The content of AFB1 after 0, 21 and 56 days treated with SCB was less than 5 μg kg^{-1} for all WRF species (Table 3).

Figure 1 shows the rate and extent of IVGP (mL/g OM) during 72 h of incubation in buffered-rumen fluid of SCB bioprocessed with *A. bisporus* (Panel A), *P. ostreatus* (Panel B), *C. indica* (Panel C) and *P. djamor* (Panel D) for 0, 21 and 56 days. Irrespective of the fungal species, the rate and extent of IVGP increased with the pre-treatment of SCB for 21 and 56 days.

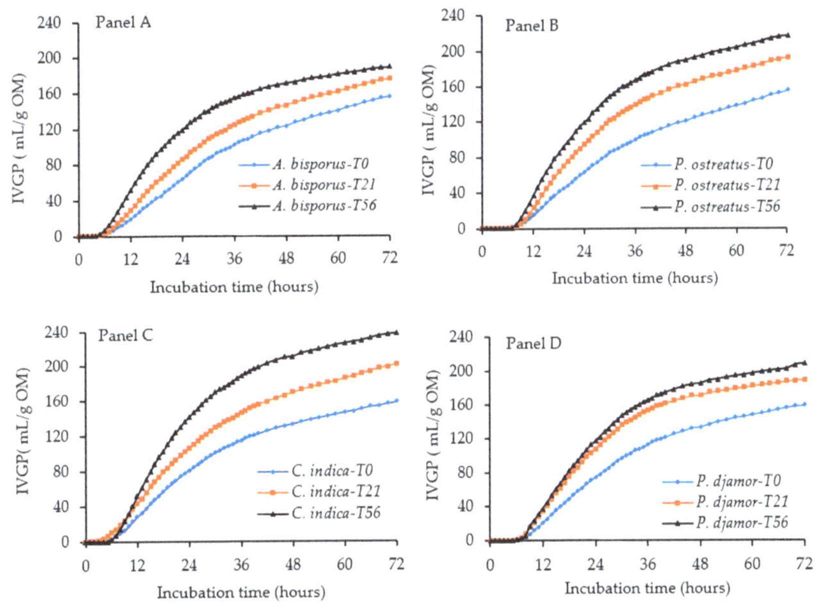

Figure 1. In vitro total gas production (IVGP; mL/g organic matter (OM)) for 72 h of incubation in buffered-rumen fluid of sugarcane bagasse treated with Panel (**A**) *Agaricus bisporus* (*A. bisporus*), Panel (**B**) *Pleurotus ostreatus* (*P. ostreatus*), Panel (**C**) *Calocybe indica* (*C. indica*) and Panel (**D**) *Pleurotus djamor* (*P. djamor*) for 0 (control; T0), 21 (T21) and 56 (T56) days.

Figure 2 clearly shows that the degradation of lignin during the bioprocessing of SCB was strongly associated with the increase in IVDMD ($R^2 = 0.72$; Panel A) and IVGP ($R^2 = 0.93$; Panel B). Moreover, the decrease in lignin-to-cellulose ratio (LCR) during the bioprocessing of SCB was strongly associated with the increase in IVDMD ($R^2 = 0.71$; Figure 3A) and IVGP ($R^2 = 0.95$; Figure 3B).

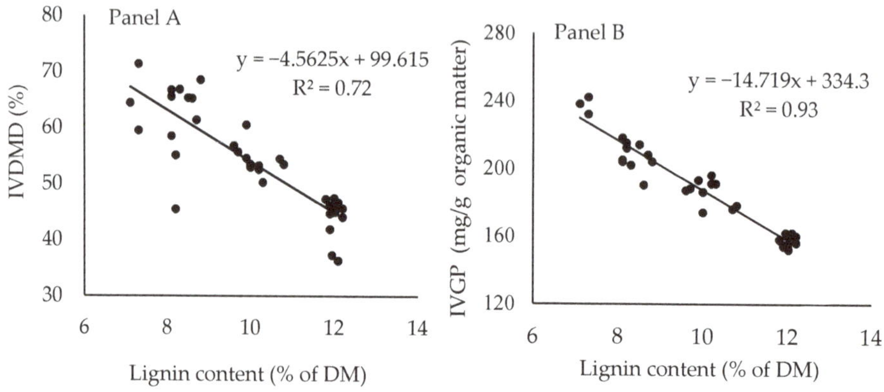

Figure 2. Relationship between lignin content and in vitro dry matter digestibility (IVDMD; Panel (**A**)) and in vitro total gas production (IVGP mL/g organic matter (OM); Panel (**B**)) of sugarcane bagasse pretreated with *Agaricus bisporus*, *Pleurotus ostreatus*, *Calocybe indica* and *Pleurotus djamor* for 0, 21 and 56 days.

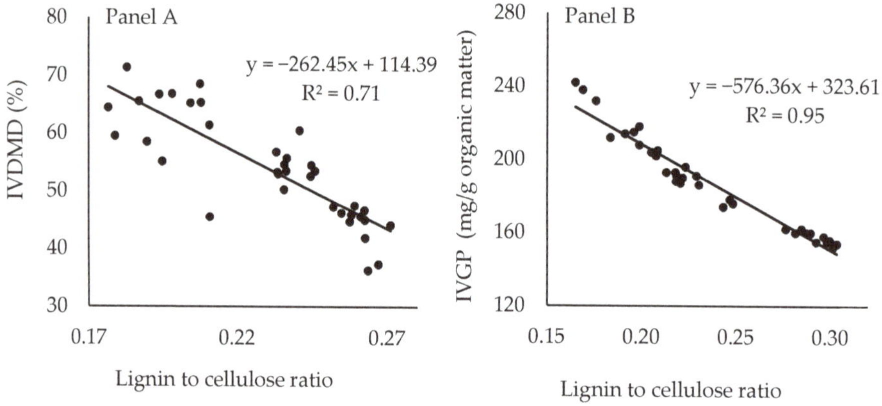

Figure 3. Relationship between lignin-to-cellulose ratio and in vitro dry matter digestibility (IVDMD; Panel (**A**)) and in vitro total gas production (IVGP; Panel (**B**)) of sugarcane bagasse pretreated with *Agaricus bisporus*, *Pleurotus ostreatus*, *Calocybe indica* and *Pleurotus djamor* for 0, 21 and 56 days.

4. Discussion

As in nature, during the mycelial colonization phase, WRF preferentially produces a complex mixture of oxidative ligninolytic enzymes and small radicals [30,31] to degrade lignin and increase the availability of polysaccharides for fruit body production. The fungal treatment is stopped before fruit body production, ideally at the stage when most of the lignin is degraded, and the maximum accessibility of cellulose is achieved for microbial fermentation in the rumen [13,14]. Lignin degradation during fungal treatment is the primary index for the improvement of the ruminal degradability of the recalcitrant structures of LCB. Nevertheless, a variable fraction of hemicellulose and cellulose is also degraded by WRF as an energy and growth substrate for mycelial colonization. Recent research has shown that the efficiency of fungal treatment of LCB not only depends on the delignification and the consequent increase in ruminal degradability but also on the availability of polysaccharides as an energy source for microbial fermentation [11,13,15]. When more polysaccharides are degraded, then the delignification process may not improve the fermentable energy supply for ruminant nutrition [20]. Therefore, the extent of lignin and polysaccharides degradation, as reflected by the changes in LCR, is a major determinant for the availability of fermentable energy and the overall efficiency of the fungal treatment of LCB [11,14].

The losses of structural components during the treatment of SCB for 21 and 56 days in current study could be related to the natural ability of WRF to mineralize fractions of lignin, hemicellulose and cellulose into CO_2 and H_2O [13,32]. Our results are consistent with the findings of previous studies, which were conducted on the other substrates [11,17,19]. Overall, lignin degradation in the present study ranged from 18.1 to 49.3%, with the greatest value being observed for *C. indica* after 56 days, followed by *P. ostreatus*, *P. djamor* and *A. bisporus*. On the other hand, cellulose degradation ranged from 10.0 to 25.6%, with the greatest values being observed for *C. indica* followed by *P. ostreatus*, *A. bisporus* and *P. djamor* (Table 2). Overall, cellulose degradation was lower than the lignin degradation for all WRF species tested in this study. As discussed earlier, the balance of polysaccharides in the treated biomass is very important, and as such, maximum lignin degradation with minimal cellulose degradation in the residual biomass is desirable for ruminant nutrition [11,33]. In the current study, the lowest degradation of cellulose was recorded for *P. djamor* compared to other WRF species. However, the comparison of the relative degradation of lignin and cellulose as expressed by LCR (calculated from Table 1) revealed that *C. indica* had lowest LCR (0.18) as compared to other WRF species (LCR ranged from 0.20 to 0.28). Next, in terms of the extent and selectivity for lignin degradation, *C. indica* also degraded lignin more quickly. In the present study, *C. indica* degraded more lignin after 21 days than those degraded by other WRF species after 56 days. Speed, selectivity and extensive delignification are the perquisites for fungal treatments of greater efficiency [34,35]. Nayan et al. [25] reported that *Ceriporiopsis subvermispora* strains degraded lignin more rapidly and selectively during the treatment of wheat straw, resulting in the greater improvement in cellulose availability and IVGP as compared to other fungal species. Datta et al. [36] reported the faster mycelial colonization of *C. indica* on SCB, as reflected by the earliest formation of pinheads (end of vegetative growth; at day 24), as compared to the other substrates (ranging from day 28 to 40). These findings highlight the potential of *C. indica* for the greater and more robust synthesis of mycelia when grown on SCB. The rapid colonization of mycelium in the substrate is a prerequisite for an effective fungal treatment. Fungi degrade lignin by releasing extracellular enzymes that have very limited ability to diffuse through the dense lignocellulosic structure. Therefore, the rate and extent of mycelial colonization and the subsequent penetration deep into the cell wall matrix are closely associated with the rate and extent of lignin degradation and the efficiency of the fungal treatment [25,37,38]. Moreover, greater biological efficiency for mushroom production has been reported for *C. indica* when grown on SCB [36], which indirectly support our findings. The greater extent of lignin degradation and lower LCR demonstrated that *C. indica* is a new potential candidate for the bioprocessing of SCB.

Notably, all WRF species consistently degraded lignin in SCB during the 56-day treatment period. However, the greater degradation of lignin was recorded during the initial phase (0–21 days) as compared to the later phase (22–56 days). In agreement with our findings, Mao et al. [39] and Van Kuijk et al. [35] reported greater lignin degradation during the initial phase (0–28 days) of the treatment of wheat straw with *C. subvermispora* as compared to the later phase (28–56 days). The substantially higher losses of hemicellulose were also observed during the initial phase (25.5%) as compared to the final phase (7.4%), which demonstrates that the fungal mycelia consumed the easily fermentable carbohydrates for its initial growth. The initial robust growth of fungal mycelia is a prerequisite to ensure desirable fermentation in the substrate and to inhibit the growth of contamination-causing microbes [39]. Such patterns of hemicellulose consumption by WRF have also been reported in recent studies [11,40].

The marked increase in the CP content was observed after 21 and 56 days of treatment of SCB for all WRF species. During fungal treatment, the CP content of the substrate (indirectly) increases due to the degradation of cell wall components into CO_2 and H_2O, which causes significant losses of OM, while N remains intact, increasing the concentration of CP in the residual mass. WRF assimilate nitrogenous compounds from the degraded substrate [41], which reduces losses and contributes to the increase in the CP content of the treated substrate. The greater increase in the CP contents of SCB pretreated with *C. indica* (104.1%) can be related to the greater degradation of lignin and structural polysaccharides by *C. indica* as compared to the other WRF species. In addition to the indirect proportional increase, higher increases in CP contents by *C. indica* may also be attributed to the greater mycelial growth and laccase activity of this fungus as compared to other fungal species [42,43], which contributes to a higher CP content in the residual mass of SCB.

In the current study, all tested WRF species exhibited markedly increased IVDMD and IVGP after 21 and 56 days of treatment of SCB under SSF. The increase in IVDMD and IVGP with the fungal treatment could be related to the decrease in lignin and fiber contents and to the faster and more extensive degradation of the treated biomass. During fungal treatment, the soluble, easily degradable, non-fiber carbohydrate fractions increased, whereas the recalcitrant lignin and least degradable fiber fractions decreased, resulting in the greater degradability of treated biomass [3,33]. Moreover, fungal treatment changes the physical structure of the biomass in such a way that it becomes more brittle, causing a more rapid reduction in particle size and ruminal degradability. Finally, the improvement in the CP content of the treated biomass is also expected to increase the IVDMD and IVGP of treated biomass [20]. In agreement with our findings, earlier studies have reported similar increases in IVDMD and IVGP for different LCB after fungal treatment [11,16,35].

Irrespective of WRF species and treatment periods, the increase in IVDMD and IVGP was strongly associated with the degradation of lignin during the fungal treatment of SCB (Figure 2). Notably, the increase in IVGP was strongly related to the decrease in LCR during the treatment of SCB, irrespective of fungal species and treatment periods (Figure 3). It may be noted that IVDMD represents the solubility of SCB, whereas IVGP represents the ruminal fermentation of OM in SCB, which could explain the stronger relationship between LCR and IVGP. It is possible that some of the fungal-degraded products, such as the small monomers of lignin, may not be fermentable in the rumen. Therefore, IVGP truly represents the fermentation of the substrate in the rumen and the supply of fermentable organic matter to ruminant animals [14]. This further demonstrates that in addition to lignin degradation, other factors, such as the availability of cell wall polysaccharides, also contribute to the increase in fermentable OM supply for the animals. LCR and IVGP appear to be the most important indices for the efficiency of fungal treatment.

Of the fungal species, the greatest improvements in IVDMD and IVGP were through *C. indica*, which may be related to the greatest lignin degradation and the greatest decrease in LCR caused by the same fungus. Such a relationship between ruminal degradation characteristics and lignin contents for WRF-treated wheat straw, rice straw and corncobs has been demonstrated in previous studies [11,33,37]. Moreover, a persistent increase in IVGP from 21 to 56 days of treatment was observed with *C. indica* as compared to the other fungal species, which may be due to the more selective and persistent lignin degradation by this species, as discussed previously. On the other hand, after 21-day treatment, *P. djamor* was the most efficient in terms of increase in IVGP per unit of DM loss (for each gram of DM loss, values were 58 mL and 38.9 mL for day 21 and day 56, respectively), as compared to the other fungal species (14.3 to 26.3 mL/g DM loss) (calculated in Tables 1 and 2). The greater increase in IVGP with smaller losses of lignin after 21-days treatment of SCB with *P. djamor* is a question for further research. Details on the production of extracellular enzymes and changes in lignin–cellulose complexes and the physical structures of SCB and their influences on rumen microbes and ecosystems could explain the underlying mechanisms involved.

A significant outcome of this study is the decrease in the CH_4 fraction of total IVGP after the treatment of SCB by all fungi, and this reduction in CH_4 was more pronounced for *C. indica*. The obvious reason for the reduction in CH_4 production was the decrease in cellulose and hemicellulose contents and the greater rate and extent of the ruminal fermentation of the treated SCB in this study. Sun et al. [44], in a comprehensive review, demonstrated that the fermentation capability of methanogenic microbes in the rumen is higher for β-1,4-linked structural polysaccharides when compared to non-structural polysaccharides. In agreement with this study, Huyen et al. [45] also reported a decrease of 11.4% in CH_4 production, after 24 h of in vitro incubation, in SCB treated with *P. eryngii* for 4 weeks.

Sugarcane bagasse naturally contains a very minute amount of AFB1 (1.55 μg kg^{-1} DM) [45] which was not detected in the screening test in this study. Moreover, some WRF degrades the AFB1 by a range of 40 to 94.7% [46,47]. Therefore, in the present study, WRF were less likely to have detectable levels of AFB1 in treated SCB regardless of the type of fungi, and it is inferred that SCB treated with all fungi used in this study was safe to feed to ruminant animals.

A significant interaction between the WRF species and treatment period for an improvement in the nutritional value and ruminal fermentation characteristics of SCB demonstrates the scope for the optimization of this biotechnology to achieve a more desirable nutritional upgradation in less time. For example, *C. indica* can be used to fast-track the delignification process during fungal treatment. In the present study, the 21-day treatment with *C. indica* was more or comparably effective than the 56-day treatment with the other fungal species in terms of the delignification and improvement of the nutritional value and ruminal fermentability of the treated SCB. Further studies are suggested with more treatment periods (like obtaining data on a weekly basis) and involving other factors like increasing the amount of inoculum, optimizing the chop size of the substrate and supplementing essential nutrients for a more precise optimization for the treatment period and the nutritional upgradation of SCB using WRF. The extraction and application of bacterial and fungal ligninolytic enzymes can also be used to improve the degradability of LCB [48,49].

Despite the promising results of fungal treatment on lignin-rich biomass for ruminant feeding at the laboratory level in this and previous studies [11,16,17], there are questions regarding the practical feasibility of this technique at a farm or commercial level. To replicate the technique at a larger scale, it is important to consider certain challenging aspects, which include the large-scale disinfection of the substrate and inoculation process, the shortening of the treatment period and the drying of mass quantities of fungal treated biomass to stop the fermentation process and for storage. In one of our recent studies [3], we addressed the issue of mass-scales and the feasible disinfection of the substrate for fungal treatment. In another study on the feeding of fungal-treated biomass to dairy cows [50], which is in the process of publication, we demonstrated that DM intake and milk production in cows was increased when they were fed WRF-treated biomass.

5. Conclusions

This study revealed that selective lignin-degrading WRF species, namely *A. bisporus*, *P. ostreatus*, *C. indica* and *P. djamor*, can be used to degrade lignin and improve the nutritional value and digestibility of SCB for ruminants. Of the fungi, *C. indica* not only caused the highest degradation of lignin but also resulted in the smallest lignin-to-cellulose ratio in the treated SCB—therefore sparing more cellulose for ruminal microbial fermentation, which is evident by the highest in vitro digestibility and gas production of SCB treated with *C. indica*. Except for *A. bisporus*, all fungi reduced in vitro CH_4 production with the highest reduction caused by *C. indica*. The increase in gas production was strongly associated with lignin degradation ($R^2 = 0.72$), and the decrease in the lignin-to-cellulose ratio ($R^2 = 0.95$) during the bioprocessing of SCB. *C. indica* presents great prospects for the rapid, selective and more extensive degradation of lignin and, as such, for the improvement in nutritional value and digestibility of SCB for ruminant nutrition.

Author Contributions: Conceptualization, N.A.K., Z.T. and S.T.; methodology, M.K., A.S. (Abubakar Sufyan) and A.S. (Ashmal Saeed); software, N.A.K.; formal analysis, N.A.K., A.S. and M.N.; Investigation, M.K., A.S. (Abubakar Sufyan) and A.S. (Ashmal Saeed); writing—original draft preparation, M.K. and M.N.; writing—review and editing N.A.K., A.S. (Abubakar Sufyan), L.S., S.W. and Y.L.; visualization, Z.T. and S.T.; supervision, N.A.K.; project administration, N.A.K., Z.T. and S.T.; Financial support, Z.T., Y.L. and S.T. All authors have read and agreed to the published version of the manuscript.

Funding: The experimental trial was financially supported by Chinese Academy of Sciences President's International Fellowship for Postdoctoral Studies (2021BP0045), the Strategic Priority Research Program of the Chinese Academy of Sciences (grant No. XDA28020400), and the Rural Revitalization Project of Chinese Academy of Sciences (KFJ-XCZX-202303). The financial support from the Higher Education Commission of Pakistan (grant IDs 2AV3-017 and IRSIP-44-Agri-01) to carry out laboratory work at Southern Illinois University (SIU), Carbondale, USA, is highly appreciated.

Institutional Review Board Statement: Not applicable.

Informed Consent Statement: Not applicable.

Data Availability Statement: Data available on reasonable demand from the corresponding author.

Acknowledgments: The technical guidance and support from Amer AbuGhazaleh and Mohamed Galal Embaby (PhD scholar) at SIU is acknowledged and greatly appreciated.

Conflicts of Interest: The authors declare no conflicts of interest.

References

1. Liang, J.; Nabi, M.; Zhang, P.; Zhang, G.; Cai, Y.; Wang, Q.; Zhou, Z.; Ding, Y. Promising biological conversion of lignocellulosic biomass to renewable energy with rumen microorganisms: A comprehensive review. *Renew. Sust. Energ. Rev.* **2020**, *134*, 110335. [CrossRef]
2. Kainthola, J.; Podder, A.; Fechner, M.; Goel, R. An overview of fungal pretreatment processes for anaerobic digestion: Applications, bottlenecks and future needs. *Bioresour. Technol.* **2021**, *321*, 124397. [CrossRef]
3. Sufyan, A.; Khan, N.A.; AbuGhazaleh, A.; Ahmad, N.; Tang, S.; Tan, Z. Novel techniques for the mass production of nutritionally improved fungus-treated lignocellulosic biomass for ruminant nutrition. *J. Sci. Food Agric.* **2024**, *104*, 2215–2224. [CrossRef]
4. Van Zanten, H.H.E.; Van Ittersum, M.K.; De Boer, I.J.M. The role of farm animals in a circular food system. *Glob. Food Secur.* **2019**, *21*, 18–22. [CrossRef]
5. FAOSTAT. Crops and Livestock Products. Food and Agricluture Organization of the United Nations, Rome. Available online: https://www.fao.org/faostat/en/#data/QCL (accessed on 24 February 2024).
6. Ungureanu, N.; Vladut, V.; Biris, S.S. Sustainable valorization of waste and by-products from sugarcane processing. *Sustainability* **2022**, *14*, 11089. [CrossRef]
7. Parameswaran, B. Sugarcane bagasse. In *Biotechnology for Agro-Industrial Residues Utilisation: Utilisation of Agro-Residues*; Nigam, P.S., Pandey, A., Eds.; Springer: Berlin/Heidelberg, Germany, 2009; pp. 239–252.
8. da Costa, D.A.; de Souza, C.L.; Saliba, E.; Carneiro, J. By-products of sugar cane industry in ruminant nutrition. *Int. J. Adv. Agric. Res.* **2015**, *3*, 1–95.
9. Gunun, N.; Wanapat, M.; Gunun, P.; Cherdthong, A.; Khejornsart, P.; Kang, S. Effect of treating sugarcane bagasse with urea and calcium hydroxide on feed intake, digestibility, and rumen fermentation in beef cattle. *Trop. Anim. Health Prod.* **2016**, *48*, 1123–1128. [CrossRef]
10. Ding, G.; Chang, Y.; Zhou, Z.; Ren, L.; Meng, Q. Effect of *Saccharomyces cerevisiae* on rumen fermentation characteristics, nutrient degradation and cellulase activity of steers fed diets with different concentrate to forage ratios. *World J. Agric. Res.* **2014**, *2*, 303–308. [CrossRef]
11. Sufyan, A.; Ahmad, N.; Shahzad, F.; Embaby, M.G.; AbuGhazaleh, A.; Khan, N.A. Improving the nutritional value and digestibility of wheat straw, rice straw and corn cob through solid state fermentation using different *Pleurotus* species. *J. Sci. Food Agric.* **2022**, *102*, 2445–2453. [CrossRef] [PubMed]
12. Van Kuijk, S.; Sonnenberg, A.; Baars, J.; Hendriks, W.; Cone, J. Fungal treated lignocellulosic biomass as ruminant feed ingredient: A review. *Biotechnol. Adv.* **2015**, *33*, 191–202. [CrossRef]
13. Van Kuijk, S.J.A.; Sonnenberg, A.S.M.; Baars, J.J.P.; Hendriks, W.H.; Cone, J.W. Fungal treatment of lignocellulosic biomass: Importance of fungal species, colonization and time on chemical composition and in vitro rumen degradability. *Anim. Feed Sci. Technol.* **2015**, *209*, 40–50. [CrossRef]
14. Nayan, N.; Sonnenberg, A.S.; Hendriks, W.H.; Cone, J.W. Prospects and feasibility of fungal pretreatment of agricultural biomass for ruminant feeding. *Anim. Feed Sci. Technol.* **2020**, *268*, 114577. [CrossRef]

15. Tuyen, D.V.; Phuong, H.N.; Cone, J.W.; Baars, J.J.; Sonnenberg, A.S.; Hendriks, W.H. Effect of fungal treatments of fibrous agricultural by-products on chemical composition and in vitro rumen fermentation and methane production. *Bioresour. Technol.* **2013**, *129*, 256–263. [CrossRef]
16. Khan, N.A.; Hussain, S.; Ahmad, N.; Alam, S.; Bezabhi, M.; Hendriks, W.H.; Yu, P.; Cone, J.W. Improving the feeding value of straws with Pleurotus ostreatus. *Anim. Prod. Sci.* **2015**, *55*, 241–245. [CrossRef]
17. Nayan, N.; van Erven, G.; Kabel, M.A.; Sonnenberg, A.S.; Hendriks, W.H.; Cone, J.W. Improving ruminal digestibility of various wheat straw types by white-rot fungi. *J. Sci. Food Agric.* **2019**, *99*, 957–965. [CrossRef] [PubMed]
18. Niu, D.; Zuo, S.; Ren, J.; Li, C.; Zheng, M.; Jiang, D.; Xu, C. Effect of wheat straw types on biological delignification and in vitro rumen degradability of wheat straws during treatment with Irpex lacteus. *Anim. Feed Sci. Technol.* **2020**, *267*, 114558. [CrossRef]
19. Datsomor, O.; Yan, Q.; Wang, K.; Mohamed, S.; Opoku-Mensah, L.; Zhao, G.; Miao, L. Effect of ammoniated and/or basidiomycete white-rot fungi treatment on rice straw proximate composition, cell wall component, and in vitro rumen fermentation characteristics. *Fermentation* **2022**, *8*, 228. [CrossRef]
20. Khan, N.A.; Sufyan, A.; Nazar, M.; Yu, P.; Tang, S.; Liu, Y.; Tan, Z. Promising biological conversion of lignocellulosic biomass to digestible ruminant feed with white rot fungi: A comprehensive review. *Biotech. Advan.* 2024; *in press*.
21. Subbiah, K.A.; Balan, V. A comprehensive review of tropical milky white mushroom (Calocybe indica P&C). *Mycobiology* **2015**, *43*, 184–194. [PubMed]
22. Khalkho, S.; Koreti, D.; Kosre, A.; Jadhav, S.; Chandrawanshi, N.K. Review on production technique and nutritional status of Calocybe indica (P&C). *NewBioWorld* **2021**, *3*, 184–194.
23. Janusz, G.; Pawlik, A.; Sulej, J.; Świderska-Burek, U.; Jarosz-Wilkołazka, A.; Paszczyński, A. Lignin degradation: Microorganisms, enzymes involved, genomes analysis and evolution. *FEMS Microbiol. Rev.* **2017**, *41*, 941–962. [CrossRef]
24. Cohen, R.; Persky, L.; Hadar, Y. Biotechnological applications and potential of wood-degrading mushrooms of the genus *Pleurotus*. *Appl. Microbiol. Biotechnol.* **2002**, *58*, 582–594. [CrossRef]
25. Nayan, N.; Sonnenberg, A.S.M.; Hendriks, W.H.; Cone, J.W. Screening of white-rot fungi for bioprocessing of wheat straw into ruminant feed. *Appl. Microbiol.* **2018**, *125*, 468–479. [CrossRef] [PubMed]
26. Khonkhaeng, B.; Cherdthong, A. Improving nutritive value of purple field corn residue and rice straw by culturing with white-rot fungi. *J. Fungi* **2020**, *6*, 69. [CrossRef] [PubMed]
27. Van Soest, P.; Robertson, J.; Lewis, B. Symposium: Carbohydrate methodology, metabolism, and nutritional implications in dairy cattle. *J. Dairy Sci.* **1991**, *74*, 3583–3597. [CrossRef] [PubMed]
28. Zhang, W.; Tang, S.; Jin, Y.; Yang, C.; He, L.; Wang, J.; Chen, Y. Multiplex SERS-based lateral flow immunosensor for the detection of major mycotoxins in maize utilizing dual Raman labels and triple test lines. *J. Hazard. Mater.* **2020**, *393*, 122348. [CrossRef] [PubMed]
29. Günal, M.; Pinski, B.; AbuGhazaleh, A.A. Evaluating the effects of essential oils on methane production and fermentation under in vitro conditions. *Ital. J. Anim. Sci.* **2017**, *16*, 500–506. [CrossRef]
30. Hatakka, A.; Hammel, K.E. Fungal biodegradation of lignocelluloses. In *Industrial Applications*; Hofrichter, M., Ed.; Springer: Berlin/Heidelberg, Germany, 2011; pp. 319–340.
31. Pollegioni, L.; Tonin, F.; Rosini, E. Lignin-degrading enzymes. *FEBS J.* **2015**, *282*, 1190–1213. [CrossRef] [PubMed]
32. Wan, C.; Li, Y. Fungal pretreatment of lignocellulosic biomass. *Biotechnol. Adv.* **2012**, *30*, 1447–1457. [CrossRef] [PubMed]
33. Tuyen, V.D.; Cone, J.W.; Baars, J.J.; Sonnenberg, A.S.; Hendriks, W.H. Fungal strain and incubation period affect chemical composition and nutrient availability of wheat straw for rumen fermentation. *Bioresour. Technol.* **2012**, *111*, 336–342. [CrossRef]
34. Van Kuijk, S.J.; Sonnenberg, A.S.; Baars, J.J.; Hendriks, W.H.; Cone, J.W. The effect of particle size and amount of inoculum on fungal treatment of wheat straw and wood chips. *J. Anim. Sci. Biotechnol.* **2016**, *7*, 39. [CrossRef]
35. Van Kuijk, S.J.A.; José, C.; Rencoret, J.; Gutiérrez, A.; Sonnenberg, A.S.; Baars, J.J.; Hendriks, W.H.; Cone, J.W. Selective ligninolysis of wheat straw and wood chips by the white-rot fungus *Lentinula edodes* and its influence on in vitro rumen degradability. *J. Anim. Sci. Biotechnol.* **2016**, *7*, 55. [CrossRef]
36. Datta, S.; Banerjee, G.; Mitra, A.K. Effect of different agricultural waste substrates on the biological efficiency and therapeutic value of Calocybe Indica. *Int. J. Pharm. Biol. Sci.* **2019**, *9*, 1157–1168.
37. Nayan, N.; van Erven, G.; Kabel, M.A.; Sonnenberg, A.S.; Hendriks, W.H.; Cone, J.W. Evaluation of fungal degradation of wheat straw cell wall using different analytical methods from ruminant nutrition perspective. *J. Sci. Food Agric.* **2019**, *99*, 4054–4062. [CrossRef] [PubMed]
38. Van Kuijk, S.J.A.; Sonnenberg, A.S.M.; Baars, J.J.P.; Hendriks, W.H.; del Río, J.C.; Rencoret, J.; Gutiérrez, A.; de Ruijter, N.C.A.; Cone, J.W. Chemical changes and increased degradability of wheat straw and oak wood chips treated with the white rot fungi Ceriporiopsis subvermispora and Lentinula edodes. *Biomass Bioenergy* **2017**, *105*, 381–391. [CrossRef]
39. Mao, L.; van Arkel, J.; Hendriks, W.H.; Cone, J.W.; de Vos, R.C.; Sonnenberg, A.S. Assessing the nutritional quality of fungal treated wheat straw: Compounds formed after treatment with Ceriporiopsis subvermispora and Lentinula edodes. *Anim. Feed Sci. Technol.* **2021**, *276*, 114924. [CrossRef]
40. Tirado-González, D.N.; Jáuregui-Rincón, J.; Tirado-Estrada, G.G.; Martínez-Hernández, P.A.; Guevara-Lara, F.; Miranda-Romero, L.A. Production of cellulases and xylanases by white-rot fungi cultured in corn stover media for ruminant feed applications. *Anim. Feed Sci. Technol.* **2016**, *221*, 147–156. [CrossRef]

41. Davis, M.A.; Wong, K.H. Nitrogen metabolism in filamentous fungi. In *Cellular and Molecular Biology of Filamentous Fungi*; Borkovich, K.A., Ebbole, D.J., Eds.; American Society for Microbiology Press: Washington, DC, USA, 2010; pp. 325–338.
42. Periasamy, K.; Natarajan, K. Role of lignocellulosic enzymes during basidiomata production by Pleurotus djamor var. roseus. *Indian J. Biotechnol.* **2004**, *3*, 577–583.
43. Kaur, M.; Sharma, S.; Sodhi, H. White rot fungus Calocybe indica: Incredulous factory of lignocellulolytic enzymes and their potential applicability for mushroom cultivation. *Acta Aliment.* **2022**, *51*, 424–436. [CrossRef]
44. Sun, X.; Cheng, L.; Jonker, A.; Munidasa, S.; Pacheco, D. A review: Plant carbohydrate types—The potential impact on ruminant methane emissions. *Front. Vet. Sci.* **2022**, *9*, 880115. [CrossRef]
45. Huyen, N.T.; Le, N.T.T.; Tuan, B.Q. Fermented sugarcane bagasse with the fungus *Pleurotus eryngii* reduced in vitro methane production. *Livest. Res. Rural Dev.* **2021**, *33*, 156–157.
46. Alberts, J.F.; Gelderblom, W.; Botha, A.; Van Zyl, W. Degradation of aflatoxin B1 by fungal laccase enzymes. *Int. J. Food Microbiol.* **2009**, *135*, 47–52. [CrossRef] [PubMed]
47. Wang, L.; Huang, W.; Shen, Y.; Zhao, Y.; Wu, D.; Yin, H.; Yang, S.; Yuan, Q.; Liang, W.; Wang, J. Enhancing the degradation of Aflatoxin B1 by co-cultivation of two fungi strains with the improved production of detoxifying enzymes. *Food Chem.* **2022**, *371*, 131092. [CrossRef] [PubMed]
48. Nazar, M.; Xu, Q.; Ullah, M.W.; Khan, N.A.; Iqbal, B.; Zhu, D. Integrated laccase delignification with improved lignocellulose recalcitrance for enhancing enzymatic saccharification of ensiled rice straw. *Ind. Crops Prod.* **2023**, *202*, 116987. [CrossRef]
49. Liu, E.Y.; Wang, S.; Wang, S.; Khan, N.A.; Zhou, X.; Tang, S.; Zhou, C.; Tan, Z.; Liu, Y. Bacterial inoculants and enzymes based silage cocktails boost the ensiling quality of biomasses from reed, corn and rice straw. *Chem. Biolog. Technol. Agricult.* **2024**, *11*, 29. [CrossRef]
50. Sufyan, A.; Khan, N.A.; Akbar, A.; Tang, S.; Tan, Z. Scaling-up fungal pretreatment of lignocellulose biomass: Impact on nutritional value, ruminal degradability, methane production, and performance of lactating dairy cows. *Livest. Sci.* 2024, *in press*.

fermentation

Communication

Prokaryote Composition and Structure of Rumen Fluid before and after In Vitro Rumen Fermentation

Rajan Dhakal [1,*], André Luis Alves Neves [1,*], Rumakanta Sapkota [2], Prabhat Khanal [3] and Hanne Helene Hansen [1]

1 Department of Veterinary and Animal Sciences, Production, Nutrition and Health, University of Copenhagen, Grønnegårdsvej 15, DK-1870 Frederiksberg C, Denmark; hhh@sund.ku.dk
2 Department of Environmental Science, Aarhus University, DK-4000 Roskilde, Denmark; rs@envs.au.dk
3 Faculty of Biosciences and Aquaculture, Nord University, Skolegata 22, 7713 Steinkjer, Norway; prabhat.khanal@nord.no
* Correspondence: dbk638@ku.dk (R.D.); andre.neves@sund.ku.dk (A.L.A.N.)

Abstract: Background: This study aimed to investigate the impact of in vitro rumen fermentation (IVRF) on the microbiome structure and composition of rumen fluid before and after fermentation assays. Methods and Results: Six separate fermentation batches were run for 48 h using maize silage as the basal feed. Rumen fluid samples were analyzed before (RF; only rumen fluid inoculant) and after 48 h fermentation assay (MS; maize silage as the substrate) and further processed for microbiome analysis using amplicon sequencing targeting the V4 region of the bacterial 16S rRNA gene. Bacterial alpha diversity revealed that the Shannon index and observed index were similar between MS and RF fluid. The core microbiome was detected in 88.6% of the amplicon sequence variants in MS and RF. Taxonomic analysis at the phylum level showed similar abundances of *Bacteroidetes*, *Proteobacteria*, *Firmicutes*, *Verrucomicrobiota*, *Spirochaetota*, *Patescibacteria*, and *Campilobacterota* in MS and RF. The Bray–Curtis distance matrix showed similar bacterial community structure among MS and RF samples. Conclusion: Our results indicated that the in vitro procedure did not affect the bacterial community structure compared to the original rumen fluid inoculum. It should be noted that assessing the microbiome at a single endpoint (i.e., 48 h) may not provide a comprehensive understanding of the microbiome profile dynamics. However, the findings of this study provide a basis for future microbiome-based in vitro fermentation tests and confirm that the technique allows a high degree of species diversity that approximates the rumen function in vivo.

Keywords: rumen microbiome; *Bacteroidota*; *Proteobacteria*; *Firmicutes*; 16S rRNA

Citation: Dhakal, R.; Neves, A.L.A.; Sapkota, R.; Khanal, P.; Hansen, H.H. Prokaryote Composition and Structure of Rumen Fluid before and after In Vitro Rumen Fermentation. *Fermentation* **2024**, *10*, 108. https://doi.org/10.3390/fermentation10020108

Academic Editors: Wansup Kwak and Siran Wang

Received: 30 December 2023
Revised: 6 February 2024
Accepted: 9 February 2024
Published: 14 February 2024

1. Background

Feeding is an essential component of any ruminant production system, playing an important role in the long-term profitability and sustainability of the business. The chemical composition, anti-nutritional components, and digestibility of feedstuffs used for ruminants must be determined regularly to ensure the success of the enterprise both financially and environmentally. The quality of the feed used in ruminant operations can be examined in three possible ways: (1) testing the feed with live animals (in vivo), (2) performing chemical analysis of the feed [1], or (3) undertaking in vitro feed evaluations [2]. In vivo testing of feed is desirable and is considered the gold standard. However, it is costly and labor intensive and may impact animal health and well-being during the experiment. Chemical analysis of the feed is useful, but it does not provide information on how the feed interacts with the digestive tract, growth, performance, and well-being of the animal. In vitro feed evaluation is a viable alternative to the first two options because it simulates a specific segment of the digestive tract of animals [3] and is a cost-effective, non-invasive, easy-to-operate, and efficient method for quickly assessing the nutritional content of feedstuffs without the need for using live animal trials. However, in vivo studies should confirm the in vitro findings.

The in vitro method is used to simulate the rumen in vivo and can be broadly classified into three categories: dynamic, semi-dynamic, and static systems (batch cultures). In each case, a sample feed is incubated with rumen fluid and a buffer in a controlled setting, and a predetermined time interval is used to maintain the ideal physico-chemical properties in the test flask for the rumen microbes to thrive [4]. In a static system, in vitro rumen fermentation is carried out by combining a sample of feed mixed with rumen fluid and a buffer made of macro and micro minerals. In a static in vitro system, the experimental conditions remain constant throughout the duration of the experiment. This means that there is no flow or movement of buffer or feedstuff in the fermentation vessel once the experiment begins. A semi-dynamic in vitro system combines aspects of both static and dynamic systems. The dynamic and semi-dynamic systems adhere to the same principles as the static system, but they enable the influx and outflux of buffer and waste and allow the expulsion of accumulated gas at a predetermined pressure or rate. In semi-dynamic systems, the experimental conditions will change periodically, but there is no continuous flow of feed stuff. Furthermore, the dynamic in vitro systems allow the continuous flow of feedstuff through the fermentation vessel. However, it is important to note that the semi-dynamic system may not provide all the functionalities of the dynamic system. For all three systems, the feeds are incubated under anaerobic conditions in a fermentation vessel, or glass bottles equipped with an automatic gas measurement system [5], fermentation tubes, or glass syringes [3]. The choice of system depends on the specific research goals and the biological processes to be studied. The fermentation process is typically carried out by maintaining a temperature of approximately 39 °C and a pH of around 6.5–7.0, similar to the conditions in the rumen in vivo. Based on research interest and objectives, fermentation is carried out for a period of 24–96 h [6].

In vitro rumen fermentation can be used to study rumen function, including the degradation of feedstuffs, fermentation kinetics, the production of volatile fatty acids (VFAs), total gas production, the concentration and yield of methane in total gas, the activity of rumen microorganisms, and the effects of feed additives on microorganisms [7–10]. Fermentation is influenced by the feed type and inoculum source (rumen fluid) and having the same microbial composition before and after the fermentation is desirable. However, there is limited knowledge of the microbiome composition before and after a fermentation assay. We hypothesized that the procedures used during in vitro fermentation do not affect the rumen prokaryotic (bacteria and archaea) community structure and that this prokaryotic composition remains unchanged compared to that of the original rumen fluid even after a 48 h fermentation. In this work, we aimed to characterize the rumen prokaryotic profile in pre- and post-fermentation samples to verify our hypothesis.

2. Materials and Methods

2.1. In Vitro Fermentation

Maize silage was used as the basal substrate for the fermented samples and 500 mg were weighed in 100 mL Duran® bottles. Fermentations were undertaken for 48 h with triplicates in each trial for the samples incubated with maize silage. Two rumen-cannulated heifers at the Large Animal Hospital from the University of Copenhagen (Taastrup, Denmark) were donors of rumen fluid (solid and semi-solid phases) for each fermentation. The heifers fasted for 12 h before rumen fluid collection. The cannulated animal use was authorized by Danish law under the research animal license no. 2012-15-2934-00648. Animals were fed haylage (85% dry matter, 7.5 MJ/kg metabolizable energy, and 11% crude protein) ad libitum for over six weeks before the experiment. Six in vitro fermentation assays were undertaken as described by Dhakal et al. [11]. Briefly, a four-part buffer solution was prepared and flushed with CO_2 for two hours to ensure anaerobic conditions, and the temperature of the medium was maintained at 39 °C before the addition of the rumen fluid. A reduction agent composed of sodium sulfide and sodium hydroxide was added

10 min before the addition of the rumen fluid. An equal amount of rumen fluid from each cannulated heifer was filtered through a double layer of commercial cheesecloth before being added to the buffered media. This rumen fluid was added to the buffer at a 1:2 ratio, and 90 mL of the rumen–buffer inoculum was added to each bottle, flushed with CO_2, and capped with an ANKOM^{RF} (ANKOMRF Technology, Macedon, NY, USA) module. Each ANKOM^{RF} module is an automated system with a pressure sensor (pressure range: from −69 to +3447 kPa; resolution: 0.27 kPa; accuracy: ±0.1% of measured value) and radio frequency that sends and receives signals to computer software via a base station. The recording time and global release pressure for valve opening were set to 10 min and 0.75 PSI, respectively.

To analyze the microbial composition and structure of in vivo rumen fluid (fresh inoculant, 0 h, without added maize silage), 10 mL of rumen fluid samples (n = 6)) were collected and stored at −20° before each incubation. Similarly, after the end of each fermentation, rumen fluid samples incubated for 48 h with maize silage (n = 9) were collected and stored at −20° until further analysis.

2.2. DNA Extraction

Frozen samples were thawed and 2 mL of rumen fluid sample from each bottle was transferred into a new sterile tube and centrifuged at 15,000× g for 10 min to obtain cell-rich pellets for genomic DNA extraction. Following the manufacturer's protocols, DNA from the cell-rich pellets was extracted using the Bead-Beat Micro Ax Gravity (A&A Biotechnology, Gdynia, Poland). The concentration and purity of the extracted DNA were tested with a NanoDrop Lite UV-Vis spectrophotometer (Thermo Fisher Scientific, Waltham, MA, USA).

The bacterial primers 515F (GTGCCAGCMGCCGCGGTAA) and 806R (GGACTACHVGGGTWTCTAAT) along with the Illumina Nextera overhang adapters were used to amplify the V4 region of the bacterial 16S rRNA region [12]. For the first PCR, thermocycler conditions were 95 °C for 2 min, 33 cycles of 95 °C for 15 s, 55 °C for 15 s, 68 °C for 40 s, and the final elongation at 68 °C for 4 min (SimpliAmp Thermal Cycler, Applied Bio-systems, Newton Dr, Carlsbad, CA, USA). Each PCR reaction of 25 µL consisted of 5xPCRBIO HiFi Buffer (5 µL) (PCRBiosystems, UK), 2 ng of DNA template, 0.25 units of PCRBIO HiFi Polymerase (PCRBiosystems, UK), 0.5 mM of forward and reverse primers, 0.5 µL bovine serum, and 16.25 µL H20. Gel electrophoresis was run to check the amplification in each sample.

After PCR1, a second PCR (PCR2) was performed to add unique index combinations (i7and i5) and adaptors. For PCR2, thermocycler conditions were 98 °C for 1 min, 13 cycles of 98 °C for 10 s, 55 °C for 20 s, 68 °C for 40 s, and the final elongation at 68 °C for 5 min. Subsequently, the amplicon product was cleaned using HighPrep™ magnetic beads (MagBio Genomics Inc., Gaithersburg, MD, USA), according to the manufacturer's instructions. Then, gel electrophoresis was run for the first and second PCR products of each sample to confirm the results. Finally, amplicons were pooled in equimolar concentration, and sequencing was carried out using the Illumina MiSeq platform.

2.3. Bioinformatics

Using the DADA2 plugin [13] implemented in the software QIIME2, the DNA reads acquired from the Illumina MiSeq run were analyzed [14]. The 'dada2 denoise-paired' command was used to screen for chimeras after denoising, joining, dereplicating, trimming primers (forward and reverse), and building amplicon sequence variant (ASV) tables. ASVs were then given a taxonomic classification using the SILVA 138 database and the 'feature-classifier classify-sklearn' method [15]. The taxonomy files and ASV table were analyzed using R version 4.2.1 [16] for data analysis and visualization. The *vegan* (version 2.6-4) and the *phyloseq* (version 1.40.0) packages, by Oksanen et al. [17] and McMurdie and Holmes [18], were used to run the diversity-based analysis. The linear discriminant analysis effect size (LEfSe) using LDA > 2 was used to identify differentially abundant bacteria and archaea among the two groups using the R package *microbiomeMarker* [19].

Alpha diversity was estimated using richness (observed) and the Shannon diversity index. The significance of differences in relative abundance of prokaryotic phyla and alpha diversity between RF and MS incubation was evaluated using the Kruskal–Wallis test. For beta diversity-based calculations, the ASV table was transformed to relative abundance and a Bray–Curtis distance matrix was used for visualization using principal coordinate analysis (PCoA). For partitioning of variance, distance matrices were subjected to permutation analysis of variance (PERMANOVA) using the *adonis* test from the *vegan* package.

3. Results

The amplicon sequencing generated 1,844,731 total reads giving an average of 23,351 ± 8562 reads per sample. Following quality filtration and the removal of chloroplast and mitochondrial reads, the sequence count was reduced to 1,844,661. The final average number of counts per sample that were assigned to ASVs (post-filtering) was 23,350 ± 8561.

As shown in Figure 1a, no difference ($p > 0.05$) was observed in alpha diversity metrics using the Shannon index (MS: 4.79, RF 4.77). Likewise, no differences of species richness (observed) (MS: 323.39, RF: 312.30) between rumen fluid fermented with maize silage (MS) and rumen fluid inoculum (RF) were observed. The first column in the UpSet plot shows the commonality of ASVs in the two rumen fluid samples while the second and third show the unique ASV. A total of 420 ASVs (88.6%) were shared between MS and RF (core microbiome), as shown in Figure 1b. The MS had 48 (10.13%) unique ASVs and RF had 6 (1.27%) unique ASVs.

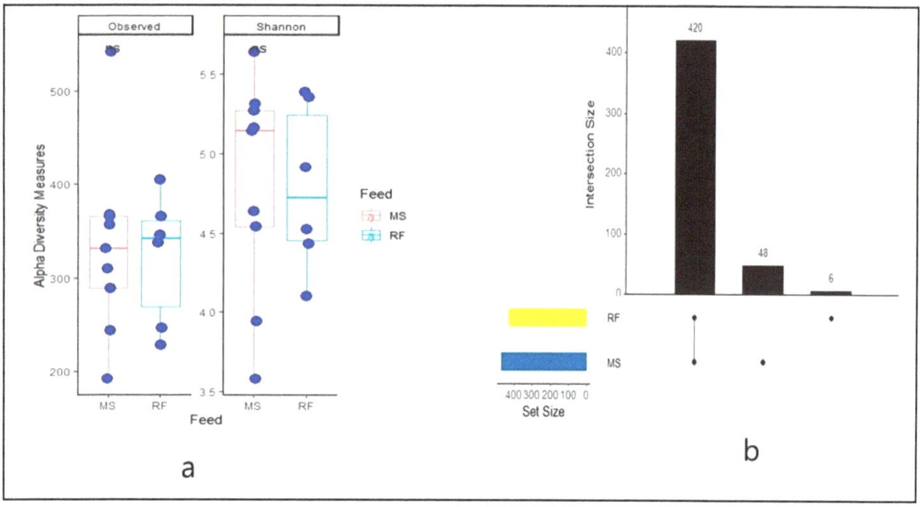

Figure 1. Box plot of alpha diversity metrics using observed (richness) and Shannon diversity index (**a**) and UPset plot (**b**) of number of ASVs common in rumen fluid (RF) and maize silage (MS).

The taxonomic analysis at the phylum level did not show statistical difference ($p > 0.05$) between RF and MS (Figure 2a). The top eight phylum were Bacteroidota (MS: 45.67%, RF: 48.52%), Proteobacteria (MS: 17.56%, RF: 18.87%), Firmicutes (MS: 17.18%, RF: 13.41%), Verrucomicrobiota (MS: 7.89%, RF: 8.78%), Spirochaetota (MS: 5.80%, RF: 4.68%), Patescibacteria (MS: 2.92%, RF: 2.36%), Campilobacterota (MS: 0.86%, RF: 0.14%), and Cyanobacteria (MS: 0.38%, RF: 1.8%). Figure 2b shows the microbiome community structure based on the fermentation batch and pre- (MS) and post-incubation (RF). PERMANOVA analysis revealed no significant difference between batches regarding the bacterial structures ($p > 0.05$). The abundance at the phylum level was similar ($p > 0.05$) between the MS and RF, as shown in Figure 3, except for Cyanobacteria ($p < 0.05$). To better understand the dominance of specific bacteria within the two groups, we used the LEfSe method (Figure 4). The genera Prevotella, Gastranaerophilales, Rhodospirillales, and Burkholderiales were more abundant in RF and the genera Methanobrevibacter, RF39, Ruminiclostridium, [Eubacterium]_ventriosum_group, Desulfovibrio, Mogibacterium, Veillonellaceae_UCG−001, Schwartzia, and Lachnospiraceae_UCG-004 were more abundant in MS.

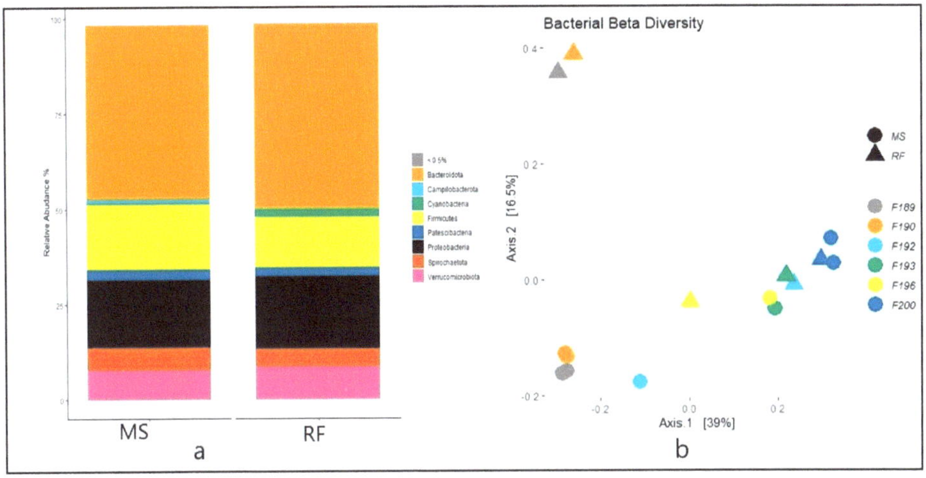

Figure 2. (**a**) The relative abundance (**a**) and principal coordinate analysis of Bray–Curtis distance beta diversity (**b**) of prokaryote phylum in rumen fluid (RF) and filtrate from in vitro rumen fluid fermented with maize silage (MS). Individual fermentation batches are shown with batch numbers and colored dots in 2b.

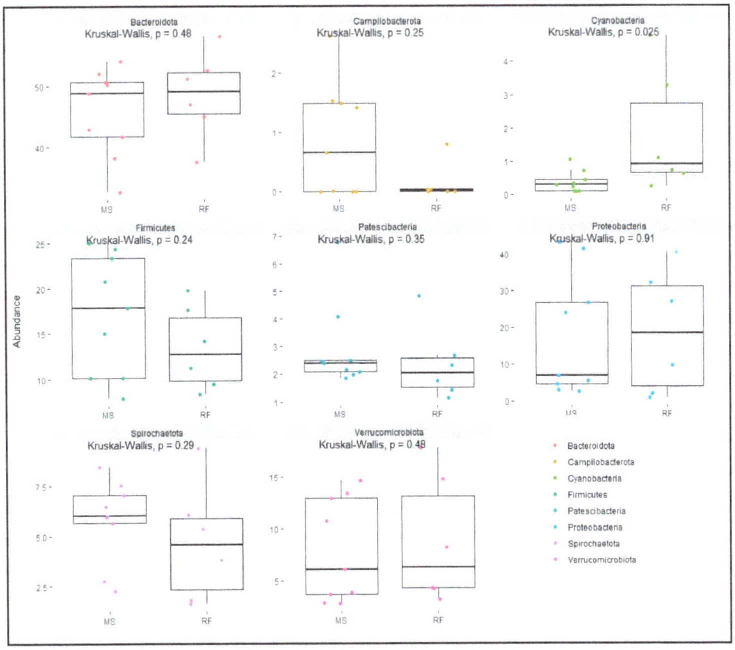

Figure 3. Relative abundance of a selected phyla of bacteria from in vitro fermented maize silage (MS) and rumen fluid (RF).

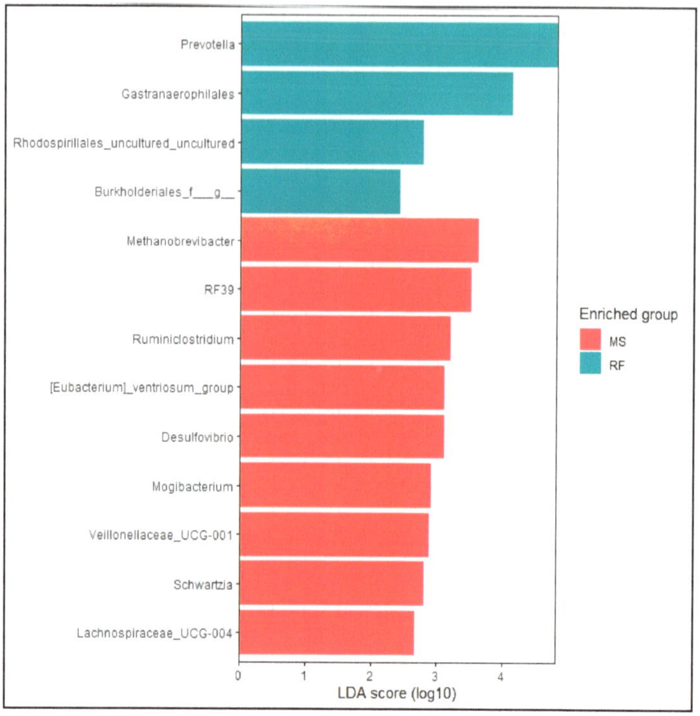

Figure 4. LEfSe analysis of rumen microbiome at genera level of prokaryote microbes. Differentially abundant ($p < 0.05$, LDA > 2) genera in rumen fluid (RF) and filtrate from in vitro rumen fluid fermented with maize silage (MS) are shown in the horizontal bars with their respective LDA score.

4. Discussion

This study examined the microbial composition of fresh rumen fluid (0 h, without addition of maize silage) and filtrate from the post-48 h fermentation stage of maize silage. We used maize silage as substrate because it is regarded as one of the most important feedstuff materials used in ruminant feeding. To our knowledge, there are no specific comparisons of microbial community structure pre- and post-fermentation. The analysis of the microbial diversity of rumen fluid (inoculant) and maize silage fermented for 48 h showed similarity in community structure and composition before and after the in vitro fermentation. In this study, the most abundant phyla were *Bacteroidota*, *Proteobacteria*, *Firmicutes*, *Verrucomicrobiota*, *Spirochaetota*, *Patescibacteria*, *Campilobacterota*, and *Cyanobacteria*, which is similar to earlier reports by Wei et al. [20] and Terry et al. [21]. The rumen microbiome consists mainly of cellulolytic and amylolytic taxa and microbes that utilize fermentation byproducts like methanogens [22]. The phyla *Bacteroidota*, *Proteobacteria*, and *Firmicutes* are responsible for fiber breakdown in the rumen [23]. The LeFSe and relative abundance analysis are in agreement. The abundant genera of MS fall within the phyla *Firmicutes*, *Desulfobacterota*, and *Euryarchaeota* and similarly abundant genera of RF fall within the phyla *Bacteroidota*, *Cyanobacteria*, and *Proteobacteria*. This study demonstrates that the in vitro system effectively replicates the in vivo rumen environment, as demonstrated by the comparable phyla levels that were observed before and after fermentation. This validation confirms the reliability of this methodology for studying the rumen microbiota.

The source of rumen fluid and replication of the in vivo conditions in in vitro fermentation are the key components to overcome the challenging steps involved in in vitro rumen fermentation. The source of rumen fluid can be from live animals fed production [24] or maintenance rations [7], rumen fluid from slaughterhouses [25], or frozen rumen fluid [26,27]. Henderson et al. [28] reported that the core microbiome in vivo is conserved across geographical regions no matter the species or feed; however, the relative abundances of the microbes vary with the diet and the ruminant species. Hence, to replicate in vivo conditions during in vitro fermentation, it is essential to maintain donor animals, their diets, and the basal feed in the bottle constant throughout the experiments to minimize any variations in the microbial community and structure.

We found similarities in the overall bacterial community estimated by the Bray–Curtis index between two fluid types. Ma et al. [26] observed similar results when fresh and frozen rumen fluids were studied for microbial community profiling. It was expected that the DNA of the inactive or dead microorganisms were retained in the fermentation flask in our research. After incubation, we assumed that the overall (most abundant) microorganisms remained active in both the inoculation source and the flask. This is supported by the lack of difference in alpha diversity metrics in both richness and evenness between MS and RF. However, any significant difference in the proportion of readily fermentable carbohydrates as substrates in the in vitro incubation would have an impact on the amylolytic microbial population that grows at a faster rate than the fibrolytic microbes, but this was not observed in the present study.

Although there was no change in the overall microbial diversity between RF and MS, the in vitro procedure affected the growth of *Cyanobacteria*. This phylum is commonly found in soil and water and its relative abundance in the rumen is less than 1% [29]. *Cyanobacteria* are aerobic microbes that can convert inert nitrogen into the organic form, but they can utilize carbohydrates in a deficient N_2 concentration [30]. The absorption of oxygen from drinking water as well as contact with the air during the sample collection may have favored the growth of these bacteria in the rumen fluid. Neves et al. [31] reported that the presence of these microbes could be related to O_2 scavenging and sugar fermentation under the restricted aerobic conditions in the rumen.

Having the same inoculum source (animals) is the goal in every in vitro rumen fermentation. However, the donor animals undergo different physiological and nutritional changes that can affect the microbiome profile [32–34] of the inoculum. Such physiological and nutritional changes make in vitro rumen fermentation difficult to replicate [34]. This research shows that these challenges can be overcome by using the same basal treatment and the same inoculum source in every fermentation batch, followed by a comparison of the microbiome of the basal treatment (control) in each batch. Such a comparison provides insight into fermentation characteristics and whether there are any differences in the microbial populations of in vivo and in vitro rumen fluid.

5. Conclusions

The study indicates that the rumen prokaryotic microbial structure and composition in both pre- and post-in vitro fermentation samples remained relatively stable despite a 48 h incubation period. This composition in the in vitro filtrate after incubation did not significantly differ from the rumen fluid before incubation (inoculum). Nevertheless, the study demonstrates that the rumen prokaryotic population remained consistent and robust even after the 48 h incubation period. It is worth considering that collecting data from a single endpoint within a 48 h fermentation may not provide a comprehensive overview of the rumen prokaryotic microbiome (structure and composition) profile dynamics. To obtain a more precise and thorough understanding, it is advisable to include multiple time points of sampling in the fermentation. In general, this study confirms that the in vitro rumen fermentation technique allows for a high degree of species diversity that closely approximates the rumen in vivo.

Author Contributions: Conceptualization, R.D., R.S. and H.H.H.; methodology, R.D. and R.S.; software, R.D., A.L.A.N. and R.S.; validation, R.D., R.S. and H.H.H.; formal analysis, R.D. and R.S.; investigation, R.D. and R.S.; resources, R.D. and H.H.H.; data curation, R.D. and R.S.; writing—original draft preparation, R.D.; writing—review and editing, R.D., R.S., P.K., A.L.A.N. and H.H.H.; supervision, A.L.A.N. and H.H.H.; project administration, R.D. and H.H.H.; funding acquisition, R.D. and H.H.H. All authors have read and agreed to the published version of the manuscript.

Funding: This research received no external funding.

Institutional Review Board Statement: The cannulated animal use was authorized by Danish law under the research animal license no. 2012-15-2934-00648.

Informed Consent Statement: Not applicable.

Data Availability Statement: Data from the sequencing raw reads can be found in the NCBI SRA database under the accession number PRJNA1020745.

Acknowledgments: We would like to acknowledge lab technicians Anni Christiansen and Lotte Ørbæk for their contributions during the experimental procedures and other technical supports in the lab.

Conflicts of Interest: The authors declare that they have no competing interests.

Abbreviations

ASV	Amplicon sequence variant
DADA2	Deficiency of adenosine deaminase 2
DNA	Deoxyribonucleic acid
IVRF	In vitro rumen fermentation
MS	Maize silage
PCoA	Principle coordinate analysis
PCR	Polymerase chain reaction
QIIME2	Quantitative insights into microbial ecology 2
RF	Rumen fluid
rRNA	Ribosomal ribonucleic acid
VFAs	Volatile fatty acids

References

1. Van Soest, P.J.; Robertson, J.B.; Lewis, B.A. Methods for Dietary Fiber, Neutral Detergent Fiber, and Nonstarch Polysaccharides in Relation to Animal Nutrition. *J. Dairy Sci.* **1991**, *74*, 3583–3597. [CrossRef]
2. Makkar, H.P.S. Applications of the In Vitro Gas Method in the evaluation of feed resources, and enhancement of nutritional value of tannin-rich tree/browse leaves and agro-industrial by-products. *Anim. Prod. Heal. Sect. Int. At. Energy Agency Vienna* **2009**, *2*, 23–41.
3. Menke, K.H.; Raab, L.; Salewski, A.; Steingass, H.; Fritz, D.; Schneider, W. The estimation of the digestibility and metabolizable energy content of ruminant feedingstuffs from the gas production when they are incubated with rumen liquor in vitro. *J. Agric. Sci.* **1979**, *93*, 217–222. [CrossRef]
4. Blümmel, M.; Makkar, H.P.S.; Becker, K. In vitro gas production: A technique revisited. *J. Anim. Physiol. Anim. Nutr.* **1997**, *77*, 24–34. [CrossRef]
5. Cornou, C.; Storm, I.M.L.D.; Hindrichsen, I.K.; Worgan, H.; Bakewell, E.; Ruiz, D.R.Y.; Abecia, L.; Tagliapietra, F.; Cattani, M.; Ritz, C.; et al. A ring test of a wireless in vitro gas production system. *Anim. Prod. Sci.* **2013**, *53*, 585–592. [CrossRef]
6. Dijkstra, J.; Kebreab, E.; Bannink, A.; France, J.; López, S. Application of the gas production technique to feed evaluation systems for ruminants. *Anim. Feed Sci. Technol.* **2005**, *123–124 Pt 1*, 561–578. [CrossRef]
7. Pandey, D.; Hansen, H.H.; Dhakal, R.; Aryal, N.; Rai, S.P.; Sapkota, R.; Nielsen, M.O.; Novoa-Garrido, M.; Khanal, P. Interspecies and seasonal variations in macroalgae from the Nordic region: Chemical composition and impacts on rumen fermentation and microbiome assembly. *J. Clean. Prod.* **2022**, *363*, 132456. [CrossRef]
8. Sembach, L.E.; Hansen, H.H.; Dhakal, R.; Eriksson, T.; Nielsen, N.I.; Nielsen, M.O. Estimates for rumen dry matter degradation of concentrates are higher, but not consistently, when evaluated based on in sacco as compared to in vitro methods. In Proceedings of the 10th Nordic Feed Science Conference, Uppsala, Sweden, 11–12 June 2019; pp. 123–128.
9. Getachew, G.; Robinson, P.H.; DePeters, E.J.; Taylor, S.J. Relationships between chemical composition, dry matter degradation and in vitro gas production of several ruminant feeds. *Anim. Feed Sci. Technol.* **2004**, *111*, 57–71. [CrossRef]
10. Vargas-Bello-Pérez, E.; Pedersen, N.C.; Khushvakov, J.; Ye, Y.; Dhakal, R.; Hansen, H.H.; Ahrné, L.; Khakimov, B. Effect of Supplementing Dairy Goat Diets with Rapeseed Oil or Sunflower Oil on Performance, Milk Composition, Milk Fatty Acid Profile, and in vitro Fermentation Kinetics. *Front. Vet. Sci.* **2022**, *9*, 614. [CrossRef] [PubMed]
11. Dhakal, R.; Ronquillo, M.G.; Vargas-Bello-Pérez, E.; Hansen, H.H. Effect of Autochthonous Nepalese Fruits on Nutrient Degradation, Fermentation Kinetics, Total Gas Production, and Methane Production in In-Vitro Rumen Fermentation. *Animals* **2022**, *12*, 2199. [CrossRef] [PubMed]

12. Caporaso, J.G.; Lauber, C.L.; Walters, W.A.; Berg-Lyons, D.; Lozupone, C.A.; Turnbaugh, P.J.; Fierer, N.; Knight, R. Global patterns of 16S rRNA diversity at a depth of millions of sequences per sample. *Proc. Natl. Acad. Sci. USA* **2011**, *108*, 4516–4522. [CrossRef] [PubMed]
13. Callahan, B.J.; McMurdie, P.J.; Rosen, M.J.; Han, A.W.; Johnson, A.J.A.; Holmes, S.P. DADA2: High-resolution sample inference from Illumina amplicon data. *Nat. Methods* **2016**, *13*, 581–583. [CrossRef] [PubMed]
14. Bolyen, E.; Rideout, J.R.; Dillon, M.R.; Bokulich, N.A.; Abnet, C.C.; Al-Ghalith, G.A.; Alexander, H.; Alm, E.J.; Arumugam, M.; Asnicar, F.; et al. Reproducible, interactive, scalable and extensible microbiome data science using QIIME 2. *Nat. Biotechnol.* **2019**, *37*, 852–857. [CrossRef] [PubMed]
15. Quast, C.; Pruesse, E.; Yilmaz, P.; Gerken, J.; Schweer, T.; Yarza, P.; Peplies, J.; Glöckner, F.O. The SILVA ribosomal RNA gene database project: Improved data processing and web-based tools. *Nucleic Acids Res.* **2013**, *41*, 590–596. [CrossRef]
16. R Core Team. R A Lang. Environ. Stat. Comput. R Found. Stat. Comput. 2022. Available online: https://www.r-project.org/ (accessed on 29 December 2023).
17. Oksanen, A.J.; Blanchet, F.G.; Friendly, M.; Kindt, R.; Legendre, P.; Mcglinn, D.; Minchin, P.R.; Hara, R.B.O.; Simpson, G.L.; Solymos, P.; et al. Package 'vegan'. 2020. Available online: https://github.com/vegandevs/vegan (accessed on 3 February 2023).
18. McMurdie, P.J.; Holmes, S. Phyloseq: An R Package for Reproducible Interactive Analysis and Graphics of Microbiome Census Data. *PLoS ONE* **2013**, *8*, e61217. [CrossRef]
19. Cao, Y.; Dong, Q.; Wang, D.; Zhang, P.; Liu, Y.; Niu, C. microbiomeMarker: An R/Bioconductor package for microbiome marker identification and visualization. *Bioinformatics* **2022**, *38*, 4027–4029. [CrossRef]
20. Wei, X.; Ouyang, K.; Long, T.; Liu, Z.; Li, Y.; Qiu, Q. Dynamic Variations in Rumen Fermentation Characteristics and Bacterial Community Composition during In Vitro Fermentation. *Fermentation* **2022**, *8*, 5–7. [CrossRef]
21. Terry, S.A.; Badhan, A.; Wang, Y.; Chaves, A.V.; McAllister, T.A. Fibre digestion by rumen microbiota—A review of recent metagenomic and metatranscriptomic studies. *Can. J. Anim. Sci.* **2019**, *99*, 678–692. [CrossRef]
22. Belanche, A.; Patra, A.K.; Morgavi, D.P.; Suen, G.; Newbold, C.J.; Yáñez-Ruiz, D.R. Editorial: Gut Microbiome Modulation in Ruminants: Enhancing Advantages and Minimizing Drawbacks. *Front. Microbiol.* **2021**, *11*, 622002. [CrossRef]
23. Mizrahi, I.; Wallace, R.J.; Moraïs, S. The rumen microbiome: Balancing food security and environmental impacts. *Nat. Rev. Microbiol.* **2021**, *19*, 553–566. [CrossRef]
24. Macome, F.M.; Pellikaan, W.F.; Hendriks, W.H.; Warner, D.; Schonewille, J.T.; Cone, J.W. In vitro gas and methane production in rumen fluid from dairy cows fed grass silages differing in plant maturity, compared to in vivo data. *J. Anim. Physiol. Anim. Nutr.* **2018**, *102*, 843–852. [CrossRef] [PubMed]
25. Fortina, R.; Glorio Patrucco, S.; Barbera, S.; Tassone, S. Rumen Fluid from Slaughtered Animals: A Standardized Procedure for Sampling, Storage and Use in Digestibility Trials. *Methods Protoc.* **2022**, *5*, 59. [CrossRef] [PubMed]
26. Ma, Z.Y.; Zhou, J.W.; Yi, S.Y.; Wang, M.; Tan, Z.L. In vitro Inoculation of Fresh or Frozen Rumen Fluid Distinguishes Contrasting Microbial Communities and Fermentation Induced by Increasing Forage to Concentrate Ratio. *Front. Nutr.* **2022**, *8*, 772645. [CrossRef] [PubMed]
27. Spanghero, M.; Chiaravalli, M.; Colombini, S.; Fabro, C.; Froldi, F.; Mason, F.; Moschini, M.; Sarnataro, C.; Schiavon, S.; Tagliapietra, F. Rumen inoculum collected from cows at slaughter or from a continuous fermenter and preserved in warm, refrigerated, chilled or freeze-dried environments for in vitro tests. *Animals* **2019**, *9*, 815. [CrossRef] [PubMed]
28. Henderson, G.; Cox, F.; Ganesh, S.; Jonker, A.; Young, W.; Janssen, P.H.; Abecia, L.; Angarita, E.; Aravena, P.; Arenas, G.N.; et al. Rumen microbial community composition varies with diet and host, but a core microbiome is found across a wide geographical range. *Sci. Rep.* **2015**, *5*, 14567. [CrossRef] [PubMed]
29. Ilina, L.A.; Filippova, V.A.; Brazhnik, E.A.; Dubrovin, A.V.; Yildirim, E.A.; Dunyashev, T.P.; Laptev, G.Y.; Novikova, N.I.; Sobolev, D.V.; Yuzhakov, A.A.; et al. The comparative analysis of the ruminal bacterial population in reindeer (*Rangifer tarandus* L.) from the Russian arctic zone: Regional and seasonal effects. *Animals* **2021**, *11*, 911. [CrossRef]
30. Nandi, R.; Sengupta, S. Microbial Production of Hydrogen: An Overview. *Crit. Rev. Microbiol.* **1998**, *24*, 61–84. [CrossRef]
31. Neves, A.L.A.; Li, F.; Ghoshal, B.; McAllister, T.; Guan, L.L. Enhancing the resolution of rumen microbial classification from metatranscriptomic data using Kraken and Mothur. *Front. Microbiol.* **2017**, *8*, 2445. [CrossRef]
32. Amin, N.; Seifert, J. Dynamic progression of the calf's microbiome and its influence on host health. *Comput. Struct. Biotechnol. J.* **2021**, *19*, 989–1001. [CrossRef]
33. Tröscher-Mußotter, J.; Saenz, J.S.; Grindler, S.; Meyer, J.; Kononov, S.U.; Mezger, B.; Borda-Molina, D.; Frahm, J.; Dänicke, S.; Camarinha-Silva, A.; et al. Microbiome Clusters Disclose Physiologic Variances in Dairy Cows Challenged by Calving and Lipopolysaccharides. *mSystems* **2021**, *6*, e00856-21. [CrossRef]
34. Kim, S.H.; Mamuad, L.L.; Kim, E.J.; Sung, H.G.; Bae, G.S.; Cho, K.K.; Lee, C.; Lee, S.S. Effect of different concentrate diet levels on rumen fluid inoculum used for determination of in vitro rumen fermentation, methane concentration, and methanogen abundance and diversity. *Ital. J. Anim. Sci.* **2018**, *17*, 359–367. [CrossRef]

fermentation

Article

Safety of Chinese Cabbage Waste and Rice Straw Mixed Silage and Its Effect on Growth and Health Performance of Hu Sheep

Zhiqi Lu [1,†], Chuang Li [1,2], Lie Hou [1,‡], Zhenbin Zhang [1], Yue Lu [1], Chun Xue [1], Ruxin Qi [1], Muhammad Hammad Zafar [1], Jun Xu [3,*] and Mengzhi Wang [1,4,*]

1 College of Animal Science and Technology, Yangzhou University, Yangzhou 225009, China; zqlu73@163.com (Z.L.); chuangliyzu@163.com (C.L.); hle19960225@163.com (L.H.); 15861273716@163.com (Y.L.); xuechun0727@163.com (C.X.); ruxinqi@126.com (R.Q.); hammadzafar075@gmail.com (M.H.Z.)
2 College of Animal Science, Xinjiang Agricultural University, Urumqi 830091, China
3 Quality & Safety and Standards of Agricultural Products Research, Jiangxi Academy of Agriculture Science, Nanchang 330200, China
4 State Key Laboratory for Sheep Genetic Improvement and Healthy Production, Xinjiang Academy of Agricultural and Reclamation Science, Shihezi 832000, China
* Correspondence: xujun0125@163.com (J.X.); mzwang@yzu.edu.cn (M.W.)
† Current address: Ningxia Dairy Science and Innovation Center of Guangming Animal Husbandry Co., Ltd., Zhongwei 751700, China.
‡ Current address: Animal Husbandry Extension Station, Yinchuan 750001, China.

Abstract: Improper disposal of vegetable waste can cause serious environmental pollution, but because they contain huge water content and organic matter, they are not suitable for disposal by methods such as incineration and landfill. However, vegetable waste contains a large amount of nutrients and have some complementary effects with rice straw in terms of physical structure, nutrients, and moisture. In this experiment, the plant feed (corn husk, peanut shells and sorghum shells) was used as the control group (CON group), and the mixed silage of Chinese cabbage waste and rice straw (mixed silage) was used as the experiment group (TRE group), and its safety performance was evaluated by testing its toxin content, pesticide residues, vitamin contents and feeding experiment of Hu sheep. In the animal experiment, 16 healthy Hu sheep (5.5 months, 39.11 ± 4.16 kg) were randomly divided into two groups of 8 each. The results of the safety performance evaluation showed that the content of mycotoxins, heavy metals, and nitrites as well as pesticide residues in the crude feeds of both groups were within the range of Chinese feed hygiene standards. In addition, the levels of deoxynivalenol (DON) and aflatoxin (AFT) in the CON group were lower, while the content of ochratoxin (OTA) and zearalenone were higher than those in the TRE group ($p < 0.05$). The levels of plumbum(Pb), chromium (Cr), cadmium (Cd), and nitrite in the CON group were lower than the mixed silage, while the levels of As were higher than the mixed silage ($p < 0.05$). It is worth noticing that the content of vitamin B_2 (VB_2) and vitamin C (VC) in the TRE group was higher than the CON group ($p < 0.05$). The results of the feeding experiment showed that the mixed silage did not affect the growth performance, nutrient digestibility, organ index, and intestinal index of Hu sheep ($p > 0.05$). In addition, the mixed silage reduced the weight of omasum, the proportion of omasum to live weight before slaughter, the amount of compound stomach, and the proportion of compound stomach to live weight before slaughter, which were higher than those in the TRE group ($p < 0.05$). The thickness of the basal layer of the rumen abdominal sac, the red blood cell count, the content of IL-10, and TNF-α in the blood, and TNF-α content in the rumen of the Hu sheep in the TRE group were higher than the CON group ($p < 0.05$). In conclusion, the feed safety index content of the mixed silage did not exceed the Chinese feed hygiene and safety standards and did not cause adverse effects on the growth performance of the Hu sheep, and it improved the immune performance of the body and digestive tract of the sheep to a certain extent and promoted the healthy development of the sheep.

Keywords: Chinese cabbage waste; growth performance; Hu sheep; rice straw; safety performance

Citation: Lu, Z.; Li, C.; Hou, L.; Zhang, Z.; Lu, Y.; Xue, C.; Qi, R.; Zafar, M.H.; Xu, J.; Wang, M. Safety of Chinese Cabbage Waste and Rice Straw Mixed Silage and Its Effect on Growth and Health Performance of Hu Sheep. *Fermentation* **2024**, *10*, 47. https://doi.org/10.3390/fermentation10010047

Academic Editors: Wansup Kwak and Siran Wang

Received: 1 December 2023
Revised: 25 December 2023
Accepted: 28 December 2023
Published: 8 January 2024

1. Introduction

Rice production is the third most important crop after wheat and maize in the world [1]. And, for each 1 kg of rice produced, 1.0 to 1.5 kg of rice straw is produced [2]. Rice straw is a very important crop by-product, which is widely used in animal silage, bioenergy, organic fertilizer, and other areas, but to a lesser extent [3–5]. Vegetables are an essential source of nutrients in people's daily life, but a large amount of vegetable tails are produced while meeting people's needs. Supermarkets, food markets, and households are the main places where vegetable tails are produced [6]. To the best of our knowledge, burning of rice straw and abandonment of tailing vegetables are still the main and effective disposal methods, and resource utilization is imminent [7,8]. The seasonal harvest and short-term concentrated outbreak of these wastes put great pressure on the collection and rapid treatment of raw materials [9,10].

There are some complementary effects between vegetable tailing and rice straw in terms of physical structure, nutrient and moisture content. Treatment by ensiling not only improves the nutritional value of rice straw, but also has less environmental impact [11]. Studies have shown that vegetable tails can be mixed with straw for ensiling, for example, mixed silage of broccoli by-products with wheat straw [12], mixed silage of corn straw and cabbage [13], co-storage of sugar beet waste with wheat straw [14]. Partovi et al. [12] replaced 20% of roughage (200 g/kg) in the diet with broccoli by-products mixed with wheat straw silage without affecting the growth performance and rumen fermentation parameters of Fashandy lambs.

Mycotoxins are inanimate, invisible, and toxic secondary metabolites produced by fungi. The presence of mycotoxins can affect crop quality, human health, and animal production, thereby affecting the global economy [15,16]. Moreover, to meet the global demand for crops, people overuse chemical fertilizers, pesticides, etc. Pesticide residues as well as heavy metal content are an increasing threat to ecosystems and human health [17,18]. Therefore, in order to ensure the safety of livestock and poultry, it is necessary to detect mycotoxins, heavy metals, and pesticide residues in the mixed silage.

We hypothesized that the mixed silage of Chinese cabbage waste and rice straw would reduce its toxins content without affecting the health performance of Hu sheep. Therefore, the objective of the research is to evaluate the safety performance of the mixed silage by measuring the vitamins content, mycotoxins content, pesticide residues content, and heavy metals content, and the feeding experiment of Hu sheep.

2. Materials and Methods

2.1. Mixed Silage Production

The rice straw and Chinese cabbage waste used for silage were collected from Suqian City, Jiangsu Province, China. *Lactobacillus plantarum* (2.0×10^{10} cfu/g) and cellulase (2.0×10^{5} U/g) were purchased from Guangzhou Greenfield Biotechnology Co., Ltd. (Guangzhou, China). Prior to silage, Chinese cabbage waste and rice straw were chopped up 2–3 cm. Chinese cabbage waste and rice straw (4:6), *lactobacillus plantarum* (0.035 g/kg) and cellulase (0.250 g/kg) were then mixed and fermented for 45 d in sealed silage bags. The weight of each silage wrap was 300 kg, resulting in a total production of 15 t.

2.2. Experimental Animals and Feeding Management

A one-way completely randomized group experimental design was used for this experiment. Sixteen healthy Hu sheep were randomly and equally divided into two groups (four rams and four ewes in each group), which were 5.5 months old and of similar weight (39.11 ± 4.16 kg). The treatment (TRE) group of Hu sheep was fed with mixed silage as roughage and the control (CON) group was fed with conventional common feed (peanut seedlings, corn husks and high grain husks) as roughage. The diets of Hu sheep were configured according to the nutritional requirements of sheep weighing 40 kg and gaining 400 g per day [19]. The ratio of concentrate to roughage was 50:50 for all groups (based on the dry matter (DM)). Their nutrient composition is shown in Table 1. The experiment consisted of a pre-feeding period (7 d) and a regular feeding period (28 d), for a total of 35 days. The sheep houses were cleaned and sterilized before the experiment. All sheep were uniformly dewormed and immunized. Hu sheep were kept in a pen. The sheep were fed twice at 7:00 and 19:00 with free access to food and water. Sheep houses were cleaned and disinfected daily to maintain cleanliness and hygiene.

Table 1. Experimental diet formula and nutrition composition (DM Basis/%).

Items	Group [1]	
	CON	TRE
	Ingredients	
Peanut seedlings	30	—
Corn husk	15	—
Sorghum shell	5	—
Mixed silage	—	50
Corn	34	34
Soybean meal	7	5.5
Bran	7.5	8
Corn protein powder	—	1
$NaHCO_3$	0.5	0.5
NaCl	0.5	0.5
Premix [2]	0.5	0.5
Total	100	100
	Nutrient levels [3]	
Digestive energy (DE) (MJ/kg)	13.52	14.73
Crud protein (CP)	15.08	15.11
Ash	4.36	12.33
Neutral Detergent Fiber (NDF)	47.64	48.23
Acid Detergent Fiber (ADF)	23.71	27.17
Ca	0.48	0.45
P	0.38	0.39

Note: [1] CON: Based on peanut seedling, corn husk and sorghum shell for roughage; TRE: Based on the mixed silage for roughage. [2] Premix contained (per kg): Vitamin A (VA) 80 kIU, Vitamin D (VD) 25 kIU, Vitamin E (VE) 130 kIU, Fe 0.6 g, Mn 0.7 g, Zn 2.3 g, Cu 0.2 g, Se 8 mg, Ca 10%, P 1%, NaCl 10%. [3] DE were estimated according to NRC (2007). The others were measured values. The experimental design of this experiment was the same as that of Li et al. [20].

2.3. Sampling

After the mixed silage was completed, six different wrapped silage samples were randomly selected for sampling and sent to the laboratory with the mill feed for crushing (40 mesh) for feed safety assessment tests. Peanut seedlings, corn husks and sorghum hulls were obtained from the mill feed of a sheep farm in Suqian, Jiangsu Province.

At the end of the feeding experiment, we randomly selected six Hu sheep (three rams and three ewes) in each of the CON group and the TRE group for slaughter. Feed and water were fasted for 24 h before slaughtering, and slaughtered after weighing. After the sheep were stunned by CO_2 gas, they were slaughtered by bloodletting from the jugular vein and their viscera were weighed. Blood, heart, liver, spleen, kidney, rumen small intestine (duodenum, jejunum, ileum), and large intestine tissues were collected from Hu sheep during slaughter.

2.4. Safety Performance Assessment

This test is to determine the content of four common mycotoxins in feed, namely DON, AFT, OTA and Zearalenone (ZEN). The four mycotoxins in the feed samples were determined using an enzyme-linked immunosorbent assay (ELISA) kit produced by Shanghai Amperexamination Technology Co., Shanghai, China. The detection of pesticide residues (hexachlorobenzene, dichlorodiphenyltrichloroethane (DDT), hexachlorobenzene, deltamethrin, dimethoate and dichlorvos) and common heavy metal ions (As, Pb, Hg, Cr, Cd and nitrite) and nitrite in the feed samples was tested by Qingdao Stander Testing Company Co., Ltd. (Qingdao, China). The contents of VA, VB_2, VC and VE in the feeds were determined by biochemical kits purchased from Shanghai Enzyme-linked Biotechnology Co., Shanghai, China.

2.5. Growth Performance Determination

At the beginning of the experiment and at the end of the experiment, each sheep was weighed on an empty stomach and the initial weight (IW) and final weight (FW) was recorded. The amount of feed fed and the amount remaining per sheep per day during the experiment were recorded, and the dry matter intake (DMI), the average daily gain (ADG) and the feed-to-weight ratio (F/W) were calculated at the end of the experiment using the following formula.

$$\text{ADG} = (\text{FW} - \text{IW})/28$$

$$\text{DMI} = (\text{Feeding rate} \times \text{DM content}) - (\text{Residual feed rate} \times \text{Residual DM content})$$

$$\text{F/W} = \text{DMI/ADG}$$

2.6. Determination of the Apparent Digestibility

One week before the end of the feeding experiment, the amount of feed and residual were accurately recorded. Feces were collected for five consecutive days by the whole feces method and weighed after collection. 20% of the daily manure sample was added to 10 mL 10% sulfuric acid and stored at −20 °C. The fresh manure samples were brought back to the laboratory for drying (65 °C), crushing, passing through 40 mesh sieve and then stored for testing.

DM, organic matter (OM), and CP of the manure samples and feed samples were determined using. The content of NDF and ADF was determined using the Van's washing method. DM digestibility (DMD), OM digestibility (OMD), CP digestibility (CPD), NDF digestibility (NDFD) and ADF digestibility (ADFD) were calculated using the following formula.

$$\text{Apparent digestibility of a nutrient} = 1 - (\text{whole manure/feed intake}) \times (\text{content of a nutrient in the manure/content of a nutrient in the ration}) \times 100\%$$

2.7. Blood Physiological and Biochemical Measurements

Blood was collected from the jugular vein of 12 (6 per group) sheep prior to slaughter. Three blood samples were collected from each sheep slaughtered. One of them was packed in EDTA anticoagulation tube, and the PE-6800VET fully automatic animal blood cell analyzer (Procan Electronics Co., Ltd, Shenzhen, China) was used to detect routine blood indexes, including white blood cell count, red blood cell count, hemoglobin concentration, platelets, absolute basophil value, basophil percentage, red blood cell pressure, mean red blood cell volume, mean hemoglobin content, mean hemoglobin concentration, erythrocyte distribution width, and standard deviation of erythrocyte distribution. The other one was divided into lithium heparin anticoagulation tubes, and the SMT-120V automatic biochemical analyzer (Seamaly, Chengdu, China)was used to test blood biochemical parameters, including total protein, albumin, globulin, albumin-to-globulin ratio, alkaline phosphatase, lactate dehydrogenase, creatinine, urea nitrogen, total cholesterol, triglycerides, high-density lipoprotein, and low-density lipoprotein. The remaining portion was divided into common tubes, placed at room temperature for 2 h, and left at 4 °C for 2–3 h. After the blood clotted and contracted, the supernatant was centrifuged for 10 min at 3000 r/min, and the supernatant was collected, and the contents of IL-1β, IL-6, IL-8, IL-10, TNF-α, IgA and IgM in the serum were determined by colorimetric method using a Huawei Delong DR-200BS enzyme standard analyzer (Beijing Huaying Biotechnology Research Institute, Beijing, China).

2.8. Determination of Organ Indices

During the slaughter process, the heart, liver, spleen, lungs, kidneys, and digestive tract were sampled and weighed in their entirety to calculate the organ index. The contents of the digestive tract were removed and cleaned before the rumen, reticulum, flap, wrinkled stomach, small intestine and large intestine were weighed and the organ indices were calculated. The organ index is calculated using the following formula.

$$\text{Organ index} = \text{organ weight (g)/live weight before slaughter (kg)} \times 100\%$$

$$\text{Complex stomach index} = \text{weight of complex stomach (g)/live weight before slaughter (kg)} \times 100\%$$

$$\text{Intestinal index} = \text{intestinal weight (g)/live weight before slaughter (kg)} \times 100\%$$

2.9. Measurement of Immune Performance

Samples of liver, spleen, kidney, rumen and small intestine (duodenum, jejunum and ileum) tissues as described above were homogenized to prepare homogenates. The levels of IL-1β, IL-6, IL-8, IL-10, TNF-α, IgA and IgM in serum as well as other tissue homogenates were determined by ELISA kits.

2.10. Data Analysis

We tested the data for normal distribution and homogeneity using SPSS Statistics V20.0 software (IBM Corporation, Armonk, NY, USA). All the data were subjected to independent samples *t*-test in SPSS. $p < 0.05$ indicates a significant difference and $p \geq 0.05$ indicates a non-significant difference. Visualization of growth performance and nutrient digestibility, development of the rumen epithelium and indicators of immune performance is done with GraphPad Prism 8.0 software.

3. Results

3.1. Safety Performance Assessment

As shown in Table 2, the contents of DON and AFT in the CON group were lower than the TRE group ($p < 0.05$), while the levels of OTA and ZEN were higher than the TRE group ($p < 0.05$). Hexachlorocyclohexane, DDT, dichlorvos and hexachlorobenzene were not detected in the diets of both groups. However, paclobutrazol was detected in the CON group and deltamethrin was detected in the TRE group diets. For heavy metals and nitrite content, the contents of Pb, Cr, Cd and nitrite in the CON feed were lower than the TRE group, while the content of As were higher than the TRE group ($p < 0.05$). In addition, the VB_2 and VC contents of the TRE group were higher than the CON group ($p < 0.05$).

Table 2. Assessment of safety performance of the mixed silage.

Items	Groups [1]		SEM [2]	p
	CON	TRE		
Mycotoxins				
DON/ppm	0.50 [b]	0.68 [a]	0.018	<0.001
AFT/ppb	3.48 [b]	4.18 [a]	0.104	0.002
OTA/ppb	16.63 [a]	11.65 [b]	1.237	0.016
ZEN/ppb	89.37 [a]	28.20 [b]	0.408	<0.001
Pesticide residues (mg/kg)				
Hexachlorocyclohexane	<0.01	<0.01	—	—
DDT	<0.01	<0.01	—	—
Paclobutrazol	0.036	<0.01	—	—

Table 2. *Cont.*

Items	Groups [1]		SEM [2]	*p*
	CON	TRE		
Dichlorvos	<0.01	<0.01	—	—
Hexachlorobenzene	<0.01	<0.01	—	—
Deltamethrin	<0.01	0.048	—	—
Heavy metals and nitrites (mg/kg)				
As	0.16 [a]	0.09 [b]	0.010	0.002
Pb	2.63 [b]	3.63 [a]	0.957	0.008
Hg	0.03	0.05	0.008	0.103
Cr	0.67 [b]	0.98 [a]	0.052	0.007
Cd	0.12 [b]	0.43 [a]	0.014	<0.001
Nitrite	3.15 [b]	6.96 [a]	1.036	<0.001
Vitamin (μg/L)				
VA	72.54	75.95	1.010	0.089
VB_2	3.69 [b]	4.10 [a]	0.100	0.028
VC	23.23 [b]	27.85 [a]	0.100	0.003
VE	8.17	7.70	0.150	0.114

Note: In the same row, values with no letter or the same letter mean no significant difference ($p \geq 0.05$), while with different small letter mean significant difference ($p < 0.05$). [1] CON: Based on peanut seedling, corn husk and sorghum shell for roughage; TRE: Based on the mixed silage for roughage. [2] SEM: Standard Error of Mean. DON: deoxynivalenol; AFT: aflatoxin; OTA: ochratoxin; ZEN: Zearalenone; DDT: Dichlorodiphenyltrichloroethane; As: arsenic; Pb: Plumbum; Hg: Hydrargyrum; Cr: Chromium; Cd: Cadmium; VA: Vitamin A; VB_2: Vitamin B_2; VC: Vitamin C; VE: Vitamin E.

3.2. Growth Performance and Nutrient Digestibility of Hu Sheep

As can be seen from Figure 1, there were no differences between the CON group and the TRE group in terms of ADG (Figure 1A), DMI (Figure 1B) and F/G (Figure 1C) ($p > 0.05$). There was no effect on DMD, OMD, CPD, NDFD and ADFD of Hu sheep between the CON and TRE group ($p > 0.05$) (Figure 1D).

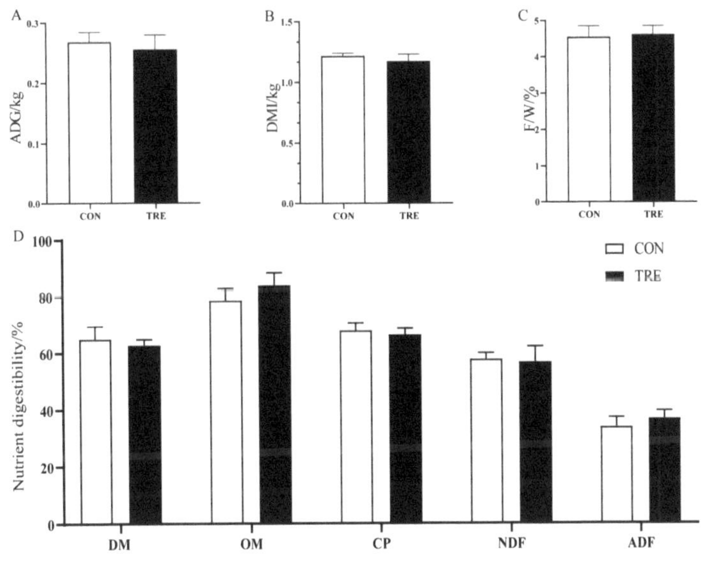

Figure 1. Effect of mixed silage on growth performance and nutrient digestibility of Hu sheep. (**A**) Average daily gain. (**B**) Dry matter intake. (**C**) the feed-to-weight ratio. (**D**) Nutrient digestibility. CON: Based on peanut seedling, corn husk and sorghum shell for roughage; TRE: Based on the mixed silage for roughage. DMI: Dry matter intake; ADG: the average daily gain; F/W: the feed-to-weight ratio; DM: Dry matter; OM: Organ matter; CP: Crude protein; NDF: Neutral Detergent Fiber; ADF: Acid Detergent Fiber.

3.3. Organs Index

As can be seen from Table 3, the organ indices of Hu sheep in the CON group were not different from those of the CON group ($p > 0.05$).

Table 3. Effects of the mixed silage on organ indices of Hu sheep.

Items	Groups [1]		SEM [2]	p
	CON	TRE		
Spleen weight (g)	78.73	67.01	20.599	0.674
Heart weight (g)	152.20	161.45	18.367	0.376
Liver weight (g)	654.38	591.99	97.661	0.143
Pancreas weight (g)	171.39	155.21	43.293	0.228
Lung weight (g)	431.20	450.56	113.679	0.682
Kidney weight (g)	112.21	100.38	20.764	0.676
Spleen index (g/kg)	1.80	1.57	0.413	0.370
Heart index (g/kg)	3.49	3.82	0.359	0.430
Liver index (g/kg)	15.03	13.85	1.570	0.312
Pancreatic index (g/kg)	3.89	3.67	0.851	0.561
Lung index (g/kg)	9.96	10.68	2.699	0.792
Kidney index (g/kg)	2.59	2.37	0.466	0.369

Note: [1] CON: Based on peanut seedling, corn husk and sorghum shell for roughage; TRE: Based on the mixed silage for roughage. [2] SEM: Standard Error of Mean.

3.4. Complex Stomach Development

The basal and granular layers of the rumen in the TRE group were higher than those in the CON group ($p < 0.05$) (Figure 2). It can be seen from Table 4 that the weight of omasum, the ratio of omasum to live weight before slaughter, the amount of compound stomach, and the ratio of compound stomach to live weight before slaughter were higher in the CON group ($p < 0.05$).

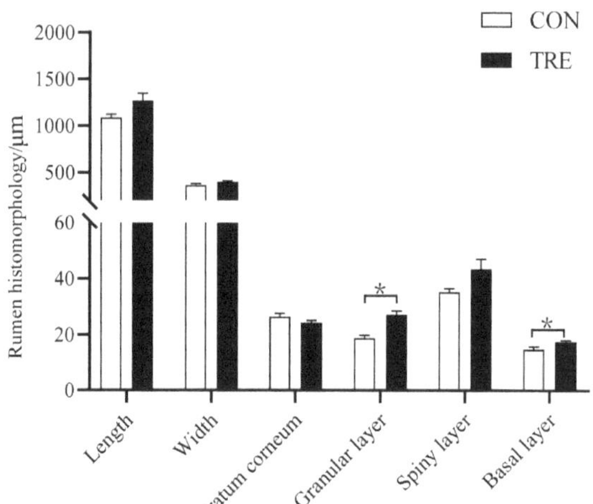

Figure 2. Effects of the mixed silage feed on rumen epithelium of Hu sheep. CON: Based on peanut seedling, corn husk, and sorghum shell for roughage; TRE: Based on the mixed silage for roughage. * indicates a significant difference between the two groups.

Table 4. Effects of the mixed silage on the development of compound stomach in Hu sheep.

Items	Groups [1]		SEM [2]	p
	CON	TRE		
rumen weight (g)	764.86	692.33	64.651	0.058
reticulum weight (g)	123.84	102.92	22.334	0.125
omasum weight (g)	150.09	116.91	24.259	0.014
abomasum weight (g)	216.72	195.20	20.078	0.072
complex stomachs weight (g)	1255.51	1107.36	110.821	0.017
rumen weight/complex stomachs weight (%)	0.61	0.62	0.019	0.206
reticulum weight/complex stomachs weight (%)	0.10	0.09	0.013	0.532
omasum weight/complex stomachs weight (%)	0.12	0.11	0.012	0.063
abomasum weight/complex stomachs weight (%)	0.17	0.18	0.014	0.648
rumen weight/LWBS (%)	17.65	16.36	1.389	0.131
reticulum weight/LWBS (%)	2.82	2.43	0.348	0.055
omasum weight/LWBS (%)	3.43	2.76	0.437	0.004
abomasum weight/LWBS (%)	5.03	4.65	0.694	0.392
complex stomachs weight/LWBS (%)	28.93 [a]	26.20 [b]	2.284	0.041

Note: In the same row, values with no letter or the same letter mean no significant difference ($p \geq 0.05$), while with different small letter mean significant difference ($p < 0.05$). [1] CON: Based on peanut seedling, corn husk and sorghum shell for roughage; TRE: Based on the mixed silage for roughage. [2] SEM: Standard Error of Mean.

3.5. Gut Development

The results in Table 5 showed that there was no difference between the CON group and the TRE group in the weight and index of duodenum, jejunum, ileum, cecum, colon, and rectum ($p > 0.05$).

Table 5. Effects of the mixed silage on intestinal development of Hu sheep.

Items		Group [1]		SEM [2]	p
		CON	TRE		
Gut weight (g)	Duodenum	33.93	31.06	2.591	0.061
	jejunum	63.75	59.42	3.811	0.054
	Ileum	218.07	199.46	18.471	0.095
	Cecum	204.27	193.55	18.013	0.347
	Colon	137.16	131.55	11.897	0.461
	Rectum	267.87	240.74	26.858	0.094
Gut index (g/Kg)	Duodenum	0.78	0.74	0.079	0.371
	jejunum	1.46	1.42	0.182	0.789
	Ileum	5.03	4.75	0.457	0.340
	Cecum	4.68	4.61	0.541	0.842
	Colon	3.15	3.13	0.382	0.925
	Rectum	6.15	5.73	0.817	0.418

Note: [1] CON: Based on peanut seedling, corn husk and sorghum shell for roughage; TRE: Based on the mixed silage for roughage. [2] SEM: Standard Error of Mean.

3.6. Blood Physiological and Biochemical Indicators

The effects of the mixed silage on blood physiological and biochemical indicators of Hu sheep are shown in Table 6. The results showed that the number of red blood cells in the blood of the TRE group was higher than the CON group ($p < 0.05$). The numbers of red blood cells in the CON group and the TRE group were both within the normal range.

Table 6. Effects of the mixed silage on the physicochemical indexes of the blood of Hu sheep.

Items	Groups [1]		SEM [2]	p	Reference Value
	CON	TRE			
Leukocytes/(10^{-9})	11.53	11.50	3.430	0.990	5.1–15.8
Erythrocytes/(10^{-12})	10.43 ^{b}	11.52 ^{a}	0.765	0.010	5.5–14.2
Hemoglobin/(g/L)	133.33	136.91	14.107	0.695	63–132
Platelets/(10^{-9})	348.33	329.63	138.372	0.835	178–462
Basophil absolute value/(10^{-9})	0.06	0.03	0.040	0.231	0–0.17
Basophil percentage/(%)	0.46	0.35	0.187	0.337	0–1.5
Erythrocyte pressure/(L/L)	40.74	38.96	3.296	0.398	20–39
Mean red blood cell volume/(f1)	37.62	36.23	2.314	0.342	-
Mean hemoglobin content/(pg)	12.78	12.95	0.887	0.765	9.2–11.0
Mean hemoglobin concentration/(g/L)	324.84	353.03	30.946	0.137	290.0–360.0
Erythrocyte distribution width/(%)	22.77	21.11	3.081	0.395	-
Standard deviation of erythrocyte distribution/(%)	21.75	20.50	1.359	0.131	-
Total protein/(g/L)	66.64	71.85	6.531	0.200	57–91
Albumin/(g/L)	20.70	21.08	1.148	0.602	27.0–45.5
Globulin/(g/L)	46.01	48.69	5.915	0.479	16.7–48.5
Leukocyte ratio	0.45	0.52	0.095	0.227	0.7–1.6
Alkaline phosphatase/(U/L)	243.98	190.19	63.644	0.171	69.5–125.0
Lactate dehydrogenase/(U/L)	812.64	613.77	164.866	0.038	44–112
Creatinine (CREA)	46.77	65.16	16.677	0.063	-
Urea nitrogen/(mmol/L)	8.45	8.59	0.794	0.780	-
Total cholesterol/(mmol/L)	1.58	1.62	0.250	0.805	1.3–1.9
Triglycerides/(mmol/L)	0.27	0.27	0.112	0.909	-
High-density lipoprotein/(mmol/L)	0.77	0.74	0.141	0.742	-
Low-density lipoprotein/(U/L)	0.57	0.57	0.127	0.952	2.7–3.1

Note: In the same row, values with no letter or the same letter mean no significant difference ($p \geq 0.05$), while with different small letter mean significant difference ($p < 0.05$). [1] CON: Based on peanut seedling, corn husk and sorghum shell for roughage; TRE: Based on the mixed silage for roughage. [2] SEM: Standard Error of Mean.

3.7. Immunological Performance

As can be seen from Figure 3, the levels of IL-10 and TNF-α in the serum of the TRE group were higher than the CON group ($p < 0.05$). From Figure 2, it can be seen that the content of IL-6 and IgA in the duodenum of the TRE group was higher than that of the CON group ($p < 0.05$), and the content of TNF-α in the rumen was higher than the CON group ($p < 0.05$).

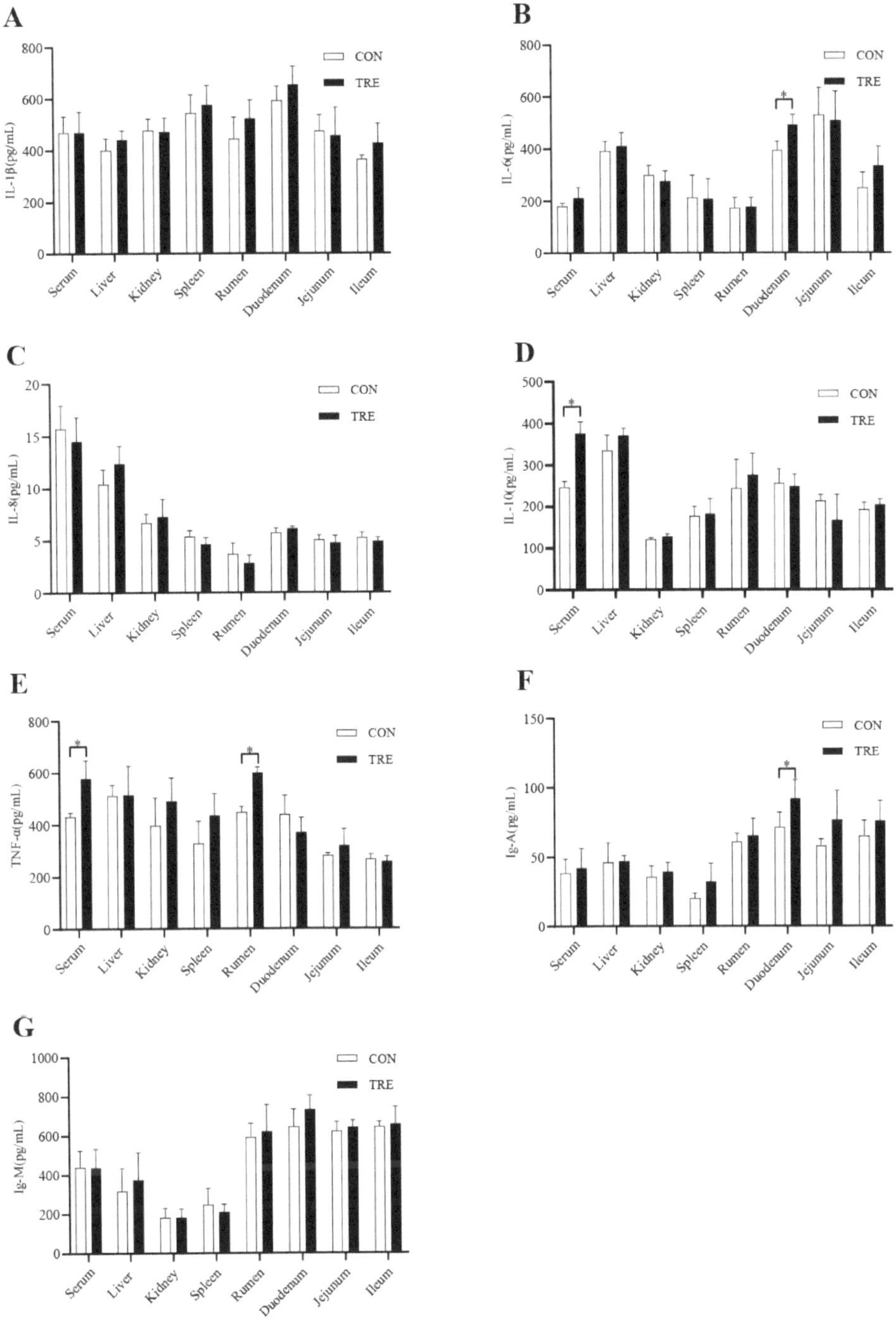

Figure 3. Effects of the mixed silage on immune performance of Hu sheep. (**A**) IL-1β. (**B**) IL-6. (**C**) IL-8. (**D**) IL-10. (**E**) TNF-α. (**F**) Ig-A. (**G**) Ig-M. CON: Based on peanut seedling, corn husk and sorghum shell for roughage; TRE: Based on the mixed silage for roughage. * indicates a significant difference between the two groups.

4. Discussion

Fungal spoilage and mycotoxin contamination are among the greatest risks of silage. Spoilage caused by fungi may lead to mold and heat, reduced palatability and loss of nutritional value of the feed [21]. Mycotoxins are anti-nutritional factors present in livestock and poultry feeds that cause mold-infected diseases and can directly harm animal health and performance [22]. In addition, in order to significantly increase the yield of crops, pesticides may be used in large quantities during cultivation, which may result in drug residues in crops [23]. Feeding livestock with high pesticide residue content may seriously endanger the lives of livestock [24]. Before silage, we also considered that Chinese cabbage has a high water content and is prone to degradation and spoilage, and its mixed storage with rice straw may be susceptible to spoilage due to high water content [25]. Therefore, our evaluation of the safety performance of mixed silage includes mycotoxins, drug residues, heavy metals, etc. The risk of fungal contamination exists during growth, before and after mowing, ensiling, transport and storage, but the level of mycotoxins can be reduced by silage [26–28]. Therefore, the mixed silage may be due to the reduction of OTA and ZEN in the feed by means of silage compared to the CON group. In addition, the levels of Pb, Cr, Cd and nitrite in the feeds of the TRE group were higher than the CON group. This may be caused by the excessive use of pesticides and nitrogen fertilizers in the process of production of rice and vegetables in the excessive pursuit of yield. The content of mycotoxins, drug residues, heavy metals and nitrite in the mixed silage were within the Chinese feed hygiene standards (GB13078-2017). It is worth noting that the VC and VB content of the mixed silage was significantly higher than the CON group, probably due to the higher vitamin content in Chinese cabbage [29,30].

Feeding experiments are necessary for the evaluation of forage resources and the response of the animal organism is the most realistic and reliable. The nutritional value of a forage is mainly determined by the digestibility of the ruminant, which depends mainly on the nutrient content of the forage [31,32]. In the present experiment, we determined the effects of the mixed silage on growth performance, digestibility, organ index, intestinal development and immune performance of Hu sheep in feeding experiment.

Complex stomach development is a major challenge for ruminants from birth, and much of this revolves around the development of the rumen [33]. The rumen epithelium (including the stratum corneum, stratum basale, stratum granulosum and stratum spinosum) has the function of absorbing fatty acids, providing the animal with animal metabolic energy, and is an important regulatory mechanism for stabilizing the rumen environment [34,35]. In this experiment, the granular and basolateral layers were higher in the CON group. The basal layer has functional mitochondria and has the metabolic properties of the rumen epithelium, which produces ketones mainly from SCFA [36]. The granular layer is characterized by a linking complex called bridging granules, which act as an osmotic barrier for the rumen epithelium [37–39]. The results from Li et al. [20] show that the mixed silage of Chinese cabbage waste and rice straw increased the rumen butyric acid. Butyric acid can effectively stimulate the proliferation and growth of rumen epithelial cells [40].

Anti-inflammatory factors (IL-10) play an important role in the termination of neuroinflammation, while pro-inflammatory cytokines (IL-1, IL-6, TNF-α, etc.) are involved in the early response to inflammation and can lead to decreased exercise capacity, loss of appetite, diminished diuretic effect, and other adverse effects [41,42]. IL-6 is a pro-inflammatory factor that promotes the activation of T and B cells and is an important effector molecule in the acute inflammatory response [43]. IL-10 is a cytokine with anti-inflammatory properties that plays a central role in infection by limiting the immune response to pathogens, thereby preventing damage to the host [44]. Our result showed that IL-6 content in the duodenum of the TRE group was higher, and IL-10 and TNF-α content in the serum of the TRE group was higher. In addition, the TNF-α content in the rumen of the TRE group was significantly higher. These results suggest that the mixed silage can reduce the damage caused by inflammatory reactions to the body of Hu sheep to some extent, probably due to the higher vitamin content of the mixed silage. It was found that the maturation, proliferation and cellular activity of lymphocytes were inhibited in mice fed a diet deficient in VB, while the addition of VB completely eliminated this inhibition [45].

Immunoglobulins, as part of the immune system, have a very important role in immune regulation and mucosal defense of the host [46]. Serum immunoglobulins can represent key information about host immunity, including IgM, IgA and IgG [47,48]. In our study, the mixed silage significantly increased the level of IgA in the duodenum, which is considered to be an important antibody isoform involved in mucosal surface protection responses [49]. In turn, the integrity of the duodenal mucosa acts as a balance between endogenous or exogenous aggressive factors and some protective mechanisms [50]. Therefore, the mixed silage may have some protective effect on the duodenal mucosa and improve the immune performance of the organism.

5. Conclusions

The levels of pesticide residues, heavy metals, and nitrites were generally lower than mill feed in line with Chinese feed safety production standards. In addition, the VC and VB contents was higher. Feeding experiment have found that the mixed silage do not adversely affect the growth performance and nutrient digestibility of Hu sheep and are beneficial to the development of rumen epithelial cells and improve the immune performance of the body. In conclusion, Chinese cabbage waste as a mixed silage ingredient does not have any impact on feed hygiene and safety and can have a beneficial effect on the healthy development of Hu sheep (Figure 4).

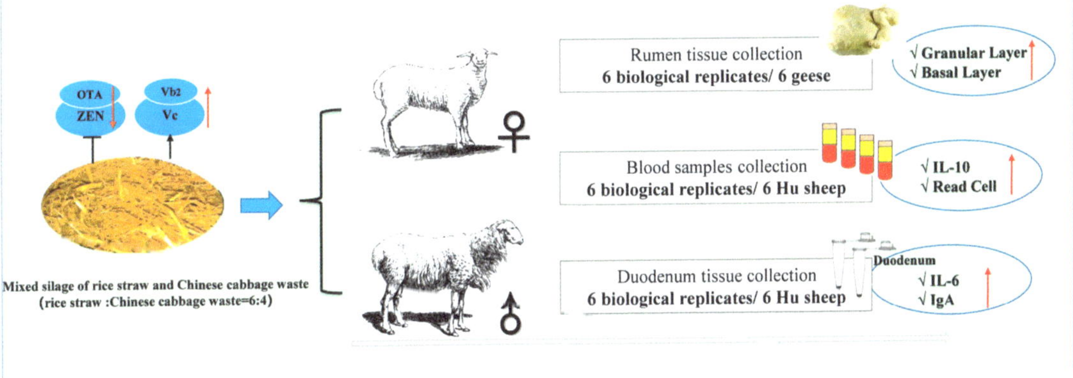

Figure 4. Safety performance evaluation and effects on growth and health performance of the mixed silage.

Author Contributions: Conceptualization, M.W., J.X., Z.L. and C.L.; methodology, J.X., Z.L., Z.Z., Y.L. and L.H.; software, Z.L., Z.Z., C.X. and C.L.; validation, Z.L., C.L., L.H., Z.Z., Y.L., C.X., R.Q. and M.H.Z.; formal analysis, Z.L., Z.Z., C.X. and L.H.; investigation, Z.L., C.L., Y.L. and M.H.Z.; resources, Z.L., R.Q. and L.H.; data curation, Z.L., C.L., R.Q. and M.H.Z.; writing—original draft preparation, Z.L., J.X. and C.L.; writing—review and editing, Z.L., J.X. and M.W.; visualization, J.X. and M.W.; supervision, J.X. and M.W.; project administration, M.W.; funding acquisition, M.W. All authors have read and agreed to the published version of the manuscript.

Funding: This research is funded by the national corps of major scientific and technological "unveiling the list of commanding project" (2023AB078), the key program of state key Laboratory of Sheep Genetic Improvement and Healthy Production (NCG202232), the High End Foreign Expert Project of the Ministry of Science and Technology (G2023014066L), and the Priority Academic Program Development of Jiangsu Higher Education Institution (PAPD), P. R. China.

Institutional Review Board Statement: The animal study was reviewed and approved by the Animal Welfare Committee of Yangzhou Veterinarians of the Agriculture Ministry of China (Yangzhou, China, license no. syxk (Su)2019-0029). Written informed consent was obtained from the owners for the participation of their animals in this study.

Informed Consent Statement: Not applicable.

Data Availability Statement: Data are contained within the article.

Acknowledgments: The author hereby expresses his gratitude to the Siyang County Huifeng Goat Breeding Professional Cooperative (Suqian, Jiangsu) for the Hu sheep and breeding sites provided.

Conflicts of Interest: The authors declare that this study was conducted without any business or financial relationships that could be considered a potential conflict of interest. Lu Zhiqi now works for Ningxia Dairy Science and Innovation Center of Guangming Animal Husbandry Co., Ltd.

References

1. Binod, P.; Sindhu, R.; Singhania, R.R.; Vikram, S.; Devi, L.; Nagalakshmi, S.; Kurien, N.; Sukumaran, R.K.; Pandey, A. Bioethanol production from rice straw: An overview. *Bioresour. Technol.* **2010**, *101*, 4767–4774. [CrossRef] [PubMed]
2. Maiorella, B.L. Ethanol Fermentation. In *Comprehensive Biotechnology*; Moo-Young, M., Ed.; Pergamon Press: Oxford, UK, 1985; Volume III, pp. 861–914.
3. Wang, X.; Song, J.; Liu, Z.; Zhang, G.; Zhang, Y. Fermentation Quality and Microbial Community of Corn Stover or Rice Straw Silage Mixed with Soybean Curd Residue. *Animals* **2022**, *12*, 919. [CrossRef] [PubMed]
4. Pal, D.B.; Tiwari, A.K.; Mohammad, A.; Prasad, N.; Srivastava, N.; Srivastava, K.; Singh, R.; Yoon, T.; Syed, A.; Bahkali, A.H. Enhanced Biogas Production Potential Analysis of Rice Straw: Biomass characterization, Kinetics and Anaerobic Co-Digestion investigations. *Bioresour. Technol.* **2022**, *358*, 127391. [CrossRef] [PubMed]
5. Abraham, A.; Mathew, A.K.; Sindhu, R.; Pandey, A.; Binod, P. Potential of rice straw for bio-refining: An overview. *Bioresour. Technol.* **2016**, *215*, 29–36. [CrossRef]
6. Ma, Y.; Yin, Y.; Liu, Y. A holistic approach for food waste management towards zero-solid disposal and energy/resource recovery. *Bioresour. Technol.* **2017**, *228*, 56–61. [CrossRef]
7. Singh, R.; Patel, M. Effective utilization of rice straw in value-added by-products: A systematic review of state of art and future perspectives. *Biomass Bioenergy* **2022**, *159*, 106411. [CrossRef]
8. Esparza, I.; Jiménez-Moreno, N.; Bimbela, F.; Ancín-Azpilicueta, C.; Gandía, L.M. Fruit and vegetable waste management: Conventional and emerging approaches. *J. Environ. Manag.* **2020**, *265*, 110510. [CrossRef]
9. Idan, F.; Adogla-Bessa, T.; Amaning-Kwarteng, K. Preference, voluntary feed intake, and digestibility of sheep fed untreated rice straw and supplemented with sole or combined fodder tree leaves. *Eur. J. Agric. Food Sci.* **2020**, *2*, 1–8. [CrossRef]
10. Tejaswini, G.S.; Mahadevakumar, S.; Sowmya, R.; Deepika, Y.S.; Meghavarshinigowda, B.R.; Nuthan, B.R.; Sharvani, K.A.; Amruthesh, K.N.; Sridhar, K.R. Molecular detection and pathological investigations on southern blight disease caused by *Sclerotium rolfsii* on cabbage (*Brassica oleracea* var. capitata): A new record in India. *J. Phytopathol.* **2022**, *170*, 363–372. [CrossRef]
11. Cherdthong, A.; Suntara, C.; Khota, W. Lactobacillus casei TH14 and additives could modulate the quality, gas kinetics and the in vitro digestibility of ensilaged rice straw. *J. Anim. Physiol. Anim. Nutr.* **2020**, *104*, 1690–1703. [CrossRef]
12. Partovi, E.; Rouzbehan, Y.; Fazaeli, H.; Rezaei, J. Broccoli byproduct-wheat straw silage as a feed resource for fattening lambs. *Transl. Anim. Sci.* **2020**, *4*, txaa078. [CrossRef] [PubMed]
13. Ren, H.W.; Wang, C.; Fan, W.G.; Zhang, B.Y.; Li, Z.Z.; Li, D. Effects of Formic or Acetic Acid on the Storage Quality of Mixed Air-Dried Corn Stover and Cabbage Waste, and Microbial Community Analysis. *Food Technol. Biotechnol.* **2018**, *56*, 71–82. [CrossRef] [PubMed]
14. Hillion, M.L.; Moscoviz, R.; Trably, E.; Leblanc, Y.; Bernet, N.; Torrijos, M.; Escudie, R. Co-ensiling as a new technique for long-term storage of agro-industrial waste with low sugar content prior to anaerobic digestion. *Waste Manag.* **2018**, *71*, 147–155. [CrossRef] [PubMed]
15. D'mello, J.; Placinta, C.; Macdonald, A. Fusarium mycotoxins: A review of global implications for animal health, welfare and productivity. *Anim. Feed Sci. Tech.* **1999**, *80*, 183–205. [CrossRef]
16. Bueno, D.; Istamboulie, G.; Muñoz, R.; Marty, J.L. Determination of mycotoxins in food: A review of bioanalytical to analytical methods. *Appl. Spectrosc. Rev.* **2015**, *50*, 728–774. [CrossRef]
17. Umapathi, R.; Ghoreishian, S.M.; Sonwal, S.; Rani, G.M.; Huh, Y.S. Portable electrochemical sensing methodologies for on-site detection of pesticide residues in fruits and vegetables. *Coord. Chem. Rev.* **2022**, *453*, 214305. [CrossRef]
18. Sandeep, G.; Vijayalatha, K.; Anitha, T. Heavy metals and its impact in vegetable crops. *Int. J. Chem. Stud.* **2019**, *7*, 1612–1621.
19. National Research Council (NRC). *Nutrient Requirements of Small Ruminants, Sheep, Goats, Cervids, and New World Camelids*; National Academy Press: Washington, DC, USA, 2007.
20. Li, C.; Chen, N.; Zhang, X.; Shahzad, K.; Qi, R.; Zhang, Z.; Lu, Z.; Lu, Y.; Yu, X.; Zafar, M.H. Mixed silage with Chinese cabbage waste enhances antioxidant ability by increasing ascorbate and aldarate metabolism through rumen Prevotellaceae UCG-004 in Hu sheep. *Front. Microbiol.* **2022**, *13*, 978940. [CrossRef]

21. Christensen, C.M. *Storage of Cereal Grains and Their Products*; American Association of Cereal Chemists: Saint Paul, MN, USA, 1974.
22. Sun, X.; Tiffany, D.G.; Urriola, P.E.; Shurson, G.G.; Hu, B. Nutrition upgrading of corn-ethanol co-product by fungal fermentation: Amino acids enrichment and anti-nutritional factors degradation. *Food Bioprod. Process.* **2021**, *130*, 1–13. [CrossRef]
23. Rusinamhodzi, L. *The Role of Ecosystem Services in Sustainable Food Systems*; Academic Press: Cambridge, MA, USA, 2019.
24. Yu, Z.; Fan, X.; Bai, C.; Tian, J.; Tharangani, R.; Bu, D.; Jia, T. Assessment of Forage Safety and Quality. In *Research Progress on Forage Production, Processing and Utilization in China*; Springer: Berlin/Heidelberg, Germany, 2022; pp. 145–181.
25. Chattopadhyay, A. Pre-and post-harvest losses in vegetables IVI. In *Advances in Postharvest Technologies of Vegetable Crops*; Apple Academic Press: Cambridge, MA, USA, 2018; pp. 25–87.
26. Storm, I.; Sørensen, J.L.; Rasmussen, R.R.; Nielsen, K.F.; Thrane, U. Mycotoxins in silage. *Stewart Postharvest Rev.* **2008**, *4*, 1–12.
27. Storm, I.M.L.D.; Kristensen, N.; Raun, B.; Smedsgaard, J.; Thrane, U. Dynamics in the microbiology of maize silage during whole-season storage. *J. Appl. Microbiol.* **2010**, *109*, 1017–1026. [CrossRef] [PubMed]
28. Vandicke, J.; De Visschere, K.; Ameye, M.; Croubels, S.; De Saeger, S.; Audenaert, K.; Haesaert, G. Multi-mycotoxin contamination of maize silages in Flanders, Belgium: Monitoring mycotoxin levels from seed to feed. *Toxins* **2021**, *13*, 202. [CrossRef] [PubMed]
29. Moreb, N.; Murphy, A.; Jaiswal, S.; Jaiswal, A.K. *Cabbage*; Elsevier: Amsterdam, The Netherlands, 2020; pp. 33–54.
30. Mosha, T.; Gaga, H. Nutritive value and effect of blanching on the trypsin and chymotrypsin inhibitor activities of selected leafy vegetables. *Plant Foods Hum. Nutr.* **1999**, *54*, 271–283. [CrossRef] [PubMed]
31. Tilley, J.; Terry, D.R. A two-stage technique for the in vitro digestion of forage crops. *Grass Forage Sci.* **1963**, *18*, 104–111. [CrossRef]
32. Li, R.; Zheng, M.; Jiang, D.; Tian, P.; Zheng, M.; Xu, C. Replacing alfalfa with paper mulberry in total mixed ration silages: Effects on ensiling characteristics, protein degradation, and in vitro digestibility. *Animals* **2021**, *11*, 1273. [CrossRef]
33. Vi, R.B.; McLeod, K.; Klotz, J.; Heitmann, R. Rumen development, intestinal growth and hepatic metabolism in the pre-and postweaning ruminant. *J. Dairy Sci.* **2004**, *87*, E55–E65.
34. Gäbel, G.; Aschenbach, J.; Müller, F. Transfer of energy substrates across the ruminal epithelium: Implications and limitations. *Anim. Health Res. Rev.* **2002**, *3*, 15–30. [CrossRef]
35. Steele, M.A.; Croom, J.; Kahler, M.; AlZahal, O.; Hook, S.E.; Plaizier, K.; McBride, B.W. Bovine rumen epithelium undergoes rapid structural adaptations during grain-induced subacute ruminal acidosis. *Am. J. Physiol.-Regul. Integr. Comp. Physiol.* **2011**, *300*, R1515–R1523. [CrossRef]
36. Leighton, B.; Nicholas, A.R.; Pogson, C.I. The pathway of ketogenesis in rumen epithelium of the sheep. *Biochem. J.* **1983**, *216*, 769–772. [CrossRef]
37. VI, R.B. Use of isolated ruminal epithelial cells in the study of rumen metabolism. *J. Nutr.* **1998**, *128*, 293S–296S.
38. Gaebel, G.; Martens, H.; Bell, M. The effect of low mucosal pH on sodium and chloride movement across the isolated rumen mucosa of sheep. *Q. J. Exp. Physiol. Transl. Integr.* **1989**, *74*, 35–44. [CrossRef] [PubMed]
39. Graham, C.; Simmons, N.L. Functional organization of the bovine rumen epithelium. *Am. J. Physiol.-Regul. Integr. Comp. Physiol.* **2005**, *288*, R173–R181. [CrossRef] [PubMed]
40. Górka, P.; Kowalski, Z.; Zabielski, R.; Guilloteau, P. Invited review: Use of butyrate to promote gastrointestinal tract development in calves. *J. Dairy Sci.* **2018**, *101*, 4785–4800. [CrossRef] [PubMed]
41. Arkhipov, V.I.; Pershina, E.V.; Levin, S.G. The role of anti-inflammatory cytokines in memory processing in a healthy brain. *Behav. Brain Res.* **2019**, *367*, 111–116. [CrossRef] [PubMed]
42. Moore, K.W.; de Waal Malefyt, R.; Coffman, R.L.; O'Garra, A. Interleukin-10 and the interleukin-10 receptor. *Annu. Rev. Immunol.* **2001**, *19*, 683. [CrossRef] [PubMed]
43. Choy, E. Understanding the dynamics: Pathways involved in the pathogenesis of rheumatoid arthritis. *Rheumatology* **2012**, *51*, v3–v11. [CrossRef] [PubMed]
44. Saraiva, M.; O'garra, A. The regulation of IL-10 production by immune cells. *Nat. Rev. Immunol.* **2010**, *10*, 170–181. [CrossRef]
45. Trakatellis, A.; Dimitriadou, A.; Exindari, M.; Scountzou, J.; Koliakos, G.; Christodoulou, D.; Malissiovas, N.; Antoniadis, A.; Polyzoni, T. Effect of pyridoxine deficiency on immunological phenomena. *Postgrad. Med. J.* **1992**, *68*, S70–S77.
46. Wang, H.; Liu, Z.; Huang, M.; Wang, S.; Cui, D.; Dong, S.; Li, S.; Qi, Z.; Liu, Y. Effects of long-term mineral block supplementation on antioxidants, immunity, and health of Tibetan sheep. *Biol. Trace Elem. Res.* **2016**, *172*, 326–335. [CrossRef]
47. Gonzalez-Quintela, A.; Alende, R.; Gude, F.; Campos, J.; Rey, J.; Meijide, L.; Fernandez-Merino, C.; Vidal, C. Serum levels of immunoglobulins (IgG, IgA, IgM) in a general adult population and their relationship with alcohol consumption, smoking and common metabolic abnormalities. *Clin. Exp. Immunol.* **2008**, *151*, 42–50. [CrossRef]
48. Greene, D.M.; Bondy, G.S.; Azcona-Olivera, J.I.; Pestka, J.J. Role of gender and strain in vomitoxin-induced dysregulation of IgA production and IgA nephropathy in the mouse. *J. Toxicol. Environ. Health Part A Curr. Issues* **1994**, *43*, 37–50. [CrossRef] [PubMed]
49. Ellis, J.A. Canine IgA and IgA deficiency: Implications for immunization against respiratory pathogens. *Can. Vet. J.* **2019**, *60*, 1305. [PubMed]
50. Abdel-Salam, O.M.; Czimmer, J.; Debreceni, A.; Szolcsányi, J.; Mózsik, G. Gastric mucosal integrity: Gastric mucosal blood flow and microcirculation. An overview. *J. Physiol.* **2001**, *95*, 105–127. [CrossRef] [PubMed]

fermentation

MDPI

Article

Effect of Yogurt Acid Whey on the Quality of Maize Silage

Irida Palamidi [1], Vasileios V. Paraskeuas [1], Basiliki Kotsampasi [2], Ioannis Hadjigeorgiou [1], Ioannis Politis [3] and Konstantinos C. Mountzouris [1,*]

[1] Laboratory of Nutritional Physiology and Feeding, Department of Animal Science, Agricultural University of Athens, 11855 Athens, Greece; inpalamidi@aua.gr (I.P.); v.paraskeuas@aua.gr (V.V.P.); ihadjig@aua.gr (I.H.)
[2] Research Institute of Animal Science, ELGO 'DIMITRA', Paralimni Giannitsa, 58100 Pella, Greece; vkotsampasi@elgo.gr
[3] Laboratory of Animal Breeding & Husbandry, Department of Animal Science, Agricultural University of Athens, 11855 Athens, Greece; i.politis@aua.gr
* Correspondence: kmountzouris@aua.gr

Abstract: The increasing popularity of Greek yogurt generates large amounts of acid whey worldwide. The use of yogurt acid whey in animal nutrition is limited. The aim of this study was to determine the effect of a yogurt acid whey powder (YAWP) addition to maize forage prior to ensiling on the nutritional, microbial and fermentation quality of maize silage. Depending on the addition level of the YAWP to maize forage, there were the following four experimental treatments: YAWP 0, 2.5, 5 and 10% *w*/*w*. An increasing YAWP inclusion level linearly increased the maize silage dry matter, crude protein and ash concentrations, whereas it reduced the crude fiber, neutral-detergent fiber and acid-detergent fiber concentrations. The silage pH decreased quadratically with the increasing YAWP level, with the lower plateau noted for the YAWP 5% addition. Concentrations of total bacteria in the silage and *Lactobacillus* spp. decreased linearly with the YAWP increase. The silage acetic acid content decreased linearly, whereas propionic acid, lactic acid and the ratio of lactic to acetic acid increased linearly with the increasing YAWP level. The ammonia-N content decreased linearly with the increasing YAWP level. In conclusion, the incorporation of the 5 and 10% YAWP addition in silage preparation improved the nutritional and fermentative quality of the produced silage.

Keywords: maize silage; yogurt acid whey; ammonia-N; pH; lactic acid; volatile fatty acids

Citation: Palamidi, I.; Paraskeuas, V.V.; Kotsampasi, B.; Hadjigeorgiou, I.; Politis, I.; Mountzouris, K.C. Effect of Yogurt Acid Whey on the Quality of Maize Silage. *Fermentation* **2023**, *9*, 994. https://doi.org/10.3390/fermentation9120994

Academic Editors: Wansup Kwak and Siran Wang

Received: 5 October 2023
Revised: 9 November 2023
Accepted: 14 November 2023
Published: 22 November 2023

1. Introduction

Ensiling is widely used as a fresh-forage-preservation method all around the world. Silages are commonly used as livestock feed because they are nutritious and can be available during the whole year. Silage ensures a high milk yield in ruminants [1,2], the beneficial modification of the milk fatty acid profile [3,4] and meat quality [5,6]. Moreover, silages keep the animals healthy when pasture growth is inadequate in relation to animal requirements [7]. Among the types of silages used in ruminant feeding, maize silage is the most common, especially in dairy cattle diets, since, despite its relatively low protein content, it is a low-cost fiber and energy feed [2,8].

Silage quality depends on various preharvest and postharvest factors [9,10] that can largely affect the final product both nutritionally and/or microbiologically. Due to this, microbial inoculants composed of lactic acid bacteria [11–13] or acids, such as formic, propionic or citric acid [7,14,15], are also incorporated into the forage biomass prior to the ensiling process for the preparation of the fermentation mixture, aiming to ensure the proper development of silage fermentation, the reduction of nutrient losses and the inhibition of the growth of undesirable microorganisms.

Byproducts from the food industry (e.g., molasses or cheese whey) have also served as alternative additives in ensiling processes [16,17]. Dairy byproducts, such as whey (produced by cheese), have been applied in alfalfa silage [17] and maize silage [18], improving the quality of silages, such as their chemical composition. Acid whey (produced from cheese) was used as a rehydration agent of maize grains, which resulted in the enhancement of the fermentation process and the aerobic stability of the feed after 30 days of ensiling [19].

Acid whey is a byproduct from the production of acid-coagulated cheeses, but also from strained (Greek) yogurt. In Greece, strained yogurt is a high-nutritional and consumer-desirable product with growing consumption [20]. The increased demand for strained yogurt creates a considerable volume of acid whey, as for every 100 L of milk used, 70 L of yogurt acid whey is produced [20] that must be either disposed of or repurposed. Recent research has focused on alternative uses for yogurt acid whey in food processing [21–23] and in microbial cultivation [24]. Given the main ingredients of yogurt acid whey, including lactose (40 g/kg), organic acids (lactic acid, 6.5 g/kg, citric acid 1.8 g/kg) functional proteins (e.g., β-lactoglobulin, α-lactalbumin, lactoferrin) and minerals (potassium, >1500 mg/kg, calcium, >1200 mg/kg and phosphorus, >600 mg/kg), with pHs of 3.5–4.5 [21,25], its use as a silage additive/component merits investigation.

The aim of this study was to determine the effect of a yogurt acid whey powder (YAWP) addition on the quality of maize silage. For this purpose, four YAWP addition levels (0, 2.5, 5 and 10% *w*/*w*) to maize forage material prior to ensilation were used, and the nutritional, microbiological and fermentation quality of the plant material and YAWP mix before and after ensiling were conducted.

2. Materials and Methods

2.1. Preparation of Crop Maize Silage In Silo Bags and Sampling

Maize (Zea mays) forage material was provided by the Research Institute of Animal Science, ELGO 'DIMITRA' (Paralimni Giannitsa, 58100 Pella, Greece, 40°44′ N, 22°27′ E), in early September 2022. The maize hybrid Leonidas (700 FAO; 135 days maturation; KWS, Italia S.p.A.) was sown in late April in the experimental fields of the Institute according to the soil type, which were characterized as clay to sandy clay and well-drained. The applied fertilization program included 400 kg/ha of N-P-K (20-10-10) and 400 kg/ha of urea (40-0-0), and an irrigation program of 6 waterings, 7 min per meter at each watering, using a hose reel traveler. The fresh maize biomass was harvested at the one-half milk-line of kernel maturity using a farm-scale maize-forage harvester (Jaguar 900, CLAAS KGaA mbH, Harsewinkel, Germany) 20 cm above the ground level, which was simultaneously chopped to an approximate particle size of 2 cm. On the same day as the harvest, the maize forage was transported to the facilities of the Department of Nutritional Physiology and Feeding at the Agricultural University of Athens (Greece).

After receiving the maize raw material in the laboratory, the experimental treatments (0, 2.5, 5 and 10% *w*/*w*) were applied to separate batches, and 7 replicates per treatment were prepared from each batch. Then, the material of the replicates was placed in appropriate vacuum-tight airbags, thus producing 7 replicate bags per experimental treatment. For each experimental treatment, prior to ensiling, from each bag (7.5 kg of the fresh maize forage batch), a 0.5 kg sample was taken and stored at −30 °C until further analysis. Therefore, all 28 bags (7.0 kg of the fresh maize forage batch) were then stored at an ambient temperature (18 ± 2 °C) for 10 weeks. The yogurt acid whey powder composition (g/kg) is presented in Table 1. At the end of the ensiling experiment (70 days), 800 g of each ensiled batch sample was taken from the center of the bag in an aseptic way, and subsequently placed in a sterile bag and immediately stored at −30 °C until further analysis.

Table 1. Yogurt acid whey powder composition (g/kg).

Lactose	720
Ash	110
Protein	50
Lactic acid	60
Dry matter	960

2.2. Chemical Composition of the Maize Forage Samples

All maize forage samples were analyzed for dry matter (DM) using the oven-drying method (65 °C for 24 h until constant weight) according to Acosta Aragón et al. [26]. Subsequently, samples were analyzed for crude fiber (CF), ether extract (EE), crude protein (CP; determined as 6.25 × Kjeldahl nitrogen) and ash using the routine procedures [27]. In addition, neutral-detergent fiber (NDF), acid-detergent fiber (ADF) and acid-detergent lignin (ADL) were determined according to Mertens [28] and Mollers [29] using an Ankom 2000 fiber analyzer (Ankom Technology Ltd., Macedon, NY, USA).

2.3. Microbial Analyses of the Maize Forage Samples

For the microbial analyses, 5 g samples of the maize forage samples were diluted 1:10 in sterile ice-cold anoxic PBS (0.1 M; pH of 7.0) and subsequently homogenized for 3 min in a stomacher (Bagmixer 100 Minimix, Interscience, Arpents, France). The suspension was serially diluted (10^{-1} to 10^{-5}) in buffered peptone water (Neogen Heywood, Bury, UK), and each dilution was spread in duplicate onto plate count agar (PCA, Neogen Heywood, UK) for the total bacteria enumeration; de Man, Rogosa and Sharpe (MRS, Neogen Heywood, UK) for the enumeration of *Lactobacillus* spp. and reinforced clostridial agar (RCA, Neogen Heywood, UK) with the addition of neutral red (0.005%) and D-cycloserine (200 ppm) for the enumeration of *Clostridium* spp. (Jonsson 1989); violet red bile agar (Oxoid Basingstoke, Basingstoke, UK) for coliforms; Sabouraud dextrose agar (SDA, Neogen Heywood, UK) with the addition of chloramphenicol supplements (Oxoid Basingstoke, UK) for the yeast and molds. PCA and SDA were incubated at 30 °C, while MRS was incubated at 37 °C, aerobically. RCA was incubated at 30 °C anaerobically. Results were expressed as log10 colony-forming units per gram silage DM.

2.4. Determination of the Fermentation Quality

Firstly, the pH was determined by direct sample measurement. Samples were diluted (1:2 w/v) in sterile deionized water and subsequently homogenized in a stomacher (Bagmixer 100 Minimix, Interscience, Arpents, France). The pH was measured by a glass electrode pH/ATC electrode #300729.1 (Denver Instrument GmbH, Göttingen, Germany).

For the determination of the ammonia-N and lactic acid contents, a fresh silage sample was diluted in sterile deionized water and subsequently homogenized in a stomacher (Bagmixer 100 Minimix, Interscience, Arpents, France). The resulting supernatants were used for the enzymatic determination of the ammonia-N and lactic acid contents using commercial kits, (kit K-AMIAR and K-DLATE, respectively; Megazyme, Bray, Ireland).

For the determination of the volatile fatty acid (VFA) concentration, samples were homogenized following a 2-fold dilution (i.e., 1:2 w/v) in sterile ice-cold phosphate-buffered saline (0.1 mol/L, pH of 7.0). Homogenates were subsequently centrifuged at 12,000× g for 10 min at 4 °C, and the resulting supernatants were stored at 80 °C until their analysis by capillary gas chromatography (GC) using an Agilent 6890 GC system, equipped with a 30 m 0.25 mm i.d. Nukol column (Supelco, Sigma-Adrich, St Louis, MO, USA) and a flame ionization detector (FID). The analysis was isothermal (185 °C), and the temperatures of the injector and FID were set at 185 and 200 °C, according to Mountzouris et al. [30]. The VFAs determined were acetic, propionic, isobutyric, butyric and heptanoic acids. Results were expressed as g/kg of the silage DM.

2.5. Statistical Analysis

Data for the nutritional, microbiological and fermentation quality were analyzed using the SPSS statistical package (version 17.0). Prior to analysis, data were tested for normality using Kolmogorov–Smirnov's test. Dependent variables that were not normally distributed were analyzed following appropriate transformation. The effect of four YAWP addition levels (0, 2.5, 5 and 10% w/w) to the quality of maize forage material ensilation was evaluated by one-way ANOVA, comparing the means of the 4 independent groups. The statistical model of the above described analysis is as follows:

$$Y_{ij} = \mu + T_i + \varepsilon_{ij}$$

where Y_{ij} is observation j in treatment i, μ is the overall mean, T_i is the fixed effect of treatment (YAWP addition), and ε_{ij} is the residual error. The vacuum-tight airbag was considered as the experimental unit. Statistically significant effects were further analyzed, and means were compared using Tukey's honestly significant difference (HSD) multiple comparison procedure. Statistical significance was determined at $p < 0.05$. The linear and quadratic effects of the dietary YAWP inclusion level were studied using polynomial contrasts.

3. Results

3.1. Chemical Composition and Epiphytic Microflora of the Maize Forage before Ensiling

The chemical composition and epiphytic microflora of the freshly chopped maize forage are shown in Table 2. Before ensiling, the counts of epiphytic microbes in the fresh maize, such as total bacteria, *Lactobacillus* spp., *Clostridium* spp. and yeasts–molds, were 7.27, 7.29, 3.90, 5.64 and 2.65 log cfu/g DM, respectively.

Table 2. Analyzed chemical and microbial composition of the fresh maize forage.

Nutritional Quality	**Average ± Std**
Dry matter (g/kg)	261.7 ± 4.87
Crude protein (g/kg DM)	81.0 ± 1.61
Ether Extract (g/kg DM)	22.5 ± 2.30
Crude fiber (g/kg DM)	211.0 ± 9.93
Neutral-detergent fiber (g/kg DM)	429.1 ± 19.36
Acid-detergent fiber (g/kg DM)	244.6 ± 1.41
Acid-detergent lignin (g/kg DM)	14.5 ± 1.11
Microbial Quality (log cfu/g DM)	
Total bacteria	7.27 ± 0.612
Lactobacillus spp.	7.29 ± 0.680
Clostridium sp.	3.90 ± 0.286
Yeast, Molds	5.64 ± 1.084
Coliforms	2.65 ± 1.169

DM: Dry matter.

3.2. Nutritional Quality of the Maize Forage Silages Prepared with the YAWP

The nutritional quality of the maize silage was affected significantly by the YAWP inclusion (Table 3). The DM content was significantly higher ($P_{anova} < 0.001$) in the 10% YAWP addition compared with the 0% and 2.5% YAWP additions, whereas the 5% YAWP addition received intermediate values. The CP content was significantly higher ($P_{anova} \leq 0.05$) in the 10% YAWP addition in comparison with the 0% addition, whereas the 2.5 and 5% additions were intermediate. Ether extract content was not affected ($P_{anova} > 0.05$) by the addition of the YAWP. Addition of the 10% YAWP significantly reduced ($P_{anova} > 0.01$) the silage CF and NDF contents compared with the other treatments, whereas the ADF content was significantly lower ($P_{anova} > 0.05$) in the 10% YAWP addition compared with the 0% and 2.5% YAWP additions, with the 5% addition being intermediate. The YAWP addition did not affect the ADL content ($P_{anova} > 0.05$). The ash content was significantly higher ($P_{anova} < 0.01$) in the 10% YAWP addition, followed by the 5% addition, compared with the 0% addition, while the 2.5% YAWP-addition silage received intermediate values and differed only from the 10% YAWP addition.

Table 3. Nutritional quality of the maize silages prepared with the yogurt acid whey powder (YAWP).

	Inclusion Level of YAWP [a]				Statistics			
	0%	2.5%	5%	10%	SEM [b]	P_{anova}	Polynomial Contrast [c]	
							P_{linear}	$P_{quadratic}$
Dry matter (DM) %	26.09 [a]	26.62 [ab]	27.30 [b]	28.73 [c]	0.388	***	***	NS
Crude protein (% DM)	7.85 [a]	7.96 [ab]	7.96 [ab]	8.08 [b]	0.769	*	*	NS
Ether extract (% DM)	2.29	2.27	2.30	2.07	0.144	NS	NS	NS
Crude fiber (% DM)	21.93 [b]	22.32 [b]	22.77 [b]	19.81 [a]	0.837	**	*	**
Neutral-detergent fiber (% DM)	44.53 [b]	42.84 [b]	42.14 [b]	38.31 [a]	1.478	**	***	NS
Acid-detergent fiber (% DM)	24.43 [b]	24.23 [b]	23.38 [ab]	21.13 [a]	1.148	*	**	NS
Acid-detergent lignin (% DM)	1.74	1.80	1.75	1.68	0.205	NS	NS	NS
Ash (% DM)	4.66 [a]	5.62 [ab]	6.29 [bc]	7.53 [c]	0.532	**	***	NS

[a] Data represent treatment means from seven replicates per treatment. Means within a row followed by the same letter are not significantly different from each other ($p > 0.05$). *** $p < 0.001$; ** $p < 0.01$; * $p \leq 0.05$; NS, not significant. [b] SEM, pooled standard error of means. [c] Polynomial contrasts test the linear and quadratic effects of the YAWP inclusion levels in the silage preparation.

Polynomial contrast results showed that the DM, CP and ash increased linearly ($P_{linear} < 0.001$, $P_{linear} < 0.05$ and $P_{linear} < 0.001$, respectively) and the NDF and ADF decreased linearly ($P_{linear} < 0.001$ and $P_{linear} < 0.01$, respectively) with the increasing YAWP addition level in the maize forage material. In addition, the CF decreased in both a linear and quadratic manner ($P_{linear} < 0.05$ and $P_{quadratic} < 0.01$, respectively) with the increasing YAWP level in the silage.

3.3. Microbial Quality of the Maize Silages Prepared with the YAWP

The effects of the YAWP addition on the microbial populations of the silage are presented in Table 4. Total bacteria were significantly higher in the 0% and 2.5% YAWP additions compared with the 5% and 10% YAWP additions ($P_{anova} < 0.001$). The addition of 5% and 10% YAWP in the silage decreased the *Lactobacillus* spp. concentration in comparison with the 0% and 2.5% YAWP. The *Clostridium* spp. concentration was not affected by the YAWP addition in the silage. Yeast and mold values were not affected ($P_{anova} > 0.05$) either by the YAWP addition.

Table 4. Microbial quality (log cfu/g dry matter) of the maize silages prepared with the yogurt acid whey powder (YAWP).

	Inclusion Level of YAWP [a]				Statistics			
	0%	2.5%	5%	10%	SEM [b]	P_{anova}	Polynomial Contrast [c]	
							P_{linear}	$P_{quadratic}$
Total bacteria	7.95 [b]	7.71 [b]	6.93 [a]	6.73 [a]	0.160	***	***	NS
Lactobacillus spp.	7.90 [b]	7.90 [b]	7.23 [a]	7.09 [a]	0.191	***	***	NS
Clostridium spp.	3.36	3.56	3.47	3.42	0.241	NS	NS	NS
Yeast, Molds	4.03	3.60	3.80	4.98	0.863	NS	NS	*
Coliforms	ND	ND	ND	ND				

[a] Data represent treatment means from seven replicates per treatment. Means within a row followed by the same letter are not significantly different from each other ($p > 0.05$). ND, not detected, i.e., below detection limits. *** $p < 0.001$; * $p \leq 0.05$; NS, not significant. [b] SEM, pooled standard error of means. [c] Polynomial contrasts test the linear and quadratic effects of the YAWP inclusion levels in the silage preparation.

Polynomial contrast results showed that the total bacteria and *Lactobacillus* spp. counts decreased linearly ($P_{linear} < 0.001$ and $P_{linear} < 0.001$, respectively) with the increasing YAWP in the silage. Yeast and mold concentrations increased quadratically ($P_{quadratic} \leq 0.05$)

with the increasing YAWP addition in the maize forage prior to ensiling, with the 2.5% YAWP addition having the lowest concentration. In addition, coliform counts were below detection limit in all treatments.

3.4. Fermentation Quality of the Maize Silage Prepared with the YAWP

The pH values displayed a quadratic pattern of decline ($P_{quadratic} \leq 0.05$) with the increasing YAWP level, with the YAWP 5% addition having the lowest value (trend, $P_{anova} = 0.083$).

The ammonia-N content was significantly affected ($P_{anova} < 0.001$) by the YAWP inclusion (Table 5). The ammonia-N content was significantly lower in the 5% YAWP addition compared with the 0% and 2.5% YAWP additions, whereas the 10% YAWP addition received intermediate values.

Table 5. Fermentation quality of the maize silages prepared with the yogurt acid whey powder (YAWP).

	Inclusion Level of YAWP [a]				Statistics			
							Polynomial Contrast [c]	
	0%	2.5%	5%	10%	SEM [b]	P_{anova}	P_{linear}	$P_{quadratic}$
pH direct	3.74	3.60	3.59	3.71	0.069	NS	NS	*
pH diluted	3.74	3.64	3.65	3.76	0.071	NS	NS	NS
Ammonia-N % DM [d]	4.03 [b]	4.02 [b]	3.71 [a]	3.81 [ab]	0.087	***	**	NS
D/L Lactic acid (g/kg DM)	22.68 [a]	30.24 [b]	34.51 [bc]	37.45 [c]	1.650	***	***	NS
Lactic acid/acetic acid ratio	2.05 [a]	3.32 [b]	4.91 [c]	4.97 [bc]	0.425	***	***	NS
Total VFAs (g/kg DM) [e]	20.75 [b]	21.56 [ab]	14.06 [a]	17.98 [ab]	2.370	**	*	NS
Acetic acid (g/kg DM)	11.18 [b]	10.23 [ab]	7.10 [a]	7.66 [a]	1.218	***	**	NS
Propionic acid (g/kg DM)	1.36 [a]	1.96 [ab]	1.52 [ab]	2.15 [b]	0.234	**	*	NS
Isobutyric acid (g/kg DM)	1.36	2.06	1.46	1.66	0.244	NS	NS	NS
Butyric acid (g/kg DM)	4.62	4.29	ND	ND				
Heptanoic acid (g/kg DM)	2.23 [a]	3.03 [ab]	3.98 [bc]	6.50 [c]	0.736	**	***	NS

[a] Data represent treatment means from seven replicates per treatment. Means within a row followed by the same letter are not significantly different from each other ($p > 0.05$). *** $p < 0.001$; ** $p < 0.01$; * $p \leq 0.05$; NS, not significant. ND, not detected, i.e., below detection limits. [b] SEM, pooled standard error of means. [c] Polynomial contrasts test the linear and quadratic effects of the YAWP inclusion levels in the silage preparation. [d] DM: dry matter. [e] Total VFAs; acetic + propionic + butyric + isobutyric + eptanoic.

The lactic acid concentration was significantly higher ($P_{anova} < 0.001$) in the 10% and 5% YAWP additions compared with the 0% addition, whereas the 2.5% addition was intermediate and differed from the 10% YAWP addition. The lactic acid/acetic acid ratio was significantly higher ($P_{anova} < 0.001$) in the 5 and 10% YAWP additions compared with the control. Furthermore, the 2.5% YAWP addition was intermediate and differed from the control and 5% YAWP addition.

Polynomial contrast results showed that the lactic acid content and lactic acid/acetic ratio increased linearly ($P_{linear} < 0.001$ and $P_{linear} < 0.001$, respectively) with the increasing YAWP in the silage. On the other hand, the ammonia-N content decreased linearly ($P_{linear} < 0.01$) with the increasing YAWP-incorporation level.

Total VFAs, acetic, propionic, butyric and heptanoic contents in the silages prepared with the YAWP addition are presented in Table 5. Total VFA content was significantly lower ($P_{anova} < 0.01$) in the 5% YAWP addition compared with the 0% YAWP addition, whereas the 2.5 and 10% YAWP additions received intermediate values. Acetic acid content significantly decreased ($P_{anova} < 0.001$) with the 5% and 10% YAWP additions in comparison with the 0% YAWP addition, with the 2.5% addition being intermediate. The incorporation of 10% YAWP in the silage significantly increased ($P_{anova} < 0.01$) the propionic content compared with the 0% YAWP, whereas the YAWP 2.5 and 5% treatments received intermediate values. The butyric acid content was detected only in the 0 and 2.5% YAWP additions. The addition of 5 and 10% YAWP in silage significantly increased ($P_{anova} < 0.01$) the heptanoic acid content compared with the 0% YAWP, whereas the 2.5% YAWP differed only from the 10% YAWP additions.

Polynomial contrast results showed that, of the total VFAs, acetic acid decreased linearly ($P_{linear} < 0.05$, $P_{linear} < 0.01$) by increasing the YAWP level. By increasing the YAWP level, the propionic and heptanoic acids increased linearly ($P_{linear} < 0.01$ and $P_{linear} < 0.001$, respectively).

4. Discussion

Usage of the food industry's byproducts in animal nutrition can play an important role in converting low-value materials into high-quality products, while at the same time reducing their environmental footprint, thus contributing to the circular economy [31,32]. In the last decade, strained (Greek) yogurt production has increased significantly due to the growing demand from consumers for its organoleptic characteristics and its health-promoting profile [21]. In the United States, approximately 771,000 metric tons of Greek yogurt was manufactured in 2015, representing nearly 40% of the yogurt market [33]. However, the production process results in yogurt acid whey, a byproduct of strained yogurt, which poses a serious environmental problem if left untreated. In particular, acid whey has a high biological oxygen demand, making it difficult to dispose of into the environment without costly effects on the surrounding ecosystems [20]. Given that yogurt acid whey is rich in lactose and organic acids (e.g., lactic acid and citric acid), its use as a silage additive component merits investigation. Lactose is a fermentable carbohydrate, and its addition to the incorporation of plant matter accelerates the fermentation phenomena and produces a better silage quality [18,19]. Lactic acid is classified among the alpha hydroxy acids and it is also known for its antibacterial activity [34].

The target of silage production is to preserve as much of the fresh plant material nutrients as possible. In this sense, the YAWP in maize forage improved the nutritional quality of the maize silage by increasing the DM and CP, and reducing the CF, NDF and ADF, in a dose-dependent manner. The increase in the DM could be linked with the high-DM content of the YAWP (960 g/kg, Table 1), but also the ability of the YAWP to improve silage fermentation by the promotion of lactic acid fermentation, the decrease in clostridial fermentation and proteolysis (discussed below). Similar results have been previously mentioned by the addition of whey powder in maize silage or alfalfa silage [17,18]. The increase in the DM content has been linked in higher feed intake by cattle [35,36]. The reduction of crude fiber and its fractions (NDF and ADF) was more or less expected due to the dilution effect of the YAWP. Although not directly comparable, other studies have also reported decreases in the NDF and ADF contents of silage prepared using whey powder [16]. Nevertheless, the decrease in fiber is expected to increase the nutritional value of the maize silage.

The addition of the YAWP increased the CP content of silages in a dose-dependent manner. Moreover, the linear reduction of the ammonia-N content by the YAWP level of addition highlights the YAWP properties to reduce protein degradation. Protein degradation in silage is attributed either to the action of enzymes, which are released from the plant cells at harvest, or to proteolytic epiphytic microorganisms present on the plant [37]. Regarding the latter, organic acid additives (e.g., citric acid, formic acid, maleic acid) via direct acidification and antimicrobial effects have been shown to limit proteolysis and to improve the fermentation quality [38–40]. Given that the YAWP not only contains high levels of lactic acid, but also reduced the concentration of total bacteria in the silages in a dose-dependent manner, it could be postulated that the YAWP modified beneficially the proteolytic processes during ensiling via its antimicrobial activity. Since maize silage is low in protein, the increase in the CP and the reduction of the ammonia-N content in YAWP-treated silages could be regarded as an enhancement of the silage dietary value.

In this study, using vacuum ensiling bags in a laboratory setting, the low-pH values (below 4) of the untreated silages indicated good fermentation. This can be attributed either to the low-buffering capacity of maize [41] and/or due to the good experimental ensiling conditions. In large-scale silos, it is harder to pack well and silages are more prone to air infiltration, and thus result in ensiling deterioration [42]. The addition of the YAWP in the silage preparation resulted in a drop in the pH (trend) when added up to 5% (3.59), while a further addition (10% YAWP) caused a numeric increase in the pH value. The high mineral composition of the YAWP (e.g., calcium, phosphorus, magnesium, potassium) [20] may have conferred a buffering capacity. Mineral addition by the YAWP could also enhance the dietary value of the silage [43].

The epiphytic microflora of forage significantly affects the quality of natural silage fermentation [44,45]. Therefore, an additional purpose of this study was to assess the concentration of various microbial populations, such as total bacteria, *Lactobacillus* spp., *Clostridium* spp., yeast and molds and coliforms. It appeared that increased addition levels of the YAWP acted more as an acidifier and silage fermentation inhibitor, due to its concentration of lactic and citric acid and other antimicrobial components, by lowering the concentration of total bacteria and *Lactobacillus* spp. in the maize silage. In general, the use of organic acids as silage additives decreases the concentration of bacterial populations due to their antibacterial activity, as has been seen in experiments adding formic acid, malic acid or citric acid to forages for silage [38,46,47]. Reducing the concentration of bacterial populations limits unwanted fermentations and nutrient losses [47]. Although the presence of *Lactobacillus* spp. is important for forage fermentation and, by extension, for the drop in its pH value, studies have shown that the reduction in the concentration of lactic acid bacteria does not negatively affect the quality of the produced product when an organic acidifier has been added to the forage [38,46,47].

The most important metabolic products of silage bacterial populations are lactic, acetic and propionic acid. Nevertheless, unwanted metabolic products can also be produced, such as butyric acid, ethanol and ammonia, and, depending on their concentration, the quality of the silage and its acceptance by the animals can be affected [48]. In this study, the addition of the YAWP in the silage preparation contributed to the increase in lactic acid as well as the decrease in acetic acid in a dose-dependent manner, therefore resulting in an increased lactic acid/acetic acid ratio. It has been demonstrated that the lactic acid/acetic acid ratio is an efficient indicator of homo- or heterofermentation. As the ratio increases, the fermentation leads to homofermentation, since homolactic acid bacteria use water-soluble carbohydrates for their growth and produce only lactic acid [49]. This indicates that the YAWP favored the homofermentation of the silage, and thus resulted in higher concentration of lactic acid. In general, lactic acid should be at least 65 to 70% of the total silage acids in a good silage [50].

In our study, the percentage of lactic acid out of the total silage acids was 52% in the YAWP 0% addition, 58% in the YAWP 2.5% addition, 71% in the YAWP 5% addition and 67% in the YAWP 10% addition. The latter showed that sufficient lactic acid was determined when the 5% and 10% YAWP additions were added in the silage preparation. On the other hand, the ratio of lactic acid/acetic acid is also used as a qualitative indicator of fermentation; with a good silage fermentation, the ratio should be between 2.5 and 3.0. In the present study, the addition of the YAWP 2.5% to the silage preparation led to this range, whereas the addition of the YAWP 5 and 10% resulted in higher values (4.91 and 4.97, respectively). Silages with very high levels of the lactic acid: acetic acid ratio may sometimes be more aerobically unstable than those with lower ratios, because low concentrations of acetic acid may not be sufficient to inhibit lactate-assimilating yeasts [41]. However, considering that the overall yeast–mold concentration in this work was not affected, it could be an indicator that aerobic stability was not compromised. In addition, given the vacuum application, the assessment of the silage aerobic stability was not specifically addressed by this study and merits research. The increase in the lactic acid: acetic acid ratio below 3 has been previously reported by the addition of 300 g/kg fresh Scotta whey in alfalfa silage [51].

Propionic acid was enhanced by the YAWP supplementation in the silage preparation in a dose-dependent manner. Propionic acid is a fermentation product resulting from the metabolism of bacteria, such as propionibacteria, *Lactobacillus bifermentans, Clostridium propionicum* and *Selenomonas ruminantium* [52,53]. Typically, its content in maize silages is low, <0.25% DM [12,54], which can also be confirmed by this study. Consequently, due to its low concentration (i.e., the higher value of 0.22% DM by the YAWP 10% addition), it may not be of practical significance. While, non-directly comparable, increasing levels of propionic acid have been reported by the increasing addition level of sour yogurt in maize silage preparation [54]. However, it is worth noting that the addition of propionic acid in maize silage preparation has been used to improve its aerobic stability by inhibiting the growth of molds and yeasts after the silage was exposed to air [14].

The butyric acid content is critical for evaluating the silage quality, as it can affect the silage intake and can induce severe ketosis when presented at a critical level in the diet of cows in early lactation [55,56]. In this experiment, butyric acid was detected in the control (0.46% of the DM) and in the 2.5% YAWP addition (0.42% of the DM), whereas, in the YAWP 5% and 10% treatments, it was not detected. The latter could highlight that the YAWP 5 and 10% addition properties inhibit the clostridial metabolic activity, further supporting the beneficial role of YAWP in silage preparation. The decrease in the butyric acid concentration has been previously reported by the addition of whey at 2 and 4% [17].

5. Conclusions

The YAWP addition in the maize silage preparation beneficially modified the silage nutritional and fermentative qualities. The YAWP 5 and 10% levels of addition in the maize forage were the most effective. In particular, the inclusion of the 10% YAWP positively modified the nutritional value of the maize silage by increasing the DM and CP, and by decreasing the CF, NDF and ADF contents. The incorporation of the 5 and 10% YAWP favored the homofermentation of the silage, and thus resulted in a higher concentration of lactic acid. The YAWP 5 and 10% levels of addition inhibited the clostridial metabolic activity, which was depicted by the absence of butyric acid in the silages prepared with these levels. The addition of the 5% YAWP in the silage preparation modified beneficially the proteolytic processes by lowering the ammonia-N concentration. Further research is recommended for the evaluation of the YAWP addition in other forages, as well as in large-scale silage-production units.

Author Contributions: Conceptualization, K.C.M. and I.P. (Ioannis Politis); Data curation, I.P. (Irida Palamidi) and K.C.M.; Funding acquisition, I.P. (Ioannis Politis); Investigation, I.P. (Irida Palamidi) and V.V.P.; Methodology, I.P. (Irida Palamidi) and K.C.M.; Project administration, I.P. (Ioannis Politis) and K.C.M.; Resources, B.K.; Supervision, I.P. (Ioannis Politis) and K.C.M.; Validation, I.P. (Irida Palamidi); Writing—original draft, I.P. (Irida Palamidi); Writing—review and editing, V.V.P., B.K., I.H., I.P. (Ioannis Politis) and K.C.M. All authors have read and agreed to the published version of the manuscript.

Funding: This research has been cofinanced by the European Regional Development Fund of the European Union and Greek national funds through the Operational Program Competitiveness, Entrepreneurship and Innovation, under the call RESEARCH—CREATE—INNOVATE (project code: T2EDK-00783).

Institutional Review Board Statement: Not applicable.

Informed Consent Statement: Not applicable.

Data Availability Statement: Data are contained within the article.

Conflicts of Interest: The authors declare no conflict of interest.

References

1. Randby, Å.T.; Weisbjerg, M.R.; Nørgaard, P.; Heringstad, B. Early Lactation Feed Intake and Milk Yield Responses of Dairy Cows Offered Grass Silages Harvested at Early Maturity Stages. *J. Dairy Sci.* **2012**, *95*, 304–317. [CrossRef] [PubMed]
2. Khan, N.A.; Yu, P.; Ali, M.; Cone, J.W.; Hendriks, W.H. Nutritive Value of Maize Silage in Relation to Dairy Cow Performance and Milk Quality. *J. Sci. Food Agric.* **2015**, *95*, 238–252. [CrossRef]

3. Bernes, G.; Turner, T.; Pickova, J. Sheep Fed Only Silage or Silage Supplemented with Concentrates. 2. Effects on Lamb Performance and Fatty Acid Profile of Ewe Milk and Lamb Meat. *Small Rumin. Res.* **2012**, *102*, 114–124. [CrossRef]
4. Patel, M.; Wredle, E.; Bertilsson, J. Effect of Dietary Proportion of Grass Silage on Milk Fat with Emphasis on Odd- and Branched-Chain Fatty Acids in Dairy Cows. *J. Dairy Sci.* **2013**, *96*, 390–397. [CrossRef]
5. Warren, H.E.; Scollan, N.D.; Nute, G.R.; Hughes, S.I.; Wood, J.D.; Richardson, R.I. Effects of Breed and a Concentrate or Grass Silage Diet on Beef Quality in Cattle of 3 Ages. II: Meat Stability and Flavour. *Meat Sci.* **2008**, *78*, 270–278. [CrossRef]
6. Berthiaume, R.; Lafrenière, C.; Girard, C.; Campbell, C.P.; Pivotto, L.M.; Mandell, I.B. Effects of Forage Silage Species on Yearling Growth Performance, Carcass and Meat Quality, and Nutrient Composition in a Forage Based Beef Production System. *Can. J. Anim. Sci.* **2015**, *95*, 173–187. [CrossRef]
7. Wilkinson, J.M.; Rinne, M. Highlights of Progress in Silage Conservation and Future Perspectives. *Grass Forage Sci.* **2018**, *73*, 40–52. [CrossRef]
8. Allen, M.S.; Coors, J.G.; Roth, G.W. Corn Silage. In *Silage Science and Technology*; John Wiley & Sons, Ltd.: Hoboken, NJ, USA, 2015; pp. 547–608, ISBN 9780891182344.
9. Buxton, D.R.; O'Kiely, P. Preharvest Plant Factors Affecting Ensiling. In *Silage Science and Technology*; John Wiley & Sons, Ltd.: Hoboken, NJ, USA, 2015; pp. 199–250, ISBN 9780891182344.
10. Johnson, L.; Harrison, J.H.; Hunt, C.; Shinners, K.; Doggett, C.G.; Sapienza, D. Nutritive Value of Corn Silage as Affected by Maturity and Mechanical Processing: A Contemporary Review. *J. Dairy Sci.* **1999**, *82*, 2813–2825. [CrossRef]
11. Filya, I.; Sucu, E. The Effects of Lactic Acid Bacteria on the Fermentation, Aerobic Stability and Nutritive Value of Maize Silage. *Grass Forage Sci.* **2010**, *65*, 446–455. [CrossRef]
12. Santos, A.O.; Ávila, C.L.S.; Schwan, R.F. Selection of Tropical Lactic Acid Bacteria for Enhancing the Quality of Maize Silage. *J. Dairy Sci.* **2013**, *96*, 7777–7789. [CrossRef]
13. Okoye, C.O.; Wang, Y.; Gao, L.; Wu, Y.; Li, X.; Sun, J.; Jiang, J. The Performance of Lactic Acid Bacteria in Silage Production: A Review of Modern Biotechnology for Silage Improvement. *Microbiol. Res.* **2023**, *266*, 127212. [CrossRef] [PubMed]
14. Kung, L.; Sheperd, A.C.; Smagala, A.M.; Endres, K.M.; Bessett, C.A.; Ranjit, N.K.; Glancey, J.L. The Effect of Preservatives Based on Propionic Acid on the Fermentation and Aerobic Stability of Corn Silage and a Total Mixed Ration. *J. Dairy Sci.* **1998**, *81*, 1322–1330. [CrossRef] [PubMed]
15. Lv, H.; Pian, R.; Xing, Y.; Zhou, W.; Yang, F.; Chen, X.; Zhang, Q. Effects of Citric Acid on Fermentation Characteristics and Bacterial Diversity of Amomum Villosum Silage. *Bioresour. Technol.* **2020**, *307*, 123290. [CrossRef] [PubMed]
16. Khorvash, M.; Colombatto, D.; Beauchemin, K.A.; Ghorbani, G.R.; Samei, A. Use of Absorbants and Inoculants to Enhance the Quality of Corn Silage. *Can. J. Anim. Sci.* **2006**, *86*, 97–107.
17. Özüretmen, S.; Özelçam, H.; Ipçak, H.H. Effects of Whey Powder on Fermentation Quality, Nutritive Value, and Digestibility of Alfalfa Silage. *J. Anim. Feed Sci.* **2022**, *31*, 65–72. [CrossRef]
18. Fallah, R. Effects of Adding Whey and Molasses on Corn Silage Quality, Growth Performance and Health of Simmental Fattening Calves. *J. Livest. Sci.* **2019**, *10*, 91–96. [CrossRef]
19. Rezende, A.V.; Rabelo, C.H.S.; Veiga, R.M.; Andrade, L.P.; Härter, C.J.; Rabelo, F.H.S.; Basso, F.C.; Nogueira, D.A.; Reis, R.A. Rehydration of Corn Grain with Acid Whey Improves the Silage Quality. *Anim. Feed Sci. Technol.* **2014**, *197*, 213–221. [CrossRef]
20. Rocha-Mendoza, D.; Kosmerl, E.; Krentz, A.; Zhang, L.; Badiger, S.; Miyagusuku-Cruzado, G.; Mayta-Apaza, A.; Giusti, M.; Jiménez-Flores, R.; García-Cano, I. Invited Review: Acid Whey Trends and Health Benefits. *J. Dairy Sci.* **2021**, *104*, 1262–1275. [CrossRef]
21. Andreou, V.; Chanioti, S.; Xanthou, M.Z.; Katsaros, G. Incorporation of Acid Whey Yogurt By-Product in Novel Sauces Formulation: Quality and Shelf-Life Evaluation. *Sustainability* **2022**, *14*, 15722. [CrossRef]
22. Flinois, J.C.; Dando, R.; Padilla-Zakour, O.I. Effects of Replacing Buttermilk with Yogurt Acid Whey in Ranch Dressing. *J. Dairy Sci.* **2019**, *102*, 7874–7883. [CrossRef]
23. Simitzis, P.; Zikou, F.; Progoulakis, D.; Theodorou, G.; Politis, I. A Note on the Effects of Yoghurt Acid Whey Marination on the Tenderness and Oxidative Stability of Different Meat Types. *Foods* **2021**, *10*, 2557. [CrossRef] [PubMed]
24. Fan, X.; Rivera Flores, V.K.; DeMarsh, T.A.; de Riancho, D.L.; Alcaine, S.D. Aerobic Cultivation of Mucor Species Enables the Deacidification of Yogurt Acid Whey and the Production of Fungal Oil. *Foods* **2023**, *12*, 1784. [CrossRef] [PubMed]
25. Menchik, P.; Zuber, T.; Zuber, A.; Moraru, C.I. Short Communication: Composition of Coproduct Streams from Dairy Processing: Acid Whey and Milk Permeate. *J. Dairy Sci.* **2019**, *102*, 3978–3984. [CrossRef]
26. Acosta Aragón, Y.; Jatkauskas, J.; Vrotniakiene, V. The Effect of a Silage Inoculant on Silage Quality, Aerobic Stability, and Meat Production on Farm Scale. *ISRN Vet. Sci.* **2012**, *2012*, 345927. [CrossRef]
27. Williams, S. *Official Methods of Analysis of the Association of Official Analytical Chemists*; Association of Official Analytical Chemists: Arlington, VA, USA, 1984; ISBN 0935584242.
28. Mertens, D.R. Gravimetric Determination of Amylase-Treated Neutral Detergent Fiber in Feeds with Refluxing in Beakers or Crucibles: Collaborative Study. *J. AOAC Int.* **2002**, *85*, 1217–1240.
29. Möller, J. Gravimetric Determination of Acid Detergent Fiber and Lignin in Feed: Interlaboratory Study. *J. AOAC Int.* **2009**, *92*, 74–90. [CrossRef]

30. Mountzouris, K.C.; Palamidi, I.; Tsirtsikos, P.; Mohnl, M.; Schatzmayr, G.; Fegeros, K. Effect of Dietary Inclusion Level of a Multi-Species Probiotic on Broiler Performance and Two Biomarkers of Their Caecal Ecology. *Anim. Prod. Sci.* **2015**, *55*, 484–493. [CrossRef]
31. Ominski, K.; Mcallister, T.; Stanford, K.; Mengistu, G.; Kebebe, E.G.; Omonijo, F.; Cordeiro, M.; Legesse, G.; Wittenberg, K. Utilization of By-Products and Food Waste in Livestock Production Systems: A Canadian Perspective. *Anim. Front.* **2021**, *11*, 55–63. [CrossRef]
32. Pires, A.F.; Marnotes, N.G.; Rubio, O.D.; Garcia, A.C.; Pereira, C.D. Dairy By-Products: A Review on the Valorization of Whey and Second Cheese Whey. *Foods* **2021**, *10*, 1067. [CrossRef]
33. Erickson, B.E. Acid Whey: Is the Waste Product an Untapped Goldmine. *Chem. Eng. News* **2017**, *95*, 26–30.
34. Reis, J.A.; Paula, A.T.; Casarotti, S.N.; Penna, A.L.B. Lactic Acid Bacteria Antimicrobial Compounds: Characteristics and Applications. *Food Eng. Rev.* **2012**, *4*, 124–140. [CrossRef]
35. Jackson, N.; Forbes, T.J. The Voluntary Intake by Cattle of Four Silages Differing in Dry Matter Content. *Anim. Prod.* **1970**, *12*, 591–599. [CrossRef]
36. Lahr, D.A.; Otterby, D.E.; Johnson, D.G.; Linn, J.G.; Lundquist, R.G. Effects of Moisture Content of Complete Diets on Feed Intake and Milk Production by Cows. *J. Dairy Sci.* **1983**, *66*, 1891–1900. [CrossRef] [PubMed]
37. Palma, J.M.; Sandalio, L.M.; Javier Corpas, F.; Romero-Puertas, M.C.; McCarthy, I.; Del Río, L.A. Plant Proteases, Protein Degradation, and Oxidative Stress: Role of Peroxisomes. *Plant Physiol. Biochem.* **2002**, *40*, 521–530. [CrossRef]
38. Ke, W.C.; Ding, W.R.; Xu, D.M.; Ding, L.M.; Zhang, P.; Li, F.D.; Guo, X.S. Effects of Addition of Malic or Citric Acids on Fermentation Quality and Chemical Characteristics of Alfalfa Silage. *J. Dairy Sci.* **2017**, *100*, 8958–8966. [CrossRef]
39. Tao, X.; Chen, S.; Zhao, J.; Wang, S.; Dong, Z.; Li, J.; Sun, F.; Shao, T. Effects of Citric Acid Residue and Lactic Acid Bacteria on Fermentation Quality and Aerobic Stability of Alfalfa Silage. *Ital. J. Anim. Sci.* **2020**, *19*, 744–752. [CrossRef]
40. Yitbarek, M.B.; Tamir, B. Silage Additives. *Open J. Appl. Sci.* **2014**, *4*, 258–274. [CrossRef]
41. Kung, L.; Shaver, R.D.; Grant, R.J.; Schmidt, R.J. Silage Review: Interpretation of Chemical, Microbial, and Organoleptic Components of Silages. *J. Dairy Sci.* **2018**, *101*, 4020–4033. [CrossRef]
42. Lima, L.M.; Dos Santos, J.P.; Casagrande, D.R.; Ávila, C.L.S.; Lara, M.S.; Bernardes, T.F. Lining Bunker Walls with Oxygen Barrier Film Reduces Nutrient Losses in Corn Silages. *J. Dairy Sci.* **2017**, *100*, 4565–4573. [CrossRef]
43. Ammerman, C.B.; Goodrich, R.D. Advances in Mineral Nutrition in Ruminants. *J. Anim. Sci.* **1983**, *57* (Suppl. S2), 519–533.
44. Wu, M.; Wang, Y.; Wang, Y.; Wang, X.; Yu, M.; Liu, G.; Tang, H. Study on the Diversity of Epiphytic Bacteria on Corn and Alfalfa Using Illumina MiSeq/NovaSeq High-Throughput Sequencing System. *Ann. Microbiol.* **2021**, *71*, 38. [CrossRef]
45. Wang, S.; Li, J.; Zhao, J.; Dong, Z.; Shao, T. Changes in Silage Quality, Bacterial Community Dynamics, and Metabolic Profiles in Whole-crop Maize Silage. *Agron. J.* **2022**, *114*, 976–990. [CrossRef]
46. Wei, S.N.; Li, Y.F.; Jeong, E.C.; Kim, H.J.; Kim, J.G. Effects of Formic Acid and Lactic Acid Bacteria Inoculant on Main Summer Crop Silages in Korea. *J. Anim. Sci. Technol.* **2021**, *63*, 91–103. [CrossRef]
47. Zhao, J.; Wang, S.; Dong, Z.; Li, J.; Jia, Y.; Shao, T. Effect of Storage Time and the Level of Formic Acid on Fermentation Characteristics, Epiphytic Microflora, Carbohydrate Components and in Vitro Digestibility of Rice Straw Silage. *Anim. Biosci.* **2021**, *34*, 1038–1048. [CrossRef]
48. de Oliveira, J.S.; Santos, E.M.; dos Santos, A.P.M. Intake and Digestibility of Silages. In *Advances in Silage Production and Utilization*; IntechOpen: London, UK, 2016.
49. Arriola, K.G.; Kim, S.C.; Adesogan, A.T. Effect of Applying Inoculants with Heterolactic or Homolactic and Heterolactic Bacteria on the Fermentation and Quality of Corn Silage. *J. Dairy Sci.* **2011**, *94*, 1511–1516. [CrossRef]
50. Kung, L. Understanding the Biology of Silage Preservation to Maximize Quality and Protect the Environment. In Proceedings of the California Alfalfa & Forage Symposium, Visalia, CA, USA, 30 November–2 December 2010; pp. 41–54.
51. Mariotti, M.; Fratini, F.; Cerri, D.; Andreuccetti, V.; Giglio, R.; Angeletti, F.G.S.; Turchi, B. Use of Fresh Scotta Whey as an Additive for Alfalfa Silage. *Agronomy* **2020**, *10*, 365. [CrossRef]
52. Merry, R.J.; Davies, D.R. Propionibacteria and Their Role in the Biological Control of Aerobic Spoilage in Silage. *Lait* **1999**, *79*, 149–164. [CrossRef]
53. Pahlow, G.; Muck, R.E.; Driehuis, F.; Oude Elferink, S.J.W.H.; Spoelstra, S.F. Microbiology of Ensiling. In *Silage Science and Technology*; American Society of Agronomy: Madison, WI, USA, 2015; pp. 31–93, ISBN 9780891182344.
54. KağanTelli, S.Ö. Effects of Sour Yogurt Addition to Corn Silage on Silage Fermentation, Aerobic Stability, and in Vitro Digestibility. *Ege Üniv. Ziraat Fak. Derg.* **2022**, *59*, 601–609.
55. Lingaas, F.; Tveit, B. Etiology of Acetonemia in Norwegian Cattle. 2. Effect of Butyric Acid, Valeric Acid, and Putrescine. *J. Dairy Sci.* **1992**, *75*, 2433–2439. [CrossRef]
56. Dirksen, G.; Breitner, W. A New Quick-Test for Semiquantitative Determination of Beta-Hydroxybutyric Acid in Bovine Milk. *J. Vet. Med. Ser. A* **1993**, *40*, 779–784. [CrossRef]

fermentation

MDPI

Article

An Assessment on the Fermentation Quality and Bacterial Community of Corn Straw Silage with Pineapple Residue

Dengte Li [1,2,†], Huade Xie [2,†], Fanquan Zeng [2], Xianqing Luo [2], Lijuan Peng [2], Xinwen Sun [1], Xinfeng Wang [1,*] and Chengjian Yang [2,*]

1 College of Animal Science and Technology, Shihezi University, Shihezi 832003, China; lidengte@163.com (D.L.); feed2008@126.com (X.S.)
2 Guangxi Key Laboratory of Buffalo Genetics, Reproduction and Breeding, Guangxi Buffalo Research Institute, Nanning 530001, China; xiehuade093@163.com (H.X.); 13471010613@163.com (F.Z.); 19101477179@163.com (X.L.); lijuanpeng2000@163.com (L.P.)
* Correspondence: wxf-4@163.com (X.W.); ycj0746@sina.com (C.Y.); Tel.: +86-135-6573-6005 (X.W.); +86-157-7716-2502 (C.Y.)
† These authors contributed equally to this work.

Simple Summary: Pineapple residue (PR) is a potential feed for ruminants, providing a rich source of fiber, protein, and water-soluble carbohydrates. However, PR has high water content, which may make storage difficult. For ruminants, corn straw (CS) is a common source of roughage, and it is often used for silage. Mixed ensiling of PR with CS may provide a solution to the problem of PR being difficult to preserve. The aim of this study was to evaluate the chemical composition, fermentation quality, and microbial community of CS silage mixed with PR. Mixed CS with appropriate PR silage showed a lower pH value and higher lactate acid and acetic acid content. Also, the addition of PR lead to an increase in the relative abundance of *Lactobacillus* in mixed silage.

Abstract: The effects of pineapple residue (PR) on fermentation quality, chemical composition, and bacterial community of corn straw (CS) silage were evaluated. CS was ensiled with 0% control group (CON), 15% (P1), 30% (P2), and 45% (P3) PR on a fresh matter (FM) basis for 45 days. P3 had lower dry matter (DM) and crude protein (CP) contents but higher ammonia-N (NH_3-N) content than the other three groups ($p < 0.05$). Compared with the other groups, P1 had lower a pH and higher lactic acid and acetic acid contents ($p < 0.05$). The lactic acid bacteria count in P1 was higher than in P2 and P3 ($p < 0.05$); the number of yeast in P2 was higher than in the other groups ($p < 0.05$). With the increasing proportion of PR addition, the relative abundance of *Lactobacillus* gradually increased, and the dominant genus in P3 was *Acetobacter*. In summary, the addition of PR can improve the quality of CS silage, and the optimum addition ratio for PR was 15% on a FM basis.

Keywords: pineapple residue; corn straw; silage quality; microbial diversity

Citation: Li, D.; Xie, H.; Zeng, F.; Luo, X.; Peng, L.; Sun, X.; Wang, X.; Yang, C. An Assessment on the Fermentation Quality and Bacterial Community of Corn Straw Silage with Pineapple Residue. *Fermentation* **2024**, *10*, 242. https://doi.org/10.3390/fermentation10050242

Academic Editors: Wansup Kwak and Siran Wang

Received: 9 April 2024
Revised: 28 April 2024
Accepted: 29 April 2024
Published: 30 April 2024

1. Introduction

Corn straw (CS) is a major field crop residue produced in large amounts that is rich in carbohydrates (cellulose and hemicellulose) and provides good roughage sources for ruminants [1,2]. However, large amounts of CS are discarded during the harvest season, leading to low CS utilization and a loss of a valuable feed resource. In order to preserve the quality of roughage, silage is commonly used in dairy farming. CS is one of primary crops for ensiling compared with other silage [3]. Silage of CS may prolong its storage time, increase its utilization efficiency, improve feed palatability, and enhance the production performance of animals [4]. Therefore, by incorporating corn stover into the feed system for ruminants, not only can it effectively utilize agricultural waste resources and reduce feed costs, but it is also expected to improve the production performance and quality of livestock and poultry products. In addition, this practice also helps to solve the environmental problems caused by stover burning, providing an environmentally friendly solution for a sustainable development path. The scientific utilization of corn stover is not only related to the development of agriculture and animal husbandry, but it is also an important contribution to sustainable agriculture and resource recycling.

Pineapple belongs to the *Bromeliaceae* family and is a high-quality and high-yield tropical fruit in tropical and subtropical regions, known for its aromatic, sweet, and crisp flavor, as well as its rich content of sugars, minerals, and vitamins [5–7]. In recent years, China has seen a steady increase in pineapple production, with a total output reaching 1.733 million tons in 2019 [8]. Pineapple residue (PR), which refers to the leftover pineapple peels and residual pulp after pressing juice, canning, or wine making, accounts for approximately 50–70% of the entire fresh fruit weight [9]. PR has similar nutritional content to pineapple pulp, rich in crude protein (CP), ether extract (EE), crude fiber (CF), and vitamin C, making it a good feed resource [10,11]. However, the majority of PR is discarded as waste, resulting in a severe waste of resources and serious environmental pollution [12]. If PR is scientifically and reasonably developed and utilized as animal feed, it can turn waste into treasure and reduce feed costs. Liu et al. found that PR is a high-quality feed resource for cattle with good feed intake and digestion properties [13]. Yang et al. found that the partial substitution of fermented PR for whole corn silage in the diet did not impair or affect the productive performance of goats [14]. Gowda et al. reported that fed dairy cattle with silage pineapple waste could improve lactation productivity [15]. Suksathit et al. reported that pineapple waste silage can improve nutrient digestibility in livestock compared to hay silage [16].

Several studies reported that mixed ensiling improves silage quality and promotes stability of the fermentation process compared to fermentation alone [17,18]. Denek et al. found that tomato pomace silage fermented and preserved well with the addition of wheat straw and wheat grain [19]. Ni et al. found that mixed ensiling of forage soybean with crop corn or sorghum could reduce the pH value, enhance *Lactobacillus* abundance, and improve the forage soybean silage quality [20]. Li et al. found that mixed silage of the banana pseudostem and fresh maize stover decreased the pH value, *Enterobacteriaceae* count, yeast, and mold count [21]. Thus, we hypothesized that ensiling CS and PR together could inhibit undesirable fermentation and improve silage fermentation quality. However, to our knowledge, relatively little information is available on the fermentation characteristics and microbial diversity of CS ensiled alone or in combination with PR, and they are poorly understood. These studies showed positive effects of feeding animals with fermented PR, providing strong support for the widespread use of pineapple pomace in ruminant feeding and demonstrating that PR is an ideal feed for ruminants.

Therefore, this experiment aims to assess the effects of adding different ratios of PR on CS silage quality and microbial diversity and ultimately provide a scientific basis for the rational utilization of PR as ruminant feed.

2. Materials and Methods

2.1. Ensiling Materials and Silage Preparation

CS and PR were obtained from the Pasture Research Base of Guangxi Buffalo Research Institute in Nanning, China in May 2023. PR includes pineapple epidermis, leaf crown, and part of fruit pulp. Moreover, the CS refers to the residues remaining after the corn cobs are harvested. This material predominantly includes the green stem and leaves, which are typically at a late vegetative stage. The chemical compositions of CS and PR are listed in Table 1. Fresh CS and PR were cut into 1 to 2 cm pieces. Then, they were spread out flat on the ground to dry out moisture to about 70%. For silage preparation, CS was mixed with varying percentages of PR and ensiled under anaerobic conditions. Silages were prepared using a small-scale system of silage fermentation. Specifically, mixes were prepared with 0% PR (control), 15% PR (P1), 30% PR (P2), and 45% PR (P3), based on fresh matter. Samples were weighed and mixed uniformly according to the specified addition ratios, packed into polyethylene film bags (1000 g per bag), vacuum-sealed using a vacuum packaging machine (DZ500; Gzrifu Co. Ltd., Guangzhou, China), and then stored in the dark at room temperature (20–30 °C). After 45 days, the polyethylene film bag was opened for the test.

Table 1. Chemical composition of corn straw and pineapple residue.

Items [1]	DM	CP	NDF	ADF	Ash	Ca	P
	(%)			(% DM)			
corn straw	26.92	13.10	68.39	38.39	9.36	0.73	0.23
pineapple residue	18.47	8.02	54.94	28.06	9.95	0.39	0.16

[1] D[1] DM, dry matter; CP, crude protein; NDF, neutral detergent fiber; ADF, acid detergent fiber.

2.2. Chemical Analysis

The raw material and silage samples (500 g) were subjected to a consistent drying process at 65°C in an oven (LBAO-250; STIK Instrument Equipment shanghai Co., Ltd. Shanghai, China) until they reached a stable weight. Subsequently, they underwent pulverization using a pulverizer (FS200; Guangzhou Bomin Electrical and mechanical equipment Co. Ltd., Guangzhou, China), while ensuring the particles passed through a 1 mm screen for uniformity. To assess the chemical compositions, several analyses were conducted following standard protocols outlined by the AOAC. Specifically, the dry matter (DM), CP, and Ash contents were determined using methods 934.01, 976.05, and 942.05, respectively [22]. Additionally, the neutral detergent fiber (NDF) and acid detergent fiber (ADF) contents were determined according to procedures in Van Soest [23].

2.3. Fermentation Characteristics Analysis

Silage samples (20 g) were mixed with 180 mL of distilled water and stored at 4 °C for 24 h [24]. Then, filtration was carried out using two layers of gauze, and the filtrate was used for determining pH, NH_3-N, microbial crude protein (MCP), and organic acid contents. The pH value was assessed using a portable pH meter (pH8180-0-00; Smart sensor Co., Ltd., Dongguan, China). Ammonia-nitrogen (NH_3-N) was determined with the phenol-hypochlorite procedure [25]. MCP was determined using procedures described by Bradford [26]. The organic acid contents, including those of lactic acid (LA), acetic acid (AA), propionic acid (PA), and butyric acid (BA), were measured via high-performance liquid chromatography (1260 Infinity II; Agilent Technologies, Inc., Waldbronn, Germany) according to Xie et al. [27]. The specific methods were as follows: Shodex RSpak KC–811 (8.0 mm × 300 mm; Showa Denko K.K., Tokyo, Japan); DAD detector set at 210 nm; elution process was carried out using 3 mmol/L $HClO_4$ as the eluent at a flow rate of 1.0 mL/min; temperature was 50 °C; and sample size was 5.0 μL.

Silage samples (10 g) were blended with 90 mL of sterilized water and serially diluted from 10^{-1} to 10^{-5} in sterilized water. Lactic acid bacteria (LAB) were measured by means of a plate count on De Man–Rogosa–Sharpe (MRS) agar (Qingdao Haibo Biotechnology Co., Ltd., Qingdao, China) and incubated at 37 °C for 48 h. Yeast were counted on potato dextrose agar (Qingdao Haibo Biotechnology Co., Ltd.) after incubation for 48 h at 37 °C. Numbers of colonies were considered to indicate the numbers of viable microorganisms (cfu g^{-1} of fresh matter [FM]) [28].

2.4. Microbial Analysis

Silage samples (15 g) were aseptically combined with 180 mL of sterile phosphate-buffered saline (PBS). This mixture was subsequently incubated at a constant temperature of 37 °C for 2 h under agitation at 200 r/min to ensure that the microbial population mixed into the buffer. Then, the samples were filtered through two layers of a sterile gauze to remove larger particulate matter. A volume of 70 mL of the resulting filtrate was then subjected to centrifugation at 12,000 r/min for 5 min at a temperature of 4 °C. This high-speed centrifugation facilitated the sedimentation of the microbial population, and the supernatant was discarded. Subsequently, it was washed 2 times with sterile PBS to remove any residual impurities. The precipitate was resuspended in 1–2 mL of the same buffer. Genomic DNA was extracted using the cetyltrimethy-lammonium bromide (CTAB) method [29]. DNA samples were sent to Majorbio Bio-pharm Technology Co., Ltd (Shanghai, China) for sequencing and analysis. The V3–V4 regions of the 16S rRNA gene were processed for amplification with the primers. The following primers were used: 338F (ACTCCTACGGGAGGCAGCAG) and 806R (GGACTACHVGGGTWTCTAAT). For the polymerase chain reactions, we used 4 μL of 5× FastPfu Buffer (TransGen Biotech Co., Ltd., Beijing, China), 2 μL of dNTPs (2.5 mmol·L^{-1}), 0.8 μL of Forward Primer (5 μmol·L^{-1}) (TransGen Biotech Co., Ltd., Beijing, China), 0.8 μL of Reverse Primer (5 μmol·L^{-1})(TransGen Biotech Co., Ltd., Beijing, China), 0.4 μL of FastPfu Polymerase (TransGen Biotech Co., Ltd., Beijing, China), 0.2 μL of BSA (TransGen Biotech Co., Ltd., Beijing, China), 10 ng of Template DNA (TransGen Biotech Co., Ltd., Beijing, China), and added ddH_2O to obtain 20 μL of total volume. The steps included initialization at 95 °C for 3 min, 27 cycles of denaturation at 95 °C for 30 s, annealing at 55 °C for 30 s, extension at 72 °C for 45 s, and a final elongation at 72 °C for 10 min. The PCR amplification products were detected via 2% agarose gel electrophoresis [30]. Purified DNA was sequenced on an Illumina MiSeq platform (Illumina, Inc., San Diego, CA, USA) at Majorbio Bio-pharm Technology Co., Ltd. (Shanghai, China). The sequences obtained from the MiSeq platform were processed using QIIME software (version 1.9.1). The sequence data reported in this study were archived in the Sequence Read Archive (SRA) under the accession number PRJNA1060478.

2.5. Statistical Analysis

Data on the fermentative characteristics were analyzed with one-way ANOVA using SPSS (SPSS 26.0 program, SPSS Inc., Chicago, IL, USA). The Alpha diversity index was calculated using Mothur software (version v.1.30.2). The histogram and heatmap were analyzed using R software (version v.3.3.1). There were significant differences only when the probability level was lower than 0.05 ($p < 0.05$).

3. Results

3.1. Chemical Compositions

The lowest DM content was found in P3, flowed by P2, P1, and CON ($p < 0.05$) (Table 2). The CP content in P3 was lower than those in CON and P1 ($p < 0.05$). The differences in NDF and ADF content among all silages were not significant ($p > 0.05$). The Ash content in P1 was lower than that in P2 and P3 ($p < 0.05$). The Ca content in P1 was lower than that in CON and P2 ($p < 0.05$).

Table 2. The effects of adding pineapple residue on the chemical compositions of corn straw silage.

Items [1]	CON	P1	P2	P3	*p*-Value
DM (%)	27.55 ± 0.31 ^{a}	26.40 ± 0.56 ^{b}	25.35 ± 0.63 ^{b}	23.61 ± 1.00 ^{c}	<0.001
CP (% DM)	13.21 ± 0.27 ^{a}	12.93 ± 0.14 ^{a}	12.55 ± 0.30 ^{ab}	11.71 ± 1.13 ^{b}	0.007
NDF (% DM)	61.69 ± 1.15	60.25 ± 1.65	60.76 ± 0.50	62.35 ± 2.88	0.277
ADF (% DM)	36.56 ± 0.80	35.97 ± 1.00	36.48 ± 0.40	38.16 ± 3.26	0.267
Ash (% DM)	9.28 ± 0.21 ^{ab}	9.16 ± 0.17 ^{b}	9.52 ± 0.05 ^{a}	9.53 ± 0.22 ^{a}	0.008
Ca (% DM)	0.67 ± 0.03 ^{a}	0.59 ± 0.01 ^{c}	0.64 ± 0.03 ^{ab}	0.60 ± 0.01 ^{bc}	0.001
P (% DM)	0.06 ± 0.00	0.06 ± 0.00	0.06 ± 0.00	0.06 ± 0.00	<0.001

[1] DM, dry matter; CP, crude protein; NDF, neutral detergent fiber; ADF, acid detergent fiber; CON, 100% CS silage; P1, 85% CS + 15% PR silage; P2, 70% CS + 30% PR silage; P3 55% CS + 45% PR silage. $^{a–c}$ Different letters in the same row indicate significant differences $p < 0.05$.

3.2. Fermentation Quality

P1 had a lower pH value but higher LA and AA content compare with the other group ($p < 0.05$) (Table 3). PA was not detected in any of the other three groups, except for P3. BA was not detected in any silages. NH_3-N content in P3 was higher than in the other groups ($p < 0.05$). NH_3-N content of all silage mixtures tended to increase after the addition of PR, resulting in CON silage having lower NH_3-N content than the other silages. The number of LAB in P1 was higher than in P2 and P3 ($p < 0.05$). The number of yeast in P2 was higher than the other groups ($p < 0.05$), followed by that in the P3, CON, and P1 silages.

Table 3. The effects of adding pineapple residue on the fermentation quality of corn straw silage.

Items [1]	Groups				
	CON	P1	P2	P3	*p*-Value
pH	4.28 ± 0.07 [b]	4.18 ± 0.04 [c]	4.35 ± 0.05 [ab]	4.43 ± 0.06 [a]	<0.001
Lactate acid (g/kg DM)	15.81 ± 0.01 [b]	17.15 ± 0.01 [a]	14.78 ± 0.02 [c]	13.88 ± 0.01 [d]	<0.001
Acetic acid (g/kg DM)	10.33 ± 0.01 [b]	12.54 ± 0.03 [a]	9.69 ± 0.02 [c]	9.56 ± 0.02 [d]	<0.001
Propionic acid (g/kg DM)	ND	ND	ND	0.91 ± 0.04	-
Butyric acid (g/kg DM)	ND	ND	ND	ND	-
NH_3-N (g/kg DM)	1.19 ± 0.02 [c]	1.22 ± 0.02 [c]	1.29 ± 0.04 [b]	1.47 ± 0.05 [a]	<0.001
MCP (mg/mL)	1.34 ± 0.38	1.01 ± 0.12	1.08 ± 0.25	1.38 ± 0.20	0.090
Microbial population (lg CFU/g of FM)					
Lactic acid bacteria	6.69 ± 0.02 [a]	6.73 ± 0.02 [a]	6.58 ± 0.06 [b]	6.59 ± 0.03 [b]	<0.001
Yeast	2.22 ± 0.04 [c]	2.21 ± 0.09 [c]	2.66 ± 0.06 [a]	2.48 ± 0.07 [b]	<0.001

[1] NH_3-N, ammonia-N; MCP, microbial crude protein; ND, not detected; CON, 100% CS silage; P1, 85% CS + 15% PR silage; P2, 70% CS + 30% PR silage; P3 55% CS + 45% PR silage. [a–d] Different letters in the same row indicate significant differences $p < 0.05$.

3.3. Microbial Analysis

The Shannon index and Simpson index were not significant among the four groups ($p > 0.05$) (Table 4). The Chao1 index in P3 was higher than in P2 ($p < 0.05$).

Table 4. The effects of adding pineapple residue on the microbial analysis of corn straw silage.

Items [1]	CON	P1	P2	P3	*p*-Value
Shannon	1.99 ± 0.11	1.58 ± 0.20	1.33 ± 0.73	1.87 ± 0.35	0.094
Simpson	0.23 ± 0.06	0.38 ± 0.08	0.47 ± 0.32	0.38 ± 0.10	0.233
Chao1	119.85 ± 11.35 [ab]	133.55 ± 20.54 [ab]	100.26 ± 33.09 [b]	167.37 ± 52.36 [a]	0.036
ACE	123.03 ± 13.56	133.81 ± 24.12	107.91 ± 37.23	172.56 ± 53.75	0.060
Coverage	0.9996 ± 0.00	0.9996 ± 0.00	0.9997 ± 0.00	0.9996 ± 0.00	0.788

[1] CON, 1[1] CON, 100% CS silage; P1, 85% CS + 15% PR silage; P2, 70% CS + 30% PR silage; P3 55% CS + 45% PR silage. [a,b] Different letters in the same row indicate significant differences $p < 0.05$.

At the order level, *Lactobacillales* and *Enterobacterales* were the top two dominant genera in terms of relative abundance in the CON, P1, and P2 silages (Figure 1). However, the dominant genus in P3 was *Acetobacterales*.

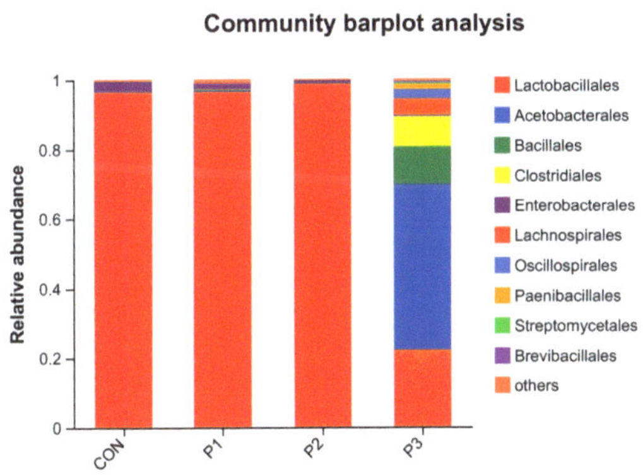

Figure 1. Effects of adding pineapple residue on the order level of microorganisms in corn straw silage. Abbreviations: CON, 100% CS silage; P1, 85% CS + 15% PR silage; P2, 70% CS + 30% PR silage; P3 55% CS + 45% PR silage.

At the genus level, the top two genera in terms of relative abundance in CON, P1, and P2 were *Lactobacillus* and *Pediococcus* (Figure 2). The relative abundance of *Lactobacillus* in CON, P1, and P2 as a proportion of all genera was 87.47%, 90.75%, and 96.21%, respectively.

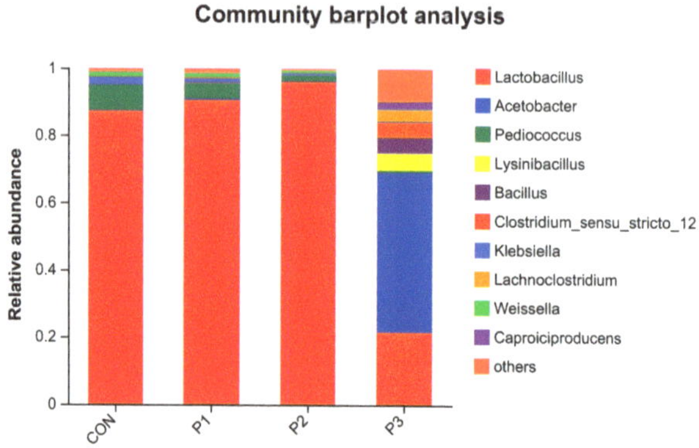

Figure 2. Effects of adding pineapple residue on the genus level of microorganisms in corn straw silage. Abbreviations: CON, 100% CS silage; P1, 85% CS + 15% PR silage; P2, 70% CS + 30% PR silage; P3 55% CS + 45% PR silage.

The pH value and NH_3-N content were negatively correlated with *un-classified_f__Enterobacteriaceae*, *Weissella*, *Klebsiella*, and *Pediococcus* ($p < 0.05$), and positively correlated *unclassified_f__Clostridiaceae*, *Caproiciproducens*, and *Lysinibacillus* ($p < 0.05$) (Figure 3). MCP content was positively correlated with *unclassified_f__Clostridiaceae* and *Bacillus* ($p < 0.05$). LA and AA contents were positively correlated with *unclassified__f__Enterobacteriaceae*, *Weissella*, *Klebsiella*, and *Pediococcus* ($p < 0.05$), and negatively correlated with *unclassified__f__Clostridiaceae*, *Caproiciproducens*, and *Lysinibacillus* ($p < 0.05$).

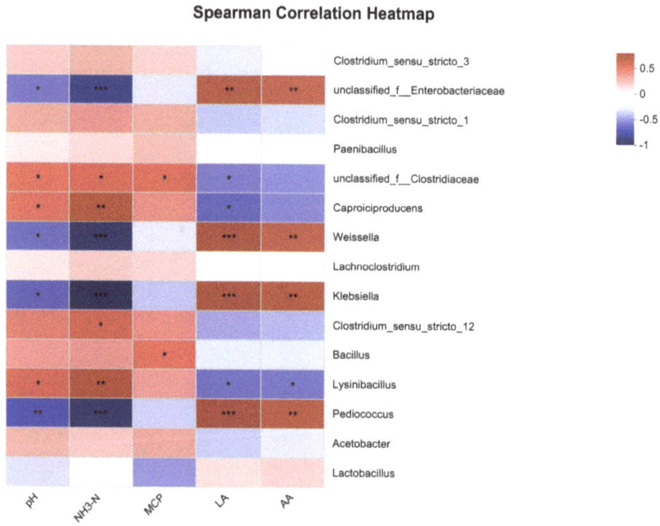

Figure 3. Correlation between relative abundance of bacteria and fermentation parameters at the genus level. Note: Columns in different colors indicate different subgroups; *p*-values are on the far right; "*" indicates $p \leq 0.05$, "**" indicates $p \leq 0.01$, and "***" indicates $p \leq 0.001$. Abbreviations: CON, 100% CS silage; P1, 85% CS + 15% PR silage; P2, 70% CS + 30% PR silage; P3 55% CS + 45% PR silage.

4. Discussion

Raw materials essential for producing high-quality silage should possess specific characteristics, including appropriate moisture levels and an environment that supports anaerobic conditions [27,28]. In this experiment, we observed that the lowest DM content occurred in P3, with sequentially higher levels in P2, P1, and CON, respectively; additionally, the CP content in P3 was noted to be lower than that in CON and P1. The reason for this result is due to the different chemical composition of the two feedstocks, with PR having lower DM content and CP content compared to CS. Consequently, an increase in the addition of PR to the silage mixture led to a decrease in both DM and CP contents. This is consistent with the findings of Muller et al., who reported that pineapple exhibits elevated levels of fiber and starch but a comparatively low content of CP [31]. According to research by Wang et al., CS can serve as a water absorbent in silage, mitigating the challenge posed by excessive water content in other materials [4]. As a result, the content of DM and CP declined concomitantly with the reduction in the proportion of CS and the increase in the proportion of the PR ratio in the silage. In the current study, the Ash content in P1 was lower than the P2 and P3, indicating that CS silage with 15% PR had the highest organic matter content. Ensiling CS with a certain proportion of PR can be modulated to produce a more nutritionally balanced feed.

Silage, an important nutritional source for herbivores, effectively guarantees a balanced supply of roughage throughout the year [32], which is the most commonly used technology for preserving forage [33]. The principle of silage fermentation is that the LAB attached to the materials, under the right temperature, moisture, and anaerobic conditions, ferments water-soluble carbohydrates into LA and other short-chain volatile fatty acids; as a result, the pH decreases, the growth of harmful microorganisms is inhibited, and the silage is preserved as long as it is not exposed to air [34–37]. The pH level of silage serves as a critical indicator for assessing its fermentation quality, where lower pH values are indicative of better fermentation quality and enhanced aerobic stability [38]. The value of pH is determined by organic acids, such as LA and AA, which are produced during the silage process. Research has shown that the content of LA is positively correlated with the overall quality of silage; similarly, AA also improves the aerobic stability of silage [39]. In this experiment, P1 exhibited a lower pH value but higher LA content and AA content compared to the other groups. This might be due to the fact that 15% PR creates a suitable environment for the growth and reproduction of benefit microorganisms, which utilize water-soluble carbohydrates to produce substantial amounts of LA under favorable conditions, causing the pH to drop. It indicates that ensiling CS with 15% PR can achieve the best fermentation quality. In this experiment, P3 had the highest pH value, the lowest lactic acid content, and poor fermentation quality, indicating that too much pineapple pomace should not be added to silage. NH_3-N content reflects the breakdown of proteins in the silage, with higher values indicating greater protein and amino acid breakdown and lower fermentation quality [40]. In this experiment, as the percentage of PR increased, the content of NH_3-N in silage increased, and P3 had the highest NH_3-N content. This is possibly due to the bromelain in PR, leading to protein breakdown and the production of NH_3-N [41].

LAB, recognized as beneficial bacteria, have the capacity to produce LA during the fermentation process. This production of LA leads to a decrease in the pH value, which effectively inhibits the growth of harmful bacteria [42]. Generally speaking, when the number of LAB reaches at least 10^5 (cfu/g FM), silage is considered to be well preserved [43]. In this experiment, the detected number of LAB in all groups reached 10^6 (cfu/g FM), which clearly indicates a successful and good fermentation process in all groups. The addition of an appropriate amount of PR enables the effective preservation of CS silage. Microorganisms play a pivotal role in silage fermentation and can affect silage fermentation through a series of metabolites [44]. Consequently, the structural composition and the abundance of different species of microorganisms present during the fermentation process are intricately linked to the overall quality of silage [45]. The Chao1 index and the Ace index are utilized to measure species richness, indicating the quantity of species [46]. In this experiment, P3 exhibited the highest values for both the Chao1 index and the ACE index. This is possible due to the higher abundance of miscellaneous bacteria in PR, suggesting that an excessive addition (45%) of PR may have adverse effects on fermentation. Furthermore, the Shannon index and the Simpson index were employed to assess species diversity within the microbial community of the silage. These indices are influenced by both the richness and evenness of species within the community. Typically, larger values of these indices indicate a higher diversity of species [47]. In this experiment, the Shannon index and Simpson index were not different among the four groups, indicating that the addition of PR did not widely change the species' diversity in CS silage.

The most common genera of LAB in silage include *Lactobacillus*, *Streptococcus*, *Pediococcus*, *Enterococcus*, *Lactococcus*, *Leuconostoc*, and *Weissella* [48]. *Lactobacillus* can increase LA content, thereby inhibiting the growth and reproduction of harmful microorganism to ensure that the quality of silage is adequate. In this experiment, the relative abundance of *Lactobacillus* in the CON group was 87.47%, the relative abundance of *Lactobacillus* increased with the addition of 15% PR to a proportion of 90.75%, and the relative abundance of *Lactobacillus* reached 96.21% with the addition of 30% PR. This indicates that the addition of an appropriate amount of PR can increase the relative abundance of *Lactobacillus* and ensure that the quality of silage is adequate. *Enterobacterales* comprises Gram-negative, facultative anaerobic bacteria that are mainly found in poor silage, and it competes with LAB for the nutrients to grow and reproduce [49]. In this experiment, the relative abundance of *Enterobacterales* in P1 and P2 was reduced compared with CON. This suggests that the addition of moderate amounts of PR can reduce the relative abundance of harmful bacteria, such as *Enterobacterales*. In general, the surface of silage materials usually has a high number of microorganisms, such as aerobic bacteria, enterobacteria, yeast, and molds, attached to it. At the beginning, there is still oxygen present in the silage environment when aerobic bacteria grow vigorously. At the same time, *Streptococcus*, *Leuconostoc*, and *Pediococcus* are also active. After lactic acid fermentation, lactic acid bacteria begin to proliferate in large quantities and produce lactic acid; subsequently, with the formation of anaerobic and acidic environments, aerobic- and acid-intolerant microorganisms are gradually reduced, and they are steadily replaced by *Lactobacillus* and *Streptococcus lamellaris* together, which are the dominant bacteria in malolactic fermentation. In this experiment, *Acetobacter* was identified as a major genus in the P3 fermentation process, despite the anaerobic vacuum conditions typically being unsuitable for aerobic bacteria, such as *Acetobacter*. It is plausible that the presence of *Acetobacter* in P3 silage could be attributed to residual oxygen trapped within the biomass at the time of sealing. Additionally, the addition of high amounts of PR, which naturally carries *Acetobacter* as part of its flora, especially from the skin of the fruit, may introduce and support a larger population of these bacteria, even under less-than-ideal conditions. Because of the presence of *Acetobacter* and a high proportion of PR in P3 silage, the PR surface carries a high number of *Acetobacter*, which grows and multiplies in large quantities during the initial aerobic fermentation phase of silage fermentation, resulting in a large relative abundance of *Acetobacter*, which in turn affects the quality of the silage.

Change of the microbial genus and its abundance will affect the generation of fermentation products during ensiling [45]. In this experiment, LA and AA contents were positively correlated with *unclassified_f__Enterobacteriaceae*, *Weissella*, *Klebsiella*, and *Pediococcus*. However, the pH value and NH_3-N content were negatively correlated with *unclassified_f__Enterobacteriaceae*, *Weissella*, *Klebsiella*, and *Pediococcus*. This suggests that LA, AA, NH_3-N content, and pH value are simultaneously affected by them and that increasing their abundance improves fermentation quality.

5. Conclusions

Mixed CS with appropriate PR silage showed a lower pH value and higher LA and AA contents. Also, the addition of PR lead to an increase in the relative abundance of *Lactobacillus* in mixed silage. Consequently, the addition of PR could enhance CS silage quality, and the optimum addition ratio for PR was 15% on a FM basis.

Author Contributions: Conceptualization, C.Y. and X.W.; methodology, H.X. and F.Z.; software, D.L.; validation, L.P., C.Y. and D.L.; formal analysis, D.L. and X.L.; investigation, C.Y.; resources, C.Y.; data curation, D.L.; writing—original draft preparation, D.L. and X.S.; writing—review and editing, D.L. and H.X.; visualization, F.Z.; supervision, C.Y.; project administration, C.Y.; funding acquisition, C.Y. All authors have read and agreed to the published version of the manuscript.

Funding: This work was financially supported by the Guangxi Natural Science Foundation (Grant No. 2023JJA130137), the National Natural Science Foundation of China (Grant No. 32361143788), the Xinjiang Production and Construction Corps. Key Areas of Technological Research and Development (2023AB008), the Guangxi Dairy Buffalo Innovation Team of National Modern Agricultural Industry Technology System (Grant No. nycytxgxcxtd-2021-21), the Guangxi Science and Technology Major Project (Grant No. GuiKe AA22068099), the 8th Group Key Areas of Technological Research and Development (2023NY03), and the Scientific Research Project of Shihezi University (KJTP202316).

Institutional Review Board Statement: Not applicable.

Informed Consent Statement: Not applicable.

Data Availability Statement: The sequence data reported in this study were archived in the Sequence Read Archive (SRA) under the accession number PRJNA1060478.

Conflicts of Interest: The authors declare no conflicts of interest.

References

1. Okoye, C.O.; Wu, Y.; Wang, Y.; Gao, L.; Li, X.; Jiang, J. Fermentation profile, aerobic stability, and microbial community dynamics of corn straw ensiled with Lactobacillus buchneri PC-C1 and Lactobacillus plantarum PC1-1. *Microbiol. Res.* **2023**, *270*, 127329. [CrossRef] [PubMed]
2. Zhao, X.; Wang, F.; Fang, Y.; Zhou, D.; Wang, S.; Wu, D.; Wang, L.; Zhong, R. High-potency white-rot fungal strains and duration of fermentation to optimize corn straw as ruminant feed. *Bioresour. Technol.* **2020**, *312*, 123512. [CrossRef] [PubMed]
3. Ma, J.; Fan, X.; Ma, Z.; Huang, X.; Tang, M.; Yin, F.; Zhao, Z.; Gan, S. Silage additives improve fermentation quality, aerobic stability and rumen degradation in mixed silage composed of amaranth and corn straw. *Front. Plant Sci.* **2023**, *14*, 1189747. [CrossRef] [PubMed]
4. Wang, X.; Song, J.; Liu, Z.; Zhang, G.; Zhang, Y. Fermentation Quality and Microbial Community of Corn Stover or Rice Straw Silage Mixed with Soybean Curd Residue. *Animals* **2022**, *12*, 919. [CrossRef] [PubMed]
5. Mohd Ali, M.; Hashim, N.; Abd Aziz, S.; Lasekan, O. Pineapple (*Ananas comosus*): A comprehensive review of nutritional values, volatile compounds, health benefits, and potential food products. *Food Res. Int.* **2020**, *137*, 109675. [CrossRef] [PubMed]
6. Cascante-Marín, A.; Núñez-Hidalgo, S. A Review of Breeding Systems in the Pineapple Family (Bromeliaceae, Poales). *Bot. Rev.* **2023**, *89*, 308–329. [CrossRef]
7. Li, D.; Jing, M.; Dai, X.; Chen, Z.; Ma, C.; Chen, J. Current status of pineapple breeding, industrial development, and genetics in China. *Euphytica* **2022**, *218*, 85. [CrossRef]
8. NBS. *China Statistical Year Book*; China Statistics Press: Beijing, China, 2020. Available online: http://www.stats.gov.cn/ (accessed on 22 December 2023).
9. Rahma Wulan, I.; Mukh, A.; Endang, P.; Agung, P. Utilization of Pineapple Waste as a Roughage Source Diets for Ruminant: A Review. In Proceedings of the International Conference on Improving Tropical Animal Production for Food Security (ITAPS 2021); Atlantis Press: Paris, France, 2022; pp. 123–130.
10. Assumi, S.; Jha, S.; Kaur, C. Valorization of pineapple waste for development of animal feed block. *Int. J. Curr. Microbiol. Appl. Sci.* **2018**, *7*, 3787–3795. [CrossRef]
11. Mello, B.L.B.; Fernandes, A.M.; de Oliveira, T.S.; Leonel, F.P.; Glória, L.S.; Silva, R.S.T. Feed intake, digestibility, and energy contents in growing bull fed pineapple crop waste silage in different planes of nutrition. *Trop. Anim. Health Prod.* **2021**, *53*, 188. [CrossRef]
12. Buliah, N.; Jamek, S.; Ajit, A.; Abu, R. Production of dairy cow pellets from pineapple leaf waste. *AIP Conf. Proc.* **2019**, *2124*, 020048.
13. Liu, C.; Asano, S.; Ogata, H.; Ito, S.; Nakase, T.; Takeda, S.; Miyoshi, K.; Numata, Y.; Takahashi, K.; Kajikawa, H. Digestive, fermentative, and physical properties of pineapple residue as a feed for cattle. *Anim. Sci. J.* **2021**, *92*, e13535. [CrossRef] [PubMed]
14. Yang, C.; Zhao, W.; Tian, H.; Wang, M.; Gao, C.; Guo, Y.; Sun, B. A preliminary study on the possibility of fermented pineapple peel residue partially replacing whole corn silage in feeding Chuanzhong black goats. *Front. Microbiol.* **2022**, *13*, 959857. [CrossRef] [PubMed]
15. Gowda, N.K.; Vallesha, N.C.; Awachat, V.B.; Anandan, S.; Pal, D.T.; Prasad, C.S. Study on evaluation of silage from pineapple (Ananas comosus) fruit residue as livestock feed. *Trop. Anim. Health Prod.* **2015**, *47*, 557–561. [CrossRef] [PubMed]

16. Suksathit, S.; Wachirapakorn, C.; Opatpatanakit, Y. Effects of levels of ensiled pineapple waste and pangola hay fed as roughage sources on feed intake, nutrient digestibility and ruminal fermentation of Southern Thai native cattle. *Songklanakarin J. Sci. Technol.* **2011**, *33*, 281–289.
17. Larsen, S.U.; Hjort-Gregersen, K.; Vazifehkhoran, A.H.; Triolo, J.M. Co-ensiling of straw with sugar beet leaves increases the methane yield from straw. *Bioresour. Technol.* **2017**, *245*, 106–115. [CrossRef] [PubMed]
18. Lee, Y.H.; Ahmadi, F.; Kim, Y.I.; Oh, Y.K.; Kwak, W.S. Co-ensiling garlic stalk with citrus pulp improves the fermentation quality and feed-nutritional value. *Asian-Australas J. Anim. Sci.* **2020**, *33*, 436–445. [CrossRef] [PubMed]
19. Denek, N.; Can, A. Feeding value of wet tomato pomace ensiled with wheat straw and wheat grain for Awassi sheep. *Small Rumin. Res.* **2006**, *65*, 260–265. [CrossRef]
20. Ni, K.; Zhao, J.; Zhu, B.; Su, R.; Pan, Y.; Ma, J.; Zhou, G.; Tao, Y.; Liu, X.; Zhong, J. Assessing the fermentation quality and microbial community of the mixed silage of forage soybean with crop corn or sorghum. *Bioresour. Technol.* **2018**, *265*, 563–567. [CrossRef]
21. Mitiku, A.A.; Vandeweyer, D.; Adriaens, I.; Kechero, Y.; Van Campenhout, L.; Aernouts, B.J.A. Mixed Silage of Banana Pseudostem and Maize Stover on Ethiopian Smallholder Farms: Effect of Fermentation Package and Location on Microbiological and Nutritional Evaluation. *Agriculture* **2023**, *13*, 2152. [CrossRef]
22. Official methods of analysis of the association of official analytical chemists: Kenneth Helrich, 15th ed., AOAC, Arlington, VA, 1990. Vol. I: xxiv + 684 + 62 (index) pp. Vol. II: xviii + 614 + 62 (index) pp. *Anal. Chim. Acta* **1991**, *242*, 302. [CrossRef]
23. Van Soest, P.J.; Robertson, J.B.; Lewis, B.A. Methods for Dietary Fiber, Neutral Detergent Fiber, and Nonstarch Polysaccharides in Relation to Animal Nutrition. *J. Dairy Sci.* **1991**, *74*, 3583–3597. [CrossRef]
24. Cai, Y. Analysis method for silage. In *Field and Laboratory Methods for Grassland Science*; Japanese Society of Grassland Science, Ed.; Tosho Printing: Tokyo, Japan, 2004; pp. 279–282.
25. Broderick, G.A.; Kang, J.H. Automated Simultaneous Determination of Ammonia and Total Amino Acids in Ruminal Fluid and In Vitro Media1. *J. Dairy Sci.* **1980**, *63*, 64–75. [CrossRef]
26. Bradford, M.M. A rapid and sensitive method for the quantitation of microgram quantities of protein utilizing the principle of protein-dye binding. *Anal. Biochem.* **1976**, *72*, 248–254. [CrossRef] [PubMed]
27. Xie, H.; Xie, F.; Guo, Y.; Liang, X.; Peng, L.; Li, M.; Tang, Z.; Peng, K.; Yang, C. Fermentation quality, nutritive value and in vitro ruminal digestion of Napier grass, sugarcane top and their mixed silages prepared using lactic acid bacteria and formic acid. *Grassl. Sci.* **2023**, *69*, 23–32. [CrossRef]
28. Cao, Y.; Cai, Y.; Takahashi, T.; Yoshida, N.; Tohno, M.; Uegaki, R.; Nonaka, K.; Terada, F. Effect of lactic acid bacteria inoculant and beet pulp addition on fermentation characteristics and in vitro ruminal digestion of vegetable residue silage. *J. Dairy Sci.* **2011**, *94*, 3902–3912. [CrossRef]
29. Denman, S.E.; McSweeney, C.S. Development of a real-time PCR assay for monitoring anaerobic fungal and cellulolytic bacterial populations within the rumen. *FEMS Microbiol. Ecol.* **2006**, *58*, 572–582. [CrossRef]
30. Ren, Y.; Yu, G.; Shi, C.; Liu, L.; Guo, Q.; Han, C.; Zhang, D.; Zhang, L.; Liu, B.; Gao, H.; et al. Majorbio Cloud: A one-stop, comprehensive bioinformatic platform for multiomics analyses. *IMeta* **2022**, *1*, e12. [CrossRef]
31. Muller, Z.O. Feeding potential of pineapple waste for cattle. *World Anim. Rev.* **1978**, *25*, 25–29.
32. Ren, H.; Feng, Y.; Liu, T.; Li, J.; Wang, Z.; Fu, S.; Zheng, Y.; Peng, Z. Effects of different simulated seasonal temperatures on the fermentation characteristics and microbial community diversities of the maize straw and cabbage waste co-ensiling system. *Sci. Total Environ.* **2020**, *708*, 135113. [CrossRef]
33. Wilkinson, J.M.; Bolsen, K.K.; Lin, C.J. History of Silage. *Silage Sci. Technol.* **2003**, *42*, 1–30. [CrossRef]
34. Mugabe, W.; Shao, T.; Li, J.; Dong, Z.; Yuan, X. Effect of hexanoic acid, Lactobacillus plantarum and their combination on the aerobic stability of napier grass silage. *J. Appl. Microbiol.* **2020**, *129*, 823–831. [CrossRef]
35. Javed, A.; Ajmal, M.; Hanif, N.Q.; Akram, A. Effects of inoculation of corn silage with Saccharomyces cerevisiae on silage fermentation characteristics, nutrient digestibility, mycoflora and aflatoxin production. *Nat. Prod. Res.* **2023**, *32*, 1–10. [CrossRef] [PubMed]
36. Ridwan, R.; Abdelbagi, M.; Sofyan, A.; Fidriyanto, R.; Astuti, W.D.; Fitri, A.; Sholikin, M.M.; Rohmatussolihat; Sarwono, K.A.; Jayanegara, A.; et al. A meta-analysis to observe silage microbiome differentiated by the use of inoculant and type of raw material. *Front. Microbiol.* **2023**, *14*, 1063333. [CrossRef] [PubMed]
37. Kim, D.H.; Lee, K.D.; Choi, K.C. Role of LAB in silage fermentation: Effect on nutritional quality and organic acid production—An overview. *AIMS Agric. Food* **2021**, *6*, 216–234. [CrossRef]
38. He, L.; Zhou, W.; Wang, Y.; Wang, C.; Chen, X.; Zhang, Q. Effect of applying lactic acid bacteria and cellulase on the fermentation quality, nutritive value, tannins profile and in vitro digestibility of Neolamarckia cadamba leaves silage. *J. Anim. Physiol. Anim. Nutr.* **2018**, *102*, 1429–1436. [CrossRef] [PubMed]
39. Lara, E.C.; Basso, F.C.; de Assis, F.B.; Souza, F.A.; Berchielli, T.T.; Reis, R.A. Changes in the nutritive value and aerobic stability of corn silages inoculated with *Bacillus subtilis* alone or combined with *Lactobacillus plantarum*. *Anim. Prod. Sci.* **2016**, *56*, 1867–1874. [CrossRef]
40. Zhao, G.; Wu, H.; Li, L.; He, J.; Hu, Z.; Yang, X.; Xie, X. Effects of applying cellulase and starch on the fermentation characteristics and microbial communities of Napier grass (*Pennisetum purpureum* Schum.) silage. *J. Anim. Sci. Technol.* **2021**, *63*, 1301–1313. [CrossRef] [PubMed]
41. Upadhyay, A.; Lama, J.P.; Tawata, S. Utilization of pineapple waste: A review. *J. Food Sci. Technol.* **2010**, *6*, 10–18. [CrossRef]

42. Soundharrajan, I.; Park, H.S.; Rengasamy, S.; Sivanesan, R.; Choi, K.C. Application and Future Prospective of Lactic Acid Bacteria as Natural Additives for Silage Production—A Review. *Appl. Sci.* **2021**, *11*, 8127. [CrossRef]
43. Cai, Y.; Benno, Y.; Ogawa, M.; Kumai, S. Effect of Applying Lactic Acid Bacteria Isolated from Forage Crops on Fermentation Characteristics and Aerobic Deterioration of Silage. *J. Dairy Sci.* **1999**, *82*, 520–526. [CrossRef]
44. Cai, Y.; Du, Z.; Yamasaki, S.; Nguluve, D.; Tinga, B.; Macome, F.; Oya, T. Community of natural lactic acid bacteria and silage fermentation of corn stover and sugarcane tops in Africa. *Asian-Australas. J. Anim. Sci.* **2020**, *33*, 1252–1264. [CrossRef]
45. Du, Z.; Yamasaki, S.; Oya, T.; Nguluve, D.; Euridse, D.; Tinga, B.; Macome, F.; Cai, Y. Microbial Co-occurrence Network and Fermentation Information of Natural Woody-Plant Silage Prepared with Grass and Crop By-Product in Southern Africa. *Front. Microbiol.* **2022**, *13*, 756209. [CrossRef] [PubMed]
46. Li, B.; Zhang, X.; Guo, F.; Wu, W.; Zhang, T. Characterization of tetracycline resistant bacterial community in saline activated sludge using batch stress incubation with high-throughput sequencing analysis. *Water Res.* **2013**, *47*, 4207–4216. [CrossRef] [PubMed]
47. Wang, X.; Wang, J.; Zhu, Y.; Zhu, X.; Qin, H.; Bian, K.; Tang, X. Comparison in structure and predicted function of epiphytic bacteria on Neopyropia yezoensis and Neopyropia katadae. *J. Oceanol. Limnol.* **2023**, *41*, 2232–2248. [CrossRef]
48. Santos, A.O.; Ávila, C.L.S.; Schwan, R.F. Selection of tropical lactic acid bacteria for enhancing the quality of maize silage. *J. Dairy Sci.* **2013**, *96*, 7777–7789. [CrossRef]
49. Ogunade, I.M.; Martinez-Tuppia, C.; Queiroz, O.C.M.; Jiang, Y.; Drouin, P.; Wu, F.; Vyas, D.; Adesogan, A.T. Silage review: Mycotoxins in silage: Occurrence, effects, prevention, and mitigation. *J. Dairy Sci.* **2018**, *101*, 4034–4059. [CrossRef]

fermentation

Article

Illumina Sequencing and Metabolomic Analysis Explored the Effects of the Mixed Silage of Rice Straw and Chinese Cabbage Waste on Fecal Microorganisms and Metabolites in Hu Sheep

Chuang Li [1,2,†], Zhiqi Lu [2,†], Ruxin Qi [2], Zhenbin Zhang [2], Yue Lu [2], Muhammad Hammad Zafar [2], Kailun Yang [1,*] and Mengzhi Wang [2,3,*]

1 College of Animal Science, Xinjiang Agricultural University, Urumqi 830091, China; chuangliyzu@163.com
2 Laboratory of Metabolic Manipulation of Herbivorous Animal Nutrition, College of Animal Science and Technology, Yangzhou University, Yangzhou 225009, China; zqlu73@163.com (Z.L.); ruxinqi@126.com (R.Q.); 15861273716@163.com (Y.L.); hammadzafar075@gmail.com (M.H.Z.)
3 State Key Laboratory of Sheep Genetic Improvement and Healthy Production, Xinjiang Academy of Agricultural Reclamation Sciences, Shihezi 832000, China
* Correspondence: ykl@xjau.edu.cn (K.Y.); mzwang@yzu.edu.cn (M.W.)
† These authors contributed equally to this work.

Citation: Li, C.; Lu, Z.; Qi, R.; Zhang, Z.; Lu, Y.; Zafar, M.H.; Yang, K.; Wang, M. Illumina Sequencing and Metabolomic Analysis Explored the Effects of the Mixed Silage of Rice Straw and Chinese Cabbage Waste on Fecal Microorganisms and Metabolites in Hu Sheep. *Fermentation* **2024**, *10*, 233. https://doi.org/10.3390/fermentation10050233

Academic Editors: Wansup Kwak and Siran Wang

Received: 19 February 2024
Revised: 1 April 2024
Accepted: 4 April 2024
Published: 26 April 2024

Abstract: Silage is the most important component of a ruminant diet and has important production and health significance in ruminant production. The aim of the research was to investigate how the mixed silage of Chinese cabbage waste and rice straw (mixed silage) impacts the fecal microorganisms and metabolites in Hu sheep using Illumina sequencing and metabolomic analysis. A total of 16 Hu sheep (8 rams and 8 ewes) weighing about 39 kg and 5.5 months old were used as experimental sheep and divided into two groups (4 rams and 4 ewes, $n = 8$) using the principle of randomized trials: the control group with peanut sprouts, corn husks, and sorghum husks as roughage and the silage group with the mixed silage as roughage. There were no significant differences in the average daily gain (ADG), dry matter intake (DMI), or feed conversion rate (FCR) between the control group and the mixed silage groups ($p > 0.05$). Microbiome results showed that 15 microorganisms such as *Ruminococcaceae UCG 010*, *Breznakia*, *Erysipelothrix*, *Desulfovibrio*, *Succiniclasticum*, and *Shuttleworthia* were significantly different between the two groups. In addition, metabolomics showed that the mixed silage modulated the concentrations and metabolic pathways of metabolites in the manure. Significantly different metabolites were mainly enriched in amino acid anabolism ("glycine, serine, and threonine metabolism", "valine, leucine, and isoleucine biosynthesis", "arginine biosynthesis", etc.), nucleic acid metabolism (pyrimidine metabolism). In conclusion, the addition of mixed silage to the diet of Hu sheep can alter the structure of the hindgut microflora and regulate the metabolism of amino acids and nucleotides, which affects health performance.

Keywords: Chinese cabbage waste; fecal microorganisms; growth performance; Hu sheep; metabolomics; mixed silage; rice straw

1. Introduction

Between 2005 and 2050, global demand for meat and milk will increase by 57% and 48%, respectively, which is a huge challenge for the livestock industry [1]. Forage resources of sufficient quality and quantity are necessary to meet human demand for animal products. However, this is a great challenge for most countries. Therefore, researchers are turning to unconventional feed ingredients. Most non-conventional feeds have limited nutritional value, and some modifications should be considered before feeding to ruminants for optimal performance. Silage is an effective and widely used technology for long-term preservation of feed, with the characteristics of low cost and easy operation [2].

Silage is the most important component of ruminant diets and has important production and health implications in ruminant production by maintaining good rumen functional status and reducing the risk of diseases such as rumen acidosis and crumpled stomach displacement [3].

The nutritional value of rice straw is low, with protein content only accounting for 2–5% of dry matter (DM), fiber and lignin content exceeding 50%, and DM digestibility being low, resulting in feed intake of 1.5–2.0% [4]. Therefore, rice straw can be used as a good source of roughage for ruminants only after further treatment. Typically, rice straw stalks are harvested and collected after wilting and drying to yellow stalks. During the air-drying process, the water in the rice straw is almost completely evaporated, so the free sugar in the air-dried rice is difficult to obtain. Therefore, lower water-soluble carbohydrates and a higher lignin content will seriously affect the quality of silage [5]. In this regard, vegetable waste has high water-soluble carbohydrates and moisture content, which can make up for the shortcomings of rice straw in silage [6]. Previous research results showed that when the mixed feed of broccoli byproducts and wheat straw (ratio 69:31) in the DM diet does not exceed 200 g/kg, the growth performance and apparent digestibility of nutrients and rumen fermentation parameters of Fashandy lambs will not be affected [7].

Fecal samples are not only convenient to collect, but also have a unique metabolome of nutrients from feed, gut microbiota, and host products [8]. Due to the emergence of high-throughput sequencing technology, we have a deeper understanding of the composition, structure, and interactions of microorganisms in the gastrointestinal tract of livestock and poultry [9]. The gut microbiota plays a crucial role in the regulation of carbohydrates, amino acids, lipids, vitamins, and mineral metabolism [10,11], along with influencing the immune system of the host [12,13]. Studies have shown that dietary changes can alter the microflora structure of the gut [14,15]. Studies have explored the impact of gut microbiota on metabolic function through metabolomics approaches. Metabolomics is a high-throughput sequencing technology, following genomics and proteomics, that provides new insights into the effects of diet, drugs, and disease, and aims to characterize and quantify all small molecules in a sample [16]. It has been used in recent years to detect plasma, fecal, rumen, and tissue metabolite biomarkers in humans and animals [17–20].

There is limited information on the effect of the mixed silage as an unconventional roughage on the fecal microbiota and metabolites in Hu sheep. Therefore, the purpose of this study was to explore the effects of the mixed silage on hindgut microbiota composition and fecal metabolites in Hu sheep using 16s RNA gene sequencing technology and LC-MS metabolomics.

2. Materials and Methods

2.1. Experimental Diet, Experimental Animals, and Feeding

The mixed silage was made based on previous research [15]. A total of 16 Hu sheep (8 rams and 8 ewes) weighing about 39 kg and 5.5 months old were used as experimental sheep and divided into two groups (4 rams and 4 ewes, $n = 8$) using the principle of randomized trials: the control group with peanut sprouts, corn husks, and sorghum husks as roughage and the silage group with the mixed silage as roughage. The feed formulation was designed based on the nutritional requirements of mutton sheep (NRC, 2007) [21]. The entire experiment spanned 35 days, consisting of a preliminary 7-day pre-feeding phase followed by an official 28-day trial period. All the Hu sheep in the experiment were raised together in one pen, and prior to the study, the sheep enclosure underwent cleaning and disinfection. Both control and silage groups were fed 50% roughage and 50% concentrate in equal portions at 8:00 and 18:00. Table 1 provides the dietary composition and nutrient levels for both the control and silage groups. Before entering the sheep house, all test sheep received uniform deworming and immunization. Throughout the experiment, a feeding management system with unlimited access to food and water was implemented.

Table 1. Experimental diet formula and nutrition level (DM basis/%).

Items	Treatment [1]	
	Control	Silage
Ingredients (% of DM)		
Peanut seedling	30	-
Corn husk	15	-
Sorghum shell	5	-
Mixed Silage	0	50
Corn	34	34
Soybean meal	7	5.5
Bran	7.5	8
Corn gluten meal	-	1
$NaHCO_3$	0.5	0.5
Premix contained [2]	0.5	0.5
Salt	0.5	0.5
Total	100	100
Nutrient composition (% of DM)		
Digestive energy/DE (MJ/kg) [3]	13.52	14.73
Metabolizable energy/ME (MJ/kg) [4]	18.93	20.62
Crude protein, CP	15.08	15.11
Ash	4.36	12.33
Neutral Detergent Fiber, NDF	47.64	48.23
Acid Detergent Fiber, ADF	23.71	27.17
Ca	0.48	0.45
P	0.38	0.39

Note: [1] Control: control group, the control group with peanut sprouts, corn husks, and sorghum husks as roughage; Silage: silage group, the silage group with the mixed silage as roughage. [2] Premix contained (per kg): VA 70~130 kIU, VD 315~30 kIU, VE $\geq$ 130 kIU, Fe 0.4~0.8 g, Mn 0.5~1.0 g, Zn 1.5~3.0 g, Cu 0.1~0.2 g, Se 4~8 mg, Ca 8~16%, P $\geq$ 1%, NaCl 5~10%. [3,4] DE and ME were estimated according to NRC (2007). The others were measured values. The experimental design of this study was the same as that of Li et al. [15].

2.2. *Sample Collection and Measurement*

2.2.1. Growth Performance

During the experiment, the quantity of feed and remaining feed for every sheep were documented, and DMI was calculated. Each sheep was weighed before the start of the formal trial and at the end of the experiment before the morning feed to record its body weight, and ADG was calculated. Feed efficiency is expressed as FCR, which is the ratio of DMI to ADG.

2.2.2. Fecal Sample Collection

On the eve of concluding the feeding experiment, random rectal fecal samples were gathered from six Hu sheep in each group, with each individual sample weighing approximately 30 g. The samples were rapidly placed in liquid nitrogen to avoid the influence of environmental microorganisms on the samples. Subsequently, they were transferred to a −80 °C refrigerator for storage and used for the determination of fecal microorganisms and metabolomes.

2.2.3. 16S rRNA Microbial Community Analysis

Total DNA was extracted from 12 fecal samples using the HiPure Soil DNA Kit soil DNA extraction kit (Beijing Tiangen Biochemical Technology Co., Ltd., Beijing, China). The A260/A280 values were measured by an ultramicro-volume spectrophotometer (NzanoDrop-1000, Thermo Fisher Scientific Co., Ltd., Waltham, MA, USA), and electrophoresis was carried out using a 1% agarose gel to check the integrity and purity of the extracted DNA. Primers 341F (5′-CCTACGGGGGNGGCWGCAG-3′) and 806R (5′-GGACTACHVGGGGTWTCTAAT-3′) were used for PCR amplification of the V3 to V4 region of the 16S rDNA gene. The PCR amplification system was a 30 μL reaction system, which included 15 μL of 2 × Phanta Master Mix, 1 μL of Bar-PCR primer F (10 μM), 1 μL of Primer R (10 μM), 10 ng of Genomic

DNA, and ddH_2O. Thermal cycling conditions were pre-denaturation at 95 °C for 5 min, followed by denaturation at 95 °C for 30 s, annealing at 55 °C for 30 s, and extension at 72 °C for 45 s for 27 cycles; the final extension was at 72 °C for 10 min. After the PCR is completed, the products are extracted by 2% agar gel electrophoresis and further purified and quantified. The Illumina Novaseq PE250 platform was used to sequence the amplified products (Genepioneer Biotechnologies Co., Ltd., Nanjing, China).

The raw sequenced sequences were quality controlled and spliced using fastp (V 0.20.0) and FLASH (V 1.2.7) software, respectively [22,23]. The quality sequences were clustered at a 97% similarity level according to UCLUST (V.1.2.22q) software to obtain each OTU. The taxonomic information of the species corresponding to each OTU can be obtained by comparing the OTUs with the SILVA rRNA database [24].

OTU-based dilution curves were used to evaluate sequencing coverage depth and were accomplished with the R (V 4.0.2) "vegan". Fecal microbial Venn diagrams were calculated with R (V 4.0.2) "VennDiagram" and visualized with "ggplot2". Microbial richness (ACE and Chao1 indices), microbial diversity (Shannon and Simpson indices), and microbial coverage (goods_coverage) were calculated using the R (V 4.0.2) "picante". Principal Coordinate Analysis (PCoA, weighted UniFrac distance) was used to assess the colony structure of the different microorganisms, and its visualization was done with the R "ggplot2" package [25]. The top ten dominant bacterial groups in phyla and genera were also visualized with R (V 4.0.2) "ggplot2". Analysis of similarity (ANOSIM) is used to determine the similarity of groups, with "0" indicating indistinguishable and "1" indicating distinguishable. LDA effect size (LEfSe) analysis was applied to screen for signature differential flora in feces of Hu sheep fed different roughages. The Harvard University online analysis platform "http://huttenhower.sph.harvard.edu/galaxy (accessed on 20 December 2021)" was used to complete the LEfSe analysis. We believe that this bacterial group differed between the two groups, when the linear discriminant analysis (LDA) effect value was greater than 3.

2.2.4. Untargeted Metabolomics Based on Liquid Chromatography-Mass Spectrometry and Data Processing

The untargeted metabolome of fecal samples was analyzed by liquid chromatography-mass spectrometry (Waters, UPLC; Thermo, Q Exactive) platforms. A total of 50 mg of fecal sample was weighed into a 1.5 mL centrifuge tube, and 800× µL of 80% methanol was added and ground for 90 s at 65 Hz, and ground for 90 s. After sufficient mixing, the sample was sonicated for 30 min at 4 °C and then allowed to stand at −40 °C for 1 h with vortex oscillation for 30 s. The sample was then separated from the sample for 1 h with the vortex oscillator. The sample mix needs to be centrifuged (12,000 rpm) for 15 min, but it needs to be left at 4 °C for 30 min first. All of the supernatant solution was pipetted into a centrifuge tube at −40 °C for 60 min, and then centrifuged again at 4 °C at 12,000 rpm for 15 min, next, 200 µL of the supernatant was transferred to a vial, with 0.14 mg/mL dichlorophenylalanine of the internal standard being added, and stored at −80 °C to be analyzed on the machine.

A small amount of treated fecal extract was taken and separated by chromatography through a chromatograph. The model of the HPLC column was ACQUITY UPLC HSS T3 (2.1 × 100 mm, 1.8 µm). Chromatographic separations were performed in positive ion mode (ESI+) and negative ion mode (ESI−) using a preheated SuperGold C18 column (100 × 4.6 mm, 3 µm i.d.). Water and 0.05% acetonitrile with 0.1% formic acid formed solvent A. Acetonitrile mixed with 0.1% formic acid formed solvent B. The samples were eluted in the mobile phase, consisting of solvent A and solvent B at a flow rate of 0.3 µL/min. The elution of the samples was divided into a total of three steps, with the ratios of mobile phase A to mobile phase B being 95%:5%, 5%:95%, and 95%:5%, respectively; and the elution times for each step were 1 min, 25.5 min, and 29.6 min, respectively. QC samples need to be tested along with the test samples for monitoring and evaluating the stability and reliability of the system's test data. The Q Exactive HF-X (Thermo Fisher

Scientific, Waltham, MA, USA) mass spectrometer operates in positive/negative mode, with each sample detected under positive (ESI+) and negative (ESI−) ion mode conditions. Mass spectrometry parameter conditions differ in the ESI+ and ESI− modes. The mass spectrometry parameter conditions were different in positive and negative modes. In ESI+ mode, the heater temperature was 300 °C, the sheath gas flow rate was 45 arb, the auxiliary gas flow rate was 15 arb, the tail gas flow rate was 1 arb, the electrospray voltage was 3 kV, the capillary temperature was 350 °C, and the S-Lens RF Level was 30%. In ESI− mode, only the electrospray voltage (3.2 kV) and the S-Lens RF Level (60%) were changed; all other conditions remained the same. Peak extraction, baseline correction, and peak matching based on mass spectrometry data were performed using ChromaTOF software (V4.3x, LECO), and these peaks were matched to the LECO/Fiehn Metabolomics Library database to identify metabolites.

The normalized data were analyzed and visualized by multivariate analysis in SIMCA (V 14.1) software. The horizontal coordinates of the supervised partial least squares (OPLS-DA) score plot indicate the score values of the main components of the orthogonal signal correction (OSC) process, from which intergroup differences can be seen; the vertical coordinates indicate the score values of the orthogonal components of the OSC process; and intragroup differences (differences between samples within a group) can be seen on the vertical coordinates. The R2 and Q2 values of the OPLS-DA model were used to assess the model's validity. VIP values were obtained from the OPLS-DA model, and p values were calculated by a one-way ANOVA with a t test. Differential metabolites were screened with VIP > 1 and $p < 0.05$. KEGG-enriched metabolic pathways were accomplished through the MetaboAnalyst online open-source website "https://www.metaboanalyst.ca (accessed on 7 July 2022)". $p < 0.05$ was used as the criterion for having a significant functional pathway among the differential metabolites.

2.3. Data Analysis

Growth performance data were statistically analyzed using an independent t-test in SPSS 20.0 software (IBM Corporation, Armonk, NY, USA), and differences were considered significant at $p < 0.05$. Fecal microbiota and metabolites were correlated using Spearman's rank correlation, and R and p values were calculated using the R (V 4.0.2) "Psych", and correlations were considered when $p < 0.05$ and R > $|0.8|$. Correlation heatmaps were visualized using the R (V 4.0.2) "ggcorrplot".

3. Results

3.1. Growth Performance

The effects of the mixed silage on the growth performance of Hu sheep are shown in Table 1. There were no differences in the ADG (Figure 1A), DMI (Figure 1B), or FCR (Figure 1C) between the control and silage groups ($p > 0.05$).

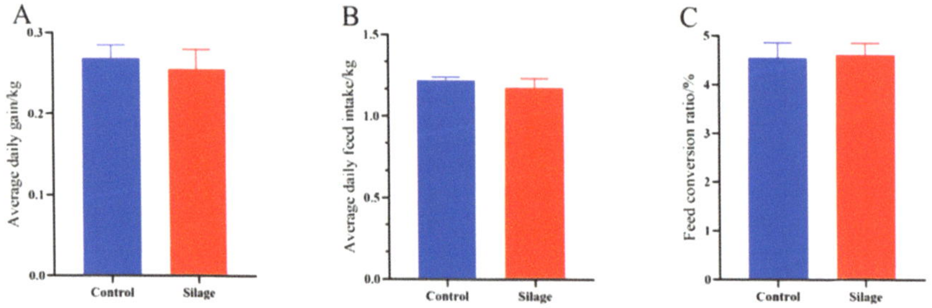

Figure 1. Effects of the mixed silage on growth performance in Hu sheep ($n = 6$). (**A**) Average daily feed intake; (**B**) average daily gain; (**C**) feed conversion ratio. Control: control group, the control group with peanut sprouts, corn husks and sorghum husks as roughage; Silage: silage group, the silage group with the mixed silage as roughage.

3.2. Fecal Microbiota Structure

In the fecal microbiome, a total of 1,111,942 reads were obtained. Based on 97% sequence similarity, an average of 894,173 valid data points were obtained after quality control, yielding 1839 OTUs. The results of the rarefaction curves (Supplementary Figure S1A) showed that with the gradual increase of sequences, the curve was in the rising stage, and a large number of microorganisms were found in the sample. As the number of measured sequences continued to increase, the dilution curve had gradually flattened, indicating that sequencing depth could cover most of the microbial groups in the sample and could be used for later experiments. The venn diagram results showed (Supplementary Figure S1B) that the number of OTUs in the control group was lower than the silage group, with 1380 OTUs shared between the two groups. In addition, the control group had exclusive access to 107 OTUs, and the silage group had exclusive access to 352 OTUs. The results of alpha diversity (Supplementary Figure S2) showed that the replacement of roughage in the control group by the mixed silage did not affect the Richness, Shannon, Simpson, Chao1, ACE, and goods_coverage indices of the microorganisms in manure ($p > 0.05$). The PCoA results (Figure 2A) showed that the contribution of the first and second principal coordinates of PCoA was 33.55% and 15.51%, respectively. Anosim analyses showed that the fecal microflora structure of Hu sheep was affected by the mixed silage ($p = 0.011$). At the phyla level (Figure 2B), *Firmicutes, Bacteroidetes,* and *Proteobacteria* were the dominant phyla in the both groups, followed by *Spirochaetes, Verrucomicrobia, Planctomycetes, Patescibacteria, Fibrobacteres, Kiritimatiellaeota,* and *Actinobacteria.* At the genera level (Figure 2C), *Ruminococcaceae UCG-005* and the *Christensenellaceae R-7 group* were the dominant genera in the control group, and the *Christensenellaceae R-7 group* and the *Rikenellaceae RC9 gut group* were the dominant genera in the silage group. The results of LEfSe analysis (Figure 2D) showed that a total of 15 differentiated microorganisms were obtained, of which four were in the control group, namely *Ruminococcaceae UCG_010, Breznakia, Erysipelothrix,* and *Desulfovibrio*, and eleven in the silage group, which were mainly distributed in the *Bacilli, Clostridiaceae1, Pseudomonadales, Succiniclasticum,* and *Shuttleworthia.*

3.3. Fecal Metabolomics

Quality control (QC) samples were used throughout the trial to assess the stability and reproducibility of the data. The quality of the data is directly proportional to the correlation of the QC samples. The values for the R-type 2 ESI+ and ESI− polarity modes are 1 and 0.79, respectively (Supplementary Figure S3). OPLS-DA with supervision was used to distinguish differences in metabolic profiles between control and silage groups, and samples from control and silage groups were significantly separated in ESI+ or ESI− mode (Figure 3A,C). In addition, there was more aggregation between samples within the silage group than in the control group. These results suggested that silage can alter fecal metabolites in Hu sheep. After 200 response replacement tests, OPLS-DA calculated the regression intercept (Q2), and we considered the model to be valid when Q2 was less than zero. From (Figure 3B,D), the Q2 value is −0.199 for the ESI+ model (Figure 3B) and −0.858 for the ESI− model (Figure 3D). Therefore, we concluded that the model was reliable and stable.

The metabolites of the feces from the control and silage groups were analyzed by LC-MS, and 582 and 612 peaks were retained after preprocessing the raw data for ESI+ and ESI− modes, resulting in the identification and quantification of 217 positive and 231 negative compounds. Among these compounds, 19 positively ionized compounds and 59 negatively ionized compounds were differentiated (VIP > 1 and $p < 0.05$), and these metabolites were mainly distributed among amino acids, peptides, and analogues, benzoic acids and derivatives, bile acids, alcohols and derivatives, carbohydrates and carbohydrate conjugates, carbonyl compounds, eicosanoids, fatty acids and conjugates, isoflav-2-enes, pyrimidine 2′-deoxyribonucleosides, tetrahydrofuran lignans, and others (Table 2).

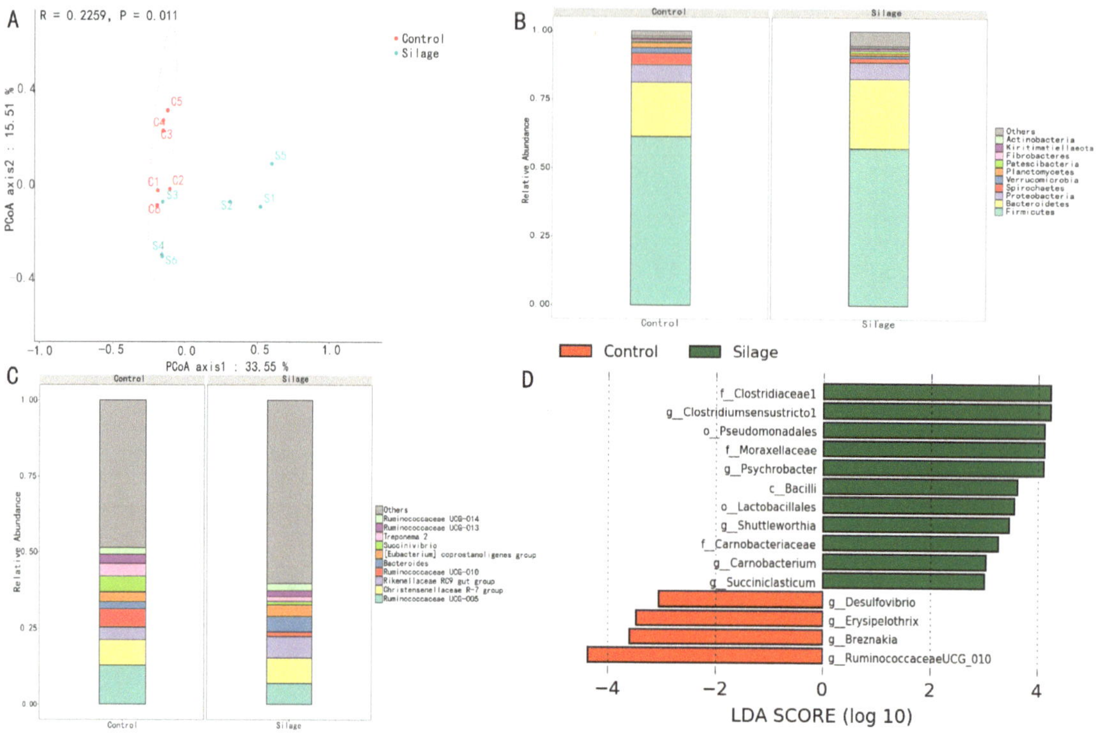

Figure 2. The structural composition of fecal microflora in the control group and silage group ($n = 6$). (**A**) Principal coordinate analysis (PCoA) demonstrated the separation of microbial communities in the feces of Hu sheep fed two diets based on the Bray–Curtis dissimilarity matrix. The top ten phyla (**B**) and genera (**C**) of fecal microbiota in the control group and silage group. (**D**) Discriminant analysis of fecal microbial OTU of Hu sheep fed with diet in the control group and silage group, with LDA score > 3. Control: control group, the control group with peanut sprouts, corn husks, and sorghum husks as roughage; Silage: silage group, the silage group with the mixed silage as roughage.

Table 2. Contents of differential metabolites in feces ($n = 6$).

Metabolites	Group [1]		VIP [2]	p	FC [3]	Trend	Mode
	Control	Silage					
Amino acids, peptides, and analogues							
DL-Alanine	0.079	0.242	1.403	0.018	3.071	Up	ESI+
Glycine	0.003	0.011	1.637	0.001	3.633	Up	ESI+
LEVODOPA	0.004	0.008	1.430	0.013	1.950	Up	ESI+
N,N-Dimethylglycine	0.012	0.003	1.277	0.042	0.268	Down	ESI+
Ala-Ile	0.065	0.027	1.119	0.045	0.407	Down	ESI−
Arginine	0.053	0.004	1.299	0.006	0.079	Down	ESI−
D-ASPARTATE	0.644	0.093	1.375	0.002	0.145	Down	ESI−
gamma-Glutamylleucine	0.083	0.027	1.407	0.001	0.328	Down	ESI−
L-Histidine	0.055	0.007	1.147	0.046	0.136	Down	ESI−
L-Phenylalanine	1.630	0.474	1.165	0.040	0.291	Down	ESI−
L-Valine	0.077	0.018	1.197	0.022	0.231	Down	ESI−
N-Acetylglutamic acid (NAG)	0.641	0.230	1.220	0.012	0.359	Down	ESI−
N-Isobutyrylglycine	1.345	0.017	1.206	0.015	0.012	Down	ESI−
N-Tigloylglycine	0.861	0.003	1.199	0.025	0.004	Down	ESI−

Table 2. *Cont.*

Metabolites	Group [1]		VIP [2]	*p*	FC [3]	Trend	Mode
	Control	Silage					
Benzoic acids and derivatives							
2-Hydroxyhippuric acid	0.544	0.007	1.199	0.028	0.013	Down	ESI−
5-Methoxysalicylic acid	0.018	0.005	1.135	0.049	0.277	Down	ESI−
Butylparaben	0.231	0.063	1.144	0.019	0.275	Down	ESI−
Bile acids, alcohols, and derivatives							
Cholic Acid	0.015	0.009	1.419	0.011	0.578	Down	ESI+
Glycochenodeoxycholate	0.014	0.001	1.274	0.045	0.040	Down	ESI+
GLYCOCHOLATE	0.005	0.002	1.411	0.011	0.441	Down	ESI+
Carbohydrates and carbohydrate conjugates							
N-Acetylneuraminic acid	0.018	0.133	1.487	0.006	7.483	Up	ESI+
D-SACCHARIC ACID	0.166	0.053	1.425	0.001	0.318	Down	ESI−
N-Acetylmannosamine	8.252	2.333	1.232	0.010	0.283	Down	ESI−
N-Acetylmuramic Acid	0.636	0.141	1.425	0.001	0.222	Down	ESI−
Carbonyl compounds							
Acetophenone	0.243	0.035	1.627	0.001	0.145	Down	ESI+
4-Hydroxybenzaldehyde	0.844	0.276	1.424	0.001	0.327	Down	ESI−
Eicosanoids							
Prostaglandin B1	0.007	0.079	1.272	0.007	11.976	Up	ESI−
Resolvin E1	0.011	0.054	1.376	0.002	4.752	Up	ESI−
Fatty acids and conjugates							
3,3-Dimethylglutaric acid	0.642	0.110	1.577	<0.001	0.172	Down	ESI−
3-Methylglutaric acid	0.050	0.224	1.154	0.032	4.455	Up	ESI−
Arachidic acid	0.322	1.498	1.126	0.016	4.655	Up	ESI−
Lauric acid	0.061	0.027	1.224	0.006	0.454	Down	ESI−
Isoflav-2-enes							
Daidzein	0.176	0.050	1.147	0.024	0.283	Down	ESI−
Genistein	2.530	0.442	1.376	0.002	0.175	Down	ESI−
Pyrimidine 2′-deoxyribonucleosides							
2′-Deoxyuridine	0.002	0.006	1.564	0.003	3.150	Up	ESI+
Thymidine	0.018	0.051	1.330	0.030	2.752	Up	ESI+
Tetrahydrofuran lignans							
Enterolactone	0.018	0.110	1.118	0.029	6.082	Up	ESI−
matairesinol	0.163	0.015	1.202	0.010	0.091	Down	ESI−
Others							
Benzothiazole	1.248	1.073	1.295	0.033	0.860	Down	ESI+
Metaxalone	0.003	0.001	1.319	0.029	0.219	Down	ESI+
9-Fluorenone	5.195	0.812	1.172	0.017	0.156	Down	ESI−
Abietic acid	0.005	0.016	1.135	0.034	3.514	Up	ESI−
Biotin	0.038	0.082	1.157	0.047	2.143	Up	ESI−
Bis(4-hydroxyphenyl)methane	0.028	0.074	1.168	0.022	2.634	Up	ESI−
cirsimaritin	0.033	0.002	1.508	<0.001	0.074	Down	ESI−
delta7-Dafachronic acid	0.090	0.047	1.104	0.048	0.525	Down	ESI−
Ecgonine	0.381	0.062	1.137	0.050	0.162	Down	ESI−
Piceatannol	0.010	0.059	1.393	0.001	6.047	Up	ESI−
Pseudouridine	1.798	0.457	1.160	0.033	0.254	Down	ESI−
Resveratrol	0.071	0.172	1.129	0.031	2.420	Up	ESI−
santin	0.019	0.085	1.286	0.009	4.525	Up	ESI−
URIDINE	1.502	0.455	1.120	0.043	0.303	Down	ESI−
3-Hydroxyphenylacetic acid	0.108	0.021	1.519	<0.001	0.199	Down	ESI−
adrenosterone	0.044	0.001	1.329	0.002	0.025	Down	ESI−
3-Indoxyl sulphate	2.911	0.004	1.160	0.042	0.001	Down	ESI−
Epinephrine	0.190	0.000	1.203	0.024	0.002	Down	ESI−
Isobutyric acid	0.085	0.019	1.208	0.020	0.220	Down	ESI−
Bisphenol A	0.003	0.045	1.409	0.001	16.963	Up	ESI−
Glabranine	1.708	0.069	1.608	<0.001	0.040	Down	ESI−
Stearamide	0.022	0.010	1.365	0.016	0.448	Down	ESI+
Chrysin	0.027	0.001	1.553	<0.001	0.045	Down	ESI−

Table 2. *Cont.*

Metabolites	Group [1]		VIP [2]	p	FC [3]	Trend	Mode
	Control	Silage					
Ferulic acid	0.005	0.012	1.375	0.023	2.257	Up	ESI+
afzelechin	0.020	0.116	1.388	0.002	5.671	Up	ESI−
3-(2-Hydroxyethyl)indole	0.073	0.001	1.171	0.035	0.019	Down	ESI−
D-(+)-Tryptophan	0.459	0.165	1.162	0.041	0.360	Down	ESI−
Equol	0.072	0.003	1.478	<0.001	0.043	Down	ESI−
Dihydrojasmonic Acid	0.306	0.128	1.269	0.011	0.419	Down	ESI−
13-HPODE	2.260	4.175	1.091	0.034	1.847	Up	ESI−
Vanillin	0.025	0.005	1.362	0.003	0.192	Down	ESI−
Formononetin	0.069	0.003	1.522	<0.001	0.045	Down	ESI−
2-Hydroxyphenylacetic acid	0.060	0.004	1.318	0.003	0.060	Down	ESI−
2′-Deoxyinosine	0.029	0.065	1.360	0.021	2.221	Up	ESI+
Uric acid	0.671	0.083	1.165	0.038	0.124	Down	ESI−
4-Pyridoxic acid	0.060	0.026	1.294	0.042	0.426	Down	ESI+
Cytosine	0.045	0.099	1.520	0.004	2.218	Up	ESI+
Acetylcholine	0.151	0.030	1.295	0.043	0.200	Down	ESI+
4-Methyl-2-oxovaleric Acid	1.376	0.415	1.162	0.013	0.302	Down	ESI−

Note: [1] Control: control group, the control group with peanut sprouts, corn husks, and sorghum husks as roughage; Silage: silage group, the silage group with the mixed silage as roughage. [2] VIP: variable importance in projection; [3] FC: fold change.

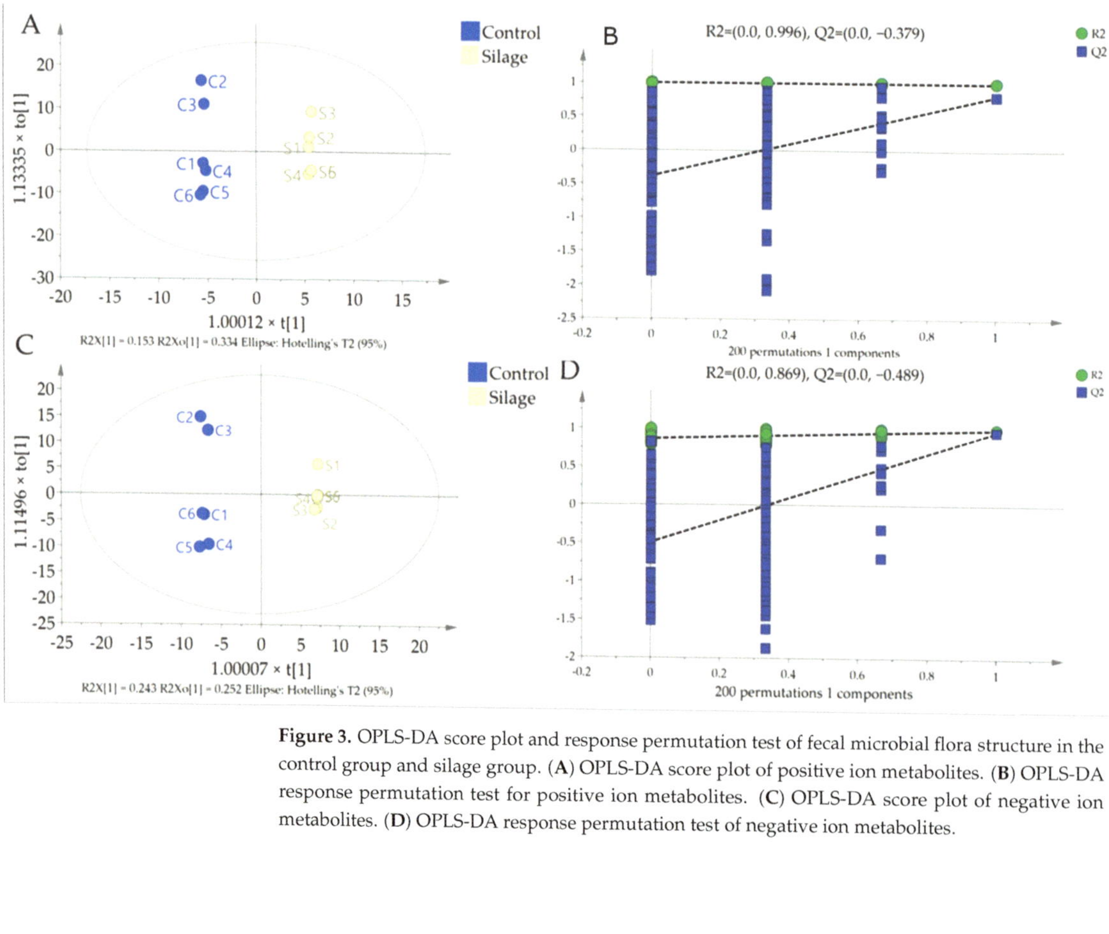

Figure 3. OPLS-DA score plot and response permutation test of fecal microbial flora structure in the control group and silage group. (**A**) OPLS-DA score plot of positive ion metabolites. (**B**) OPLS-DA response permutation test for positive ion metabolites. (**C**) OPLS-DA score plot of negative ion metabolites. (**D**) OPLS-DA response permutation test of negative ion metabolites.

3.4. KEGG Enrichment Analysis

Enrichment analysis based on fecal differential metabolism metabolites (KEGG) showed that the mixed silage altered nine metabolic pathways (Figure 4). In the ESI+ mode (Figure 4A), the mixed silage diet altered four stored metabolic pathways, namely primary bile acid biosynthesis, pyrimidine metabolism, and glycine, serine, and threonine metabolism ($p < 0.05$). In the ESI− mode (Figure 4B), five metabolic pathways were altered, namely aminoacyl-tRNA biosynthesis, valine, leucine, and isoleucine biosynthesis, phenylalanine metabolism, arginine biosynthesis, and phenylalanine, tyrosine, and tryptophan biosynthesis ($p < 0.05$).

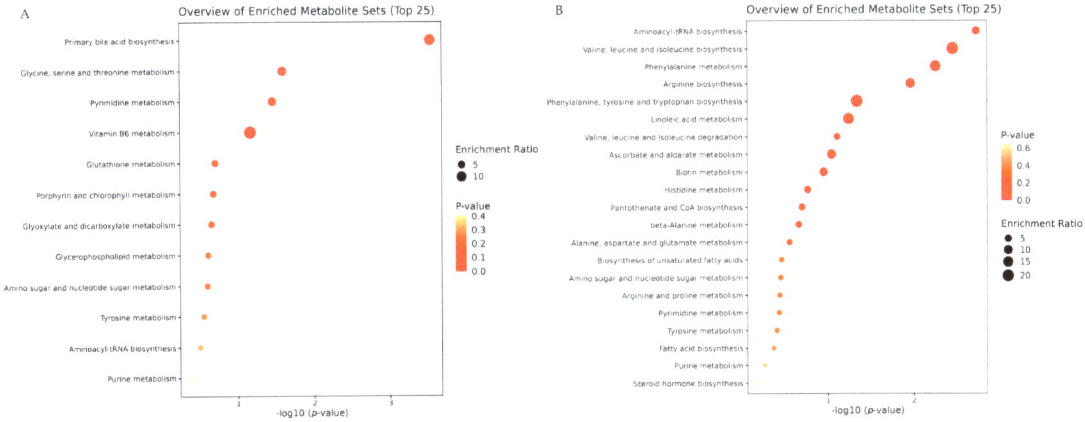

Figure 4. Enrichment analysis (KEGG) of metabolic pathways of differential metabolites in the feces of Hu sheep fed control and silage diets. (**A**) ESI+. (**B**) ESI−.

3.5. Analysis of Differential Metabolites and Fecal Microbial Correlations

The correlation analysis between differential metabolites and differential microorganisms is shown in Figure 5. *Ruminococcaceae UCG_010* and L-valine, D-saccharic acid, lauric acid, NAG, metaxalone, vanillin, 3-hydroxyphenylacetic acid, D-aspartate, 3-indoxyl sulfate, gamma-glutamylleucine, L-phenylalanine, L-histidine, arginine, 3′2-hydroxyethyl indole, and 2-hydroxyhippuric acid were positively correlated, while they were negatively correlated with ferulic acid, santin, and 13-hpode ($p < 0.05$). *Breznakia* has a positive and significant correlation with N-isobutyrylglycine, daidzein, pseudouridine, butylparaben, uridine, equol, and dihydrojasmonic acid ($p < 0.05$). *Desulfovibrio* was positively correlated with pseudouridine, uridine, dihydrojasmonic acid, 2-hydroxyhippuric acid, uric acid, 3-indoxyl sulfate, ecgonine, vanillin, N-tigloylglycine, epinephrine, 3,2-hydroxyethyl indole, L-histidine, arginine, N-isobutyrylglycine, and L-valine ($p < 0.05$). *Carnobacterium* was positively correlated with DL-alanine, N-acetylneuraminic acid, glycine, and ferulic acid. *Psychrobacter* was positively correlated with 3-methylglutaric acid ($p < 0.05$).

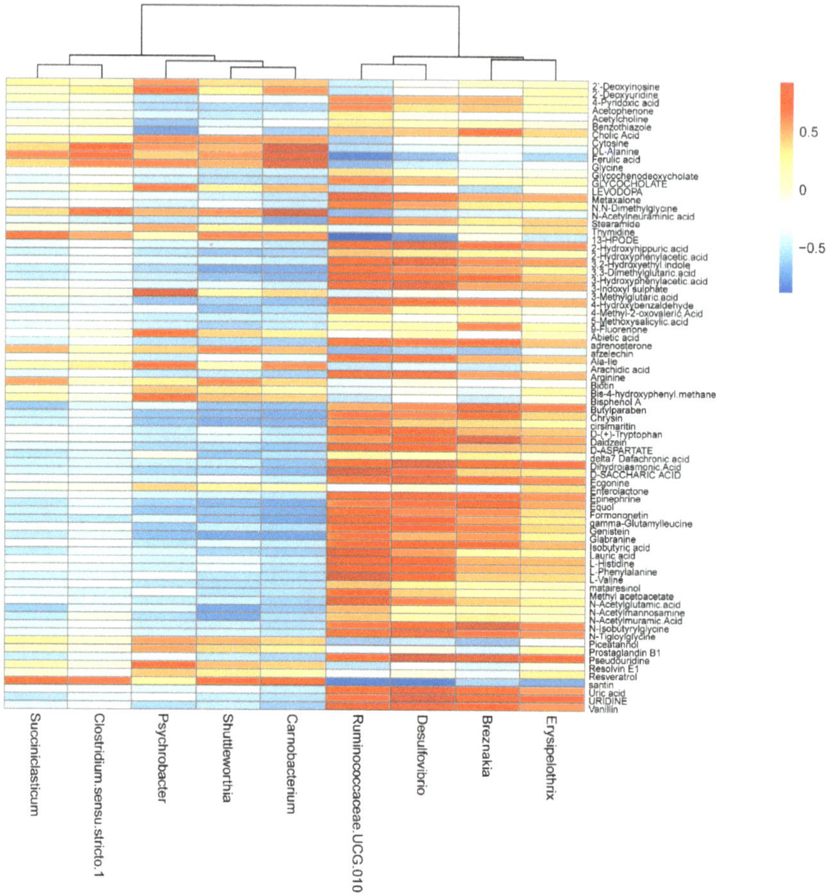

Figure 5. Spearman rank correlation between fecal microbiota and metabolites. Blue indicates negative correlations, and red indicates positive correlations.

4. Discussion

Silage can make up for the shortage of seasonal roughage and contribute to the feeding of ruminants. In this experiment, our results showed that the use of mixed silage in the diets of Hu sheep did not affect growth performance but had an effect on the structure of the microflora, metabolites, and metabolic pathways in the hindgut.

In this experiment, the difference in fecal microbial diversity and abundance between the silage and control groups was not significant, which is consistent with previous studies [26]. PCoA and Anosim analyses indicated that the mixed silage had a highly significant effect on the microflora structure of manure. At the phyla level, our results indicated that the predominant phyla in both groups was *Firmicutes*, followed by *Bacteroidetes* and *Proteobacteria*. Previous studies have shown that the main microorganisms considered to be present in sheep feces are *Firmicutes* [27]. The main phyla were shown to be *Firmicutes*, *Bacteroidetes*, and *Proteobacteria* in a study on the evolution of mammalian gut microorganisms [13]. In the intestinal tract of ruminants, *Bacteroides* plays a very important role in the degradation of cellulose [28]. It has been shown that *Firmicutes* plays a very important role in the digestion of carbohydrates and the fermentation of organic matter [29]. The role of *Proteobacteria* is not fully understood, and further studies are needed [30]. However, studies on the diversity and function of fecal microbiota in sheep have shown that the fecal microbiota is mainly associated with carbohydrate degradation and catabolism [27]. At the genera level, the main microorganisms within the silage group are dominated by the *Christensenellaceae R-7 group* and the *Rikenellaceae RC9 gut group*. The *Christensenellaceae R-7 group* is a genus in the family *Christensenaceae*, which is widespread in the human gut and animal intestinal tract and is closely related to the host's health [31]. The function of the *Rikenellaceae RC9 gut group* may be related to fiber degradation [32,33]. Furthermore, the proportion of *Rikenellaceae RC9 gut group* in the feces of musk deer is significantly reduced when they experience diarrhea and exacerbate it [34]. The *Christensenellaceae R-7 group* was the dominant genera within the silage group, which is more favorable for Hu sheep health.

The LEfSe results showed that at the genera level, the signature microorganisms in the silage group were *Psychrobacter*, *Succiniclasticum*, and *Carnobacterium*. *Psychrobacter* has a wide range of cryophilic bacteria, with members of the genera varying widely in cold suitability and genome, and can be isolated from antarctic soils, seawater, Siberian tundra, and the intestinal tracts of marine fish. *Psychrobacter* has a wide range of cryophiles, and members of the genera have widely varying cold suitability and genomes that can be isolated from Antarctic soil, seawater, Siberian tundra, and the intestinal tracts of marine fish, but their metabolisms are unknown [35–37]. The addition of *Psychrobacter* to the diet could enhance the activity of digestive enzymes in the gut and improve the digestive utilization of feed [26]. *Succiniclasticum* may be a normal genus in the intestinal tract of ruminants that catabolizes succinic acid to produce propionic acid, and also has some catabolic capacity for cellulose and starch [38]. The elevated relative abundance of Succiniclasticum may be related to higher levels of NDF and ADF in the mixed silage. *Carnobacterium* is widely distributed in the gastrointestinal tract of meat, fish, and poultry or in lakes and produces lactic acid, which in part inhibits the action of harmful microorganisms [39,40]. In addition, the mixed silage reduced the proportion of *Desulfovibrio*, which has a toxic effect on the intestinal epithelium and contributes to gastrointestinal disorders [41,42].

Metabolomics can explain phenotypic changes better than genomics and proteomics [43]. Our metabolomics data suggest that the mixed silage altered the concentrations of many fecal metabolites, which may be related to changes in fecal microbial abundance. The screened differential metabolites were enriched in glycine, serine, and threonine metabolism, valine, leucine, and isoleucine biosynthesis, phenylalanine metabolism, arginine biosynthesis, phenylalanine metabolism, arginine, and phenylalanine, tyrosine, and tryptophan biosynthesis. Amino acids are important for microbial growth and metabolism, and are key components of protein and peptide synthesis and regulate several metabolic pathways [44]. L-valine enters the tricarboxylic acid cycle to provide energy and has important roles in protein synthesis, cell proliferation, and signaling pathway activation [45,46]. NAG is an important allosteric activator of carbamoyl phosphate synthase. NAG can activate carbamoylphosphate synthase and promote its synthesis of carbamoyl phosphate, thus promoting the synthesis of arginine. NAG participates in the urea cycle in the liver and combines with ornithine transcarbamylase to generate citrulline in the intestine, and then citrulline is converted into an important pathway for the synthesis of arginine in the kidney [47]. L-arginine is involved in the tricarboxylic acid cycle as well as the urea cycle and can regulate energy metabolism, amino acid metabolism, and microbial metabolism in animals [48]. NAG, L-valine, and L-arginine were significantly down-regulated in the silage group, which may be related to the decrease in the proportion of *Ruminococcaceae UCG 010*.

It is noteworthy that we found that the concentrated metabolites of concentrated pyrimidine metabolism also changed significantly. In our experiments, the mixed silage significantly upregulated 2′-deoxyuridine and thymidine. Pyrimidine metabolism is a prominent feature of the hindgut, and different combinations of nucleosides have variable effects on animal growth performance [49]. Feeding broilers with yeast extract as a source of nucleotides does not affect the growth performance of broilers, but adding nucleotides to the diet can improve the performance of piglets [50,51]. In addition, the study of Ma et al. [52] showed that the accumulation of nucleosides functioned to enhance the growth performance of Dorper sheep. Therefore, mixed silage has the potential to improve the growth performance of Hu sheep.

5. Conclusions

In conclusion, the mixed silage increased the relative abundance of microorganisms such as *Psychrobacter*, *Succiniclasticum*, and *Carnobacterium* and decreased the relative abundance of microorganisms such as *Ruminococcaceae UCG_010* and *Breznakia*. In addition, the mixed silage alters the concentration of many fecal metabolites and enriches metabolic pathways, which is more beneficial to the healthy growth of Hu sheep.

Supplementary Materials: The following supporting information can be downloaded at: https://www.mdpi.com/article/10.3390/fermentation10050233/s1, Figure S1: Ruminal microbial OTUs with the different dietary groups. (A) OTU-based microbial sparsity profiles were used to assess the depth of coverage for each sample. (B) Venn diagram of fecal bacterial OTUs; Figure S2: Number of species, richness and diversity indices observed in fecal samples. Figure S3: Person Correlation of QC samples. (A) ESI+; (B) ESI−.

Author Contributions: Conceptualization, M.W., C.L. and Z.L.; methodology, R.Q., C.L., Y.L. and Z.L.; software, C.L., Z.Z. and Z.L.; validation, C.L., M.H.Z., R.Q., Z.Z., Z.L. and Y.L.; formal analysis, C.L., Y.L. and Z.L.; investigation, R.Q., C.L., Z.L. and M.H.Z.; resources, C.L. and Z.L.; data curation, C.L. and Z.L.; writing—original draft preparation, C.L. and M.H.Z.; writing—review and editing, M.W.; visualization, K.Y. and M.W.; supervision, K.Y. and M.W.; project administration, K.Y. and M.W.; funding acquisition, K.Y. and M.W. All authors have read and agreed to the published version of the manuscript.

Funding: This research is funded by the national 14th Five year Plan Key Research and Development Program (2023YFD1301705, 2021YFD1600702), the Key Program of State Key Laboratory of Sheep Genetic Improvement and Healthy Production (NCG202232), the High end Foreign Expert Project of the Ministry of Science and Technology (G2023014066L), and the Priority Academic Program Development of Jiangsu Higher Education Institution (PAPD), P.R. China.

Institutional Review Board Statement: All animal experiments were conducted in accordance with the Animal Protection Law, in accordance with the "Guidelines for the Care and Use of Experimental Animals" (SXXY 2015-0054) approved by the Ethics Committee of Yangzhou University.

Informed Consent Statement: The animal owners have expressed written informed consent to their animals participating in this study.

Data Availability Statement: The datasets presented in this study can be found in online repositories. The names of the reposi-tory/repositories and accession number(s) can be found below: NCBI SRA BioProject, accession no: PRJNA 1028414.

Acknowledgments: The authors sincerely thank Yangzhou University and Xinjiang Agricultural University for their support.

Conflicts of Interest: The authors declare that this study was conducted without any business or financial relationships that could be considered a potential conflict of interest.

References

1. Alexandratos, N.; Bruinsma, J. *World Agriculture towards 2030/2050: The 2012 Revision*; Food and Agriculture Organization: Rome, Italy, 2012.
2. Wang, C.; Zheng, M.; Wu, S.; Zou, X.; Chen, X.; Ge, L.; Zhang, Q. Effects of gallic acid on fermentation parameters, protein fraction, and bacterial community of whole plant soybean silage. *Front. Microbiol.* **2021**, *12*, 662966. [CrossRef]
3. Wilkinson, J.; Rinne, M. Highlights of progress in silage conservation and future perspectives. *Grass Forage Sci.* **2018**, *73*, 40–52. [CrossRef]
4. Wanapat, M.; Sundstøl, F.; Garmo, T. A comparison of alkali treatment methods to improve the nutritive value of straw. I. Digestibility and metabolizability. *Anim. Feed Sci. Technol.* **1985**, *12*, 295–309. [CrossRef]
5. Weinberg, Z.; Chen, Y. Effects of storage period on the composition of whole crop wheat and corn silages. *Anim. Feed Sci. Technol.* **2013**, *185*, 196–200. [CrossRef]
6. Ren, H.; Xu, N.; Li, J.; Li, Z.; Zhao, T.; Pei, F.; Yao, X.; Sun, Y. Effects of different mixed ratio of maize straw and cabbage wastes on silage quality. *J. Biobased Mater. Bioenergy* **2015**, *9*, 88–94. [CrossRef]

7. Partovi, E.; Rouzbehan, Y.; Fazaeli, H.; Rezaei, J. Broccoli byproduct-wheat straw silage as a feed resource for fattening lambs. *Transl. Anim. Sci.* **2020**, *4*, txaa078. [CrossRef]
8. Lu, H.; Yan, H.; Almeida, V.; Adeola, O.; Ajuwon, K. Effects of dietary resistant starch content on nutrient and energy digestibility and fecal metabolomic profile in growing pigs. *J. Anim. Sci.* **2016**, *94*, 364–368. [CrossRef]
9. Kim, J.-H.; Hong, S.W.; Park, B.-Y.; Yoo, J.G.; Oh, M.-H. Characterisation of the bacterial community in the gastrointestinal tracts of elk (*Cervus canadensis*). *Antonie Van Leeuwenhoek* **2019**, *112*, 225–235. [CrossRef]
10. Al-Lahham, S.a.H.; Roelofsen, H.; Priebe, M.; Weening, D.; Dijkstra, M.; Hoek, A.; Rezaee, F.; Venema, K.; Vonk, R.J. Regulation of adipokine production in human adipose tissue by propionic acid. *Eur. J. Clin. Investig.* **2010**, *40*, 401–407. [CrossRef]
11. Ussar, S.; Fujisaka, S.; Kahn, C.R. Interactions between host genetics and gut microbiome in diabetes and metabolic syndrome. *Mol. Metab.* **2016**, *5*, 795–803. [CrossRef]
12. Jia, W.; Li, H.; Zhao, L.; Nicholson, J.K. Gut microbiota: A potential new territory for drug targeting. *Nat. Rev. Drug Discov.* **2008**, *7*, 123–129. [CrossRef]
13. Ley, R.E.; Hamady, M.; Lozupone, C.; Turnbaugh, P.J.; Ramey, R.R.; Bircher, J.S.; Schlegel, M.L.; Tucker, T.A.; Schrenzel, M.D.; Knight, R. Evolution of mammals and their gut microbes. *Science* **2008**, *320*, 1647–1651. [CrossRef]
14. Johnson, A.J.; Vangay, P.; Al-Ghalith, G.A.; Hillmann, B.M.; Ward, T.L.; Shields-Cutler, R.R.; Kim, A.D.; Shmagel, A.K.; Syed, A.N.; Personalized Microbiome Class, S.; et al. Daily Sampling Reveals Personalized Diet-Microbiome Associations in Humans. *Cell Host Microbe* **2019**, *25*, 789–802.e785. [CrossRef]
15. Li, C.; Chen, N.; Zhang, X.; Shahzad, K.; Qi, R.; Zhang, Z.; Lu, Z.; Lu, Y.; Yu, X.; Zafar, M.H. Mixed silage with Chinese cabbage waste enhances antioxidant ability by increasing ascorbate and aldarate metabolism through rumen Prevotellaceae UCG-004 in Hu sheep. *Front. Microbiol.* **2022**, *13*, 978940. [CrossRef]
16. Nicholson, J.K.; Lindon, J.C. Metabonomics. *Nature* **2008**, *455*, 1054–1056. [CrossRef]
17. Zhang, Z.; Shahzad, K.; Shen, S.; Dai, R.; Lu, Y.; Lu, Z.; Li, C.; Chen, Y.; Qi, R.; Gao, P. Altering Dietary Soluble Protein Levels With Decreasing Crude Protein May Be a Potential Strategy to Improve Nitrogen Efficiency in Hu Sheep Based on Rumen Microbiome and Metabolomics. *Front. Nutr.* **2021**, *8*, 815358. [CrossRef]
18. Yi, H.; Yang, G.; Xiong, Y.; Wu, Q.; Xiao, H.; Wen, X.; Yang, X.; Wang, L.; Jiang, Z. Integrated metabolomic and proteomics profiling reveals the promotion of Lactobacillus reuteri LR1 on amino acid metabolism in the gut–liver axis of weaned pigs. *Food Funct.* **2019**, *10*, 7387–7396. [CrossRef]
19. Plaizier, J.C.; Li, S.; Danscher, A.M.; Derakshani, H.; Andersen, P.H.; Khafipour, E. Changes in microbiota in rumen digesta and feces due to a grain-based subacute ruminal acidosis (SARA) challenge. *Microb. Ecol.* **2017**, *74*, 485–495. [CrossRef]
20. Warner, R.D.; Jacob, R.H.; Rosenvold, K.; Rochfort, S.; Trenerry, C.; Plozza, T.; McDonagh, M.B. Altered post-mortem metabolism identified in very fast chilled lamb M. longissimus thoracis et lumborum using metabolomic analysis. *Meat Sci.* **2015**, *108*, 155–164. [CrossRef]
21. National Research Council (NRC). *Nutrient Requirements of Small Ruminants, Sheep, Goats, Cervids, and New World Camelids*; National Academy Press: Washington, DC, USA, 2007.
22. Magoč, T.; Salzberg, S.L. FLASH: Fast length adjustment of short reads to improve genome assemblies. *Bioinformatics* **2011**, *27*, 2957–2963. [CrossRef]
23. Chen, S.; Zhou, Y.; Chen, Y.; Gu, J. fastp: An ultra-fast all-in-one FASTQ preprocessor. *Bioinformatics* **2018**, *34*, i884–i890. [CrossRef] [PubMed]
24. Edgar, R.C. UPARSE: Highly accurate OTU sequences from microbial amplicon reads. *Nat. Methods* **2013**, *10*, 996–998. [CrossRef] [PubMed]
25. Lozupone, C.; Knight, R. UniFrac: A new phylogenetic method for comparing microbial communities. *Appl. Environ. Microbiol.* **2005**, *71*, 8228–8235. [CrossRef] [PubMed]
26. Sun, Y.Z.; Yang, H.L.; Ma, R.L.; Zhang, C.X.; Lin, W.Y. Effect of dietary administration of *Psychrobacter* sp. on the growth, feed utilization, digestive enzymes and immune responses of grouper *Epinephelus coioides*. *Aquacult. Nutr.* **2011**, *17*, e733–e740. [CrossRef]
27. Tanca, A.; Fraumene, C.; Manghina, V.; Palomba, A.; Abbondio, M.; Deligios, M.; Pagnozzi, D.; Addis, M.F.; Uzzau, S. Diversity and functions of the sheep faecal microbiota: A multi-omic characterization. *Microb. Biotechnol.* **2017**, *10*, 541–554. [CrossRef] [PubMed]
28. Shabana, I.; Albakri, N.; Bouqellah, N. Metagenomic investigation of faecal microbiota in sheep and goats of the same ages. *J. Taibah Univ. Sci.* **2021**, *15*, 1–9. [CrossRef]
29. Spence, C.; Wells, W.G.; Smith, C.J. Characterization of the primary starch utilization operon in the obligate anaerobe Bacteroides fragilis: Regulation by carbon source and oxygen. *J. Bacteriol.* **2006**, *188*, 4663–4672. [CrossRef] [PubMed]
30. Wang, J.; Fan, H.; Han, Y.; Zhao, J.; Zhou, Z. Characterization of the microbial communities along the gastrointestinal tract of sheep by 454 pyrosequencing analysis. *Asian Australas. J. Anim. Sci.* **2017**, *30*, 100. [CrossRef]
31. Waters, J.L.; Ley, R.E. The human gut bacteria Christensenellaceae are widespread, heritable, and associated with health. *BMC Biol.* **2019**, *17*, 83. [CrossRef]
32. Asma, Z.; Sylvie, C.; Laurent, C.; Jérôme, M.; Christophe, K.; Olivier, B.; Annabelle, T.-M.; Francis, E. Microbial ecology of the rumen evaluated by 454 GS FLX pyrosequencing is affected by starch and oil supplementation of diets. *FEMS Microbiol. Ecol.* **2013**, *83*, 504–514.

33. Huang, C.; Ge, F.; Yao, X.; Guo, X.; Bao, P.; Ma, X.; Wu, X.; Chu, M.; Yan, P.; Liang, C. Microbiome and metabolomics reveal the effects of different feeding systems on the growth and ruminal development of yaks. *Front. Microbiol.* **2021**, *12*, 682989. [CrossRef]
34. Li, Y.; Hu, X.; Yang, S.; Zhou, J.; Qi, L.; Sun, X.; Fan, M.; Xu, S.; Cha, M.; Zhang, M. Comparison between the fecal bacterial microbiota of healthy and diarrheic captive musk deer. *Front. Microbiol.* **2018**, *9*, 300. [CrossRef]
35. Zhang, S.; Song, W.; Yu, M.; Lin, X. Comparative genomics analysis of five Psychrobacter strains isolated from world-wide habitats reveal high intra-genus variations. *Extremophiles* **2017**, *21*, 581–589. [CrossRef]
36. Amato, P.; Christner, B.C. Energy metabolism response to low-temperature and frozen conditions in Psychrobacter cryohalolentis. *Appl. Environ. Microbiol.* **2009**, *75*, 711–718. [CrossRef]
37. Lin, X.; Cui, S.; Xu, G.; Wang, S.; Du, N.; Shen, J. Cloning and heterologous expression of two cold-active lipases from the Antarctic bacterium *Psychrobacter* sp. G. *Polar Res.* **2010**, *29*, 421–429.
38. Van Gylswyk, N. *Succiniclasticum ruminis* gen. nov., sp. nov., a ruminal bacterium converting succinate to propionate as the sole energy-yielding mechanism. *Int. J. Syst. Evol. Microbiol.* **1995**, *45*, 297–300. [CrossRef] [PubMed]
39. Hammes, W.P.; Hertel, C. The genera lactobacillus and carnobacterium. *Prokaryotes* **2006**, *4*, 320–403.
40. Zeineldin, M.; Barakat, R.; Elolimy, A.; Salem, A.Z.; Elghandour, M.M.; Monroy, J.C. Synergetic action between the rumen microbiota and bovine health. *Microb. Pathog.* **2018**, *124*, 106–115. [CrossRef]
41. Murros, K.E.; Huynh, V.A.; Takala, T.M.; Saris, P.E. Desulfovibrio Bacteria Are Associated with Parkinson's Disease. *Front. Cell. Infect. Microbiol.* **2021**, *11*, 378. [CrossRef]
42. Xu, Y.; Shao, M.; Fang, X.; Tang, W.; Zhou, C.; Hu, X.; Zhang, X.; Su, K.-P. Antipsychotic-induced gastrointestinal hypomotility and the alteration in gut microbiota in patients with schizophrenia. *Brain Behav. Immun.* **2022**, *99*, 119–129. [CrossRef]
43. Vinayavekhin, N.; Homan, E.A.; Saghatelian, A. Exploring disease through metabolomics. *ACS Chem. Biol.* **2010**, *5*, 91–103. [CrossRef] [PubMed]
44. Mariz, L.; Amaral, P.; Valadares Filho, S.; Santos, S.; Detmann, E.; Marcondes, M.; Pereira, J.; Silva Júnior, J.; Prados, L.; Faciola, A. Dietary protein reduction on microbial protein, amino acid digestibility, and body retention in beef cattle: 2. Amino acid intestinal absorption and their efficiency for whole-body deposition. *J. Anim. Sci.* **2018**, *96*, 670–683. [CrossRef] [PubMed]
45. Michonneau, P.; Fleurat-Lessard, P.; Cantereau, A.; Crépin, A.; Roblin, G.; Berjeaud, J.M. Implication of actin in the uptake of sucrose and valine in the tap root and leaf of sugar beet. *Physiol. Plant.* **2021**, *172*, 218–232. [CrossRef] [PubMed]
46. Rose, A.J. Amino acid nutrition and metabolism in health and disease. *Nutrients* **2019**, *11*, 2623. [CrossRef] [PubMed]
47. Shigesada, K.; Tatibana, M. Enzymatic synthesis of acetylglutamate by mammalian liver preparations and its stimulation by arginine. *Biochem. Biophys. Res. Commun.* **1971**, *44*, 1117–1124. [CrossRef] [PubMed]
48. Guo, C.; Xue, Y.; Seddik, H.-E.; Yin, Y.; Hu, F.; Mao, S. Dynamic changes of plasma metabolome in response to severe feed restriction in pregnant ewes. *Metabolites* **2019**, *9*, 112. [CrossRef] [PubMed]
49. Daneshmand, A.; Kermanshahi, H.; Mesgaran, M.D.; King, A.J.; Ibrahim, S.A.; Klasing, K.C. Combination of purine and pyrimidine nucleosides influences growth performance, gut morphology, digestive enzymes, serum biochemical indices and immune functions in broiler chickens. *Anim. Feed Sci. Technol.* **2017**, *228*, 186–193. [CrossRef]
50. Jung, B.; Batal, A. Effect of dietary nucleotide supplementation on performance and development of the gastrointestinal tract of broilers. *Brit. Poult. Sci.* **2012**, *53*, 98–105. [CrossRef]
51. Superchi, P.; Saleri, R.; Borghetti, P.; De Angelis, E.; Ferrari, L.; Cavalli, V.; Amicucci, P.; Ossiprandi, M.; Sabbioni, A. Effects of dietary nucleotide supplementation on growth performance and hormonal and immune responses of piglets. *Animal* **2012**, *6*, 902–908. [CrossRef]
52. Ma, Y.; Yang, X.; Hua, G.; Deng, X.; Xia, T.; Li, X.; Feng, D.; Deng, X. Contribution of gut microbiomes and their metabolomes to the performance of Dorper and Tan sheep. *Front. Microbiol.* **2022**, *13*, 1047744. [CrossRef]

fermentation

MDPI

Article

Evaluation of Essential Oils and Their Blends on the Fermentative Profile, Microbial Count, and Aerobic Stability of Sorghum Silage

Elon S. Aniceto [1], Tadeu S. Oliveira [1,*], José R. Meirelles, Jr. [1], Ismael N. Silva [1], Elvanio J. L. Mozelli Filho [1], Raphael S. Gomes [2], Juliana P. Arévalo [1] and Patrícia R. Moraes [1]

1 Laboratório de Zootecnia, Universidade Estadual do Norte Fluminense, Campos dos Goytacazes 28013-602, RJ, Brazil; elon1995@hotmail.com (E.S.A.)
2 Instituto Federal de Rondônia, Colorado do Oeste 76993-000, RO, Brazil
* Correspondence: tsoliveira@uenf.br

Abstract: This study aims to evaluate the effect of these essential oils and their blends on the fermentative profile, losses by gases and effluents, nutritional value, microbial count, and aerobic stability of sorghum silage. A completely randomized design was used with eight treatments and four repetitions. The evaluated treatments were the following: control (CON), without any essential oil; rosemary (Ros); tea tree (TT); citronella (Cit); Ros + TT (50% + 50%); Ros + Cit (50% + 50%); TT + Cit (50% + 50%); and Ros + TT + Cit (33% + 33% + 33%). A 1000 mg/kg dose of ensiled mass (as-fed basis) was used for each of the treatments. The addition of essential oils and their blends had a significant impact ($p < 0.05$) on the chemical composition of sorghum silage. Crude protein content increased ($p < 0.001$) with the use of essential oils and their blends. The Ros affected ($p < 0.05$) the fibrous fraction of sorghum silage. Neutral detergent fiber in vitro degradability was reduced ($p = 0.003$) when we used the blend TT + Cit compared to Ros and TT. We observed that only Ros did not reduce acetic acid concentration ($p = 0.031$) compared to the CON. The essential oils and their blends did not affect losses ($p > 0.05$). Lactic acid bacteria population increased ($p = 0.039$) when using the blend Ros + TT + Cit compared to the CON. However, the populations of entero-bacteria and fungi were not affected ($p > 0.05$) by the essential oils or their blends. For aerobic stability, we observed that Ros increased ($p < 0.001$) the air exposure time of the sorghum silage. Furthermore, the essential oils impacted the sorghum silage's pH, which affected ($p = 0.003$) its aerobic stability. In conclusion, the essential oils did not reduce sorghum silage losses. However, the Ros improved the nutritional quality and aerobic stability of sorghum silage, while the blend Ros + TT + Cit increased the lactic acid bacteria count in the silage. More in-depth studies are needed to elucidate the action of essential oils as silage additives.

Keywords: antifungal; fermentative quality; secondary plant metabolites; *Sorghum bicolor*

Citation: Aniceto, E.S.; Oliveira, T.S.; Meirelles, J.R., Jr.; Silva, I.N.; Mozelli Filho, E.J.L.; Gomes, R.S.; Arévalo, J.P.; Moraes, P.R. Evaluation of Essential Oils and Their Blends on the Fermentative Profile, Microbial Count, and Aerobic Stability of Sorghum Silage. *Fermentation* **2024**, *10*, 335. https://doi.org/10.3390/fermentation10070335

Academic Editors: Wansup Kwak and Siran Wang

Received: 29 May 2024
Revised: 20 June 2024
Accepted: 21 June 2024
Published: 27 June 2024

1. Introduction

Maintaining an anaerobic environment and low pH are the key factors for preserving stored forage [1]). Microorganisms that can cause silage to deteriorate are inhibited by a combination of acids produced during fermentation, high osmotic pressure, and the absence of oxygen [1,2]. However, when silage is exposed to air, the anaerobic environment is quickly replaced by an aerobic one. This change allows microorganisms, such as *Aspergillus* sp., *Penicillium* sp., *Clostridium* sp., and yeasts, to multiply and cause the silage to deteriorate [3].

Sorghum silage is susceptible to aerobic deterioration mainly due to the high moisture content, which can create a favorable environment for the proliferation of aerobic microorganisms when the silage is exposed to air after opening the silo [4], in addition to the greater availability of substrate for deteriorating microorganisms, producing small amounts of substances that inhibit such microorganisms [5]. With this, using additives and/or inoculants can increase silage safety, optimizing both fermentation and preservation of nutrients and the quality of the final product [6].

Antifungal additives play a crucial role in preserving silage quality, especially when it is exposed to air after the silo has been opened. These additives help prevent the growth of fungi and yeast, which can cause losses during fermentation. In this sense, essential oils (EOs) have been shown to be effective in antifungal control in silage [2,7]. Some studies have shown positive results against a range of microorganisms present in silage [8–11]. According to [12], essential oils, such as citronella, can disrupt vital metabolic processes in fungi, inhibiting fungal respiration, interfering with protein synthesis and DNA replication, and inducing oxidative stress through the antioxidant properties of essential oils. However, it is important to note that the effectiveness of essential oils may vary depending on the chemical components of each oil and the applied concentration [13]. The use of EO combinations can cause additive, synergistic, or antagonistic effects [14]. In a study conducted by [15], the authors used combinations of carvacrol and thymol, including thymol, eugenol, and a ternary compound of carvacrol; thymol and eugenol had a synergistic effect on in vitro inactivation of *Listeria innocua*.

In this context, essential oils have been studied because of their antimicrobial power in silage. The species *Rosmarinus officinalis* L., commonly known as rosemary, originates from the Mediterranean region and is cultivated almost everywhere on the planet. According to [16], the main components of rosemary essential oil are 1,8-cineole (40.55% to 45.10%), camphene (17.40% to 19.35%), and α-pinene (10.73 to [16] 15.06%). Rosemary essential oil has excellent antibacterial and antifungal characteristics. Citronella (*Cymbopogon winterianus*) is widely cultivated in the planet's tropical regions. Citronella essential oil is rich in citronellal (40%), geraniol (27.44%), and citronellol (10.45%), which are responsible for antibacterial and antifungal activities [17]. Melaleuca, commonly called tea tree, is a plant native to southern Australia. The essential oil of tea tree is composed of terpinene–4–ol (42%), γ-terpinene (19%), and α-terpinene (10%). These compounds are responsible for antibacterial and antifungal activities [18].

Thus, we hypothesized that using essential oils of rosemary, citronella, and tea tree and their blends would reduce fermentative losses in sorghum silage, improving nutritional value and aerobic stability. We aimed to evaluate the effect of these essential oils and their blends on the fermentative profile, losses by gases and effluents, nutritional value, microbial community, and aerobic stability of sorghum silage.

2. Materials and Methods

2.1. Location

The experiment took place in the municipality of Campos dos Goytacazes, RJ, Brazil (21°45′45″ S, 41°17′06″ W, and 8 m a.s.l.) between May and September 2023. The location's climate is classified as Aw, which means it is a humid tropical climate with rainy summers and dry winters, according to the Köppen–Geiger classification system [19], with an annual rainfall of 1020 mm in this area.

The Institutional Ethics Committee on the Use of Experimental Animals approved all experimental procedures under protocol 503/2021.

2.2. Harvesting, Ensiling, and Treatments

Sorghum plants (*Sorghum bicolor* (L.) Moench) were manually harvested (average dry matter content of 363.84 g/kg as-fed) and chopped in a stationary forage harvester (JF Maxxium, JF Agricultural Machinery LTDA, Itapira, Brazil) to an average particle size of 1.5 cm.

Cylindrical silos made of polyvinyl chloride (PVC), measuring 150 mm in diameter and 50 cm in height, were used in this study. These silos were equipped with a Bunsen valve for gas exhaust and contained approximately 600 g of dry sand, separated by cotton fabric to assess effluent losses. The silos were filled to a density of 600 kg/m^3 (as-fed) and stored at 25 ± 3.2 °C for 60 days.

The study was conducted using a completely randomized design, which included eight treatments and four replicates. The evaluated treatments were the following: control (CON) without any essential oil; rosemary (Ros); tea tree (TT); citronella (Cit); Ros + TT (50% rosemary + 50% tea tree); Ros + Cit (50% rosemary + 50% citronella); TT + Cit (50% rosemary + 50% citronella); and Ros + TT + Cit (33% rosemary + 33% tea tree + 33% citronella). Blends were produced from essential oils after extraction. A 1000 mg/kg dose of ensiled mass (as-fed basis) was used for each treatment.

The essential oils were commercially (FERQUIMA Essential Oils Industry, Vargem Grande Paulista, Brazil) purchased and extracted through a steam distillation process. The specifications of the rosemary essential oil are as follows: density (20 °C) = 0.9–0.93 (g/mL) and refractive index (20 °C) = 1.460–1.475 (g/cm). The main components are 1,8-cineol (40%), camphor (15%), α-pinene (13%), and β-pinene (7%). The specifications of the tea tree essential oil are as follows: density (20 °C) = 0.885–0.906 (g/mL) and refractive index (20 °C) = 1.470–1.482 (g/cm). The main components are terpinene-4-ol (41%), γ-terpinene (21%), and α-terpinene (9%). The specifications of the citronella essential oil are as follows: density (20 °C) = 0.875–0.895 (g/mL) and refractive index (20 °C) = 1.463–1.473 (g/cm). The main components are β-citronella (32.7%), geraniol (28.9%), and citronellol (9.6%).

2.3. Chemical Composition and In Vitro Assay

Plant and silage samples were dried in a forced-air oven at 55 °C for 72 h. After drying, the samples were ground in a Wiley mill (Tecnal, Piracicaba, São Paulo, Brazil) fitted with a 1 mm sieve. We analyzed dry matter (DM, method 967.03), crude fat (CF, method 2003.06), ash (method 942.05), and crude protein ([N × 6.25] CP) as described by [20]. Neutral detergent fiber (NDF) was analyzed with sodium sulfite and two additions of a standardized heat-stable amylase solution, excluding ash (aNDF, INCT-CA method F-001/1; [21], acid detergent fiber (ADF), according to INCT-CA F-003/1, as described by [21], and lignin (Lig) (INCT-CA method F-005/1; [21]). Non-fiber carbohydrate (NFC) content was estimated as follows: $NFC(\text{g/kg}) = 1000 - CP - \text{CF} - Ash - NDF$. Hemicellulose was calculated based on the difference between NDF and ADF, and cellulose was calculated based on the difference between ADF and lignin, all expressed in g/kg DM.

The in vitro degradation of DM and NDF was determined following the methodology recommended by [22]. Each sample was analyzed in triplicate by weighing approximately 200 mg of the sample and transferring it into 100 mL amber bottles. We added 20 mL of buffer solution and inoculum to the bottles and sealed them with rubber stoppers to prevent fermentation gas escape. The inoculum was made using [23] buffer and ruminal fluid from three sheep with rumen cannulas and 52 ⊥ 5.2 kg body mass. The animals were fed a total mixed ration with 180 g of crude protein kg/DM and 520 g of neutral detergent fiber kg/DM and supplemented with a mineral premix. After 48 h of incubation, the bottles were withdrawn from the water bath and immediately rinsed with hot distilled water exceeding 90 °C. Following the rinsing, the resulting material was dried (55 °C for 24 h followed by 105 °C for 16 h), and the weight was recorded, yielding the undigested residue of DM. Subsequently, this material underwent analysis for NDF content, resulting in the undigested residue of NDF. The degradability (D) of DM and NDF was calculated according to the equation:

$$D = (M - [R - B]/M) \times 1000$$

where M = incubated mass (g) of DM or NDF; R = residue of DM or NDF from incubation (g); and B = residue of DM or NDF from blanks (g).

2.4. Gas and Effluent Losses and Dry Matter Recovery

Losses were calculated according to the equations proposed by [24].

Gas losses were calculated using Equation (1):

$$GL = (WSWE - WSWO)/EDM \quad (1)$$

where GL = gas loss (g/kg DM), $WSWE$ is the whole silo weight at ensiling (g), $WSWO$ is the whole silo weight at the opening (g), and EDM is the ensiled dry matter (kg).

Effluent losses were calculated according to Equation (2):

$$EL = (ESWO - ESWE)/EDM \quad (2)$$

where EL = effluent loss (g/kg DM), $ESWO$ is the empty silo weight at the opening (g), and $ESWE$ is the empty silo weight at ensiling (g).

Dry matter recovery was calculated using Equation (3):

$$DMR = ODM/EDM \quad (3)$$

where DMR = dry matter recovery (g/kg DM) and ODM is the DM at the opening (g).

Flieg's score was calculated by assessing the DM and pH values of silages, following the equation by [25]:

$$Flieg's\ Score = 200 + (2 \times \%DM - 15) - 40 \times pH \quad (4)$$

2.5. Fermentative Profile and Microbial Count

Upon opening each silo, the contents were thoroughly mixed to ensure homogeneity. A 25 g sample of fresh silage was then collected and blended with 225 mL of saline solution (8.5 g of NaCl/L distilled water) for 1 min. The mixture was then filtered, and three separate aliquots were obtained. Two were used to determine the fermentative profile, while the third was used to determine the microbial community. The pH of the first aliquot was measured, and 0.036 N sulfuric acid was added. The mixture was then frozen to later assess the ammonia nitrogen content (NH_3-N) with magnesium oxide using the procedure described by [26]. The second aliquot was used to measure the concentration of short-chain fatty acids (SCFAs). A total of 0.5 mL of sulfuric acid solution (50%) was added to this aliquot, according to [27], and it was stored at −18 °C until analysis. The concentrations of SCFAs were determined using High-Performance Liquid Chromatography (HPLC; YL9100 HPLC System [Young Lin]) with a REZEX RCM-Monosaccharide Ca^{+2} (8%) column. Ultra-pure water served as the mobile phase with a flow rate of 0.7 mL/min, maintaining the column at 60 °C, and a refractive index detector was employed.

The third aliquot of the aqueous silage extract was filtered, 9 mL was added in a sterile falcon tube, and it was subjected to serial dilutions (10^{-1} to 10^{-6}). We used the Violet Red Bile (VRB) culture medium to count enterobacteria and incubated it at 37 °C for 24 h (h). For the fungi count, the Potato Dextrose Agar (PDA) was incubated at 30 °C for four days. The De Man, Rogosa, Sharpe (MRS) was used for lactic acid bacteria count by incubating at 37 °C for 48 h. Microbial counts were expressed as colony-forming units per gram (cfu g^{-1}) and transformed to $\log_{10}$ to obtain the lognormal distribution.

2.6. Aerobic Stability

After opening the silos, 2.0 kg of silage was placed in plastic buckets with a capacity of 5.0 kg. These buckets were then left for seven days at room temperature to evaluate their aerobic stability [28]. To monitor the temperature, data loggers not connected to the computer (Log 110 EXF Inconterm, Porto Alegre, Brazil) were inserted 10 cm deep in the center of the silage mass, and readings were taken every 6 h. Additionally, samples (200 g) were collected from silos of each treatment every 24 h to measure the pH after exposure to oxygen. The aerobic stability was calculated as the time in hours when the temperature of the silage exceeded the ambient temperature by 2 °C after air exposure [27].

2.7. Statistical Analysis

Data on chemical composition, gas and effluent losses, microbial count, fermentative profile, in vitro degradability, gross energy, and Flieg's score were compared using Tukey's test with a significance level of 0.05. The analysis was performed using the MIXED package of SAS (SAS OnDemand for Academics, SAS Institute Inc., Cary, NC, USA). A tendency was considered when the *p*-value was between 0.05 and 0.10. The Shapiro–Wilk test (PROC UNIVARIATE) was used to check for data normality.

The following statistical model was used:

$$Y_{ij} = \mu + \alpha_i + e_{ij}$$

In which Y_{ij} is the value observed for the variable under study, referring to the *j-th* replicate of the *i-th* factor level, α; μ is the mean of all experimental units for the variable under study; α_i is the addition of essential oils and their blends in silages with i = 1, 2, 3, 4, 5, 6, 7, 8; and e_{ij} is the error associated with the observation.

The aerobic stability and pH data were analyzed as repeated measures over time using regression analysis with a significance level of 0.05 using the MIXED package of SAS (SAS OnDemand for Academics, SAS Institute Inc., Cary, NC, USA, https://www.sas.com/en_us/software/on-demand-for-academics.html, accessed on 21 April 2024).

The following statistical model was used:

$$Y_{ijk} = \mu + \alpha_i + \tau_j + \alpha\tau_{ij} + e_{ijk}$$

In which Y_{ijk} is the value observed for the variable under study, referring to *the k-th* replicate of the *i-th* factor level α in the *j-th* hour; μ is the mean of all experimental units for the variable under study; α_i is the addition of essential oils and their blends in silages with i = 1, 2, 3, 4, 5, 6, 7, 8; τ_k is the random effect of the evaluation hours with j = 0.24, . . ., 144 for pH and 0, 8, 16, . . ., 162 for temperature; $\alpha\tau_{ij}$ is the interaction between essential oils and their blends and evaluation hours; and e_{ijk} is the error associated with observation Y_{ijk}.

3. Results

The addition of essential oils and their blends had a significant impact ($p < 0.05$) on the chemical composition of sorghum silage. The CP content increased ($p < 0.001$) with the use of essential oils and their blends (Table 1). The analysis of the fiber fraction revealed that the Ros increased the lignin content ($p = 0.007$) compared to the CON. The Cell content differed between essential oils and blends, e.g., Ros reduced it by 13.27% compared to the blend Ros + TT (Table 1). NDF degradability was reduced ($p = 0.003$) when we used the blend TT + Cit compared to Ros and TT. The other variables were not ($p > 0.05$) influenced by the oils (Table 1). The Ros and TT showed a tendency ($p = 0.091$) to reduce NDF content (Table 1). However, the variables NFC, Hem, IVDMD, and GE were not influenced ($p > 0.05$) by essential oils or their blends (Table 1).

When evaluating the fermentative profile, we observed that only Ros did not reduce the acetic acid concentration ($p = 0.031$) compared to the CON (Table 2). The other variables remained unchanged ($p > 0.05$) with the use of essential oils and their blends, among them lactic acid ($p = 0.533$), which is an important indicator of silage fermentation (Table 2).

The essential oils and their blends did not affect gas losses ($p = 0.240$), effluents ($p = 0.891$), or Flieg's score ($p = 0.932$) (Table 3). However, Cit and its blend TT + CIT increased dry matter recovery ($p = 0.012$) in sorghum silage (Table 3).

In terms of the microbial count, the LAB population increased ($p = 0.039$) when using the blend Ros + TT + Cit compared to the CON (Figure 1). However, the populations of enterobacteria and fungi were not affected ($p > 0.05$) by the essential oils or their blends (Figure 1).

Regarding the aerobic stability, we observed that Ros increased ($p < 0.001$) the air exposure time of sorghum silage in 138 h (Figure 2a). Furthermore, the essential oils impacted the sorghum silage's pH, which affected ($p = 0.003$) its aerobic stability (Figure 2b).

Table 1. Effects of the essential oils and their blends on the chemical composition and in vitro degradability of sorghum silage.

Variables	Sorghum	Essential Oils								SEM	*p*-Value
		CON	Ros	TT	Cit	Ros + TT	Ros + Cit	TT + Cit	Ros + TT + Cit		
DM	363.84	345.01	365.28	387.18	404.21	367.76	378.95	331.55	352.8	6.747	0.465
CP	47.79	42.28 c	53.49 a	47.77 b	49.13 ab	50.10 ab	51.01 ab	46.43 bc	49.61 ab	0.462	<0.0001
NDF	685.26	726.19	670.3	687.99	701.34	738.72	709.37	733.64	723.32	5.227	0.091
Lig	15.46	14.76 b	17.96 a	15.26 ab	14.02 b	15.82 ab	14.97 ab	15.47 ab	16.94 ab	0.233	0.007
NFC	186.74	120.53	174.04	172.52	151.09	109.63	137.45	132.63	123.5	5.189	0.129
Hem	226.94	259.8	237.0	225.63	246.22	244.02	238.04	254.52	255.6	2.586	0.135
Cell	442.86	451.62 ab	415.34 b	447.09 ab	441.1 ab	478.89 a	456.36 ab	463.65 ab	450.78 ab	3.754	0.031
IVDMD	487.67	514.96	514.35	538.88	533.68	562.99	548.81	564.42	582.31	5.444	0.201
IVNDFD	570.48	586.83 b	584.73 b	573.74 b	610.54 ab	639.32 ab	616.93 ab	655.79 a	640.36 ab	5.617	0.003
GE	16.46	16.74	16.97	16.74	17.24	17.09	17.16	16.96	17.39	0.069	0.715

Sorghum = sorghum plant before ensiling; rosemary (Ros); tea tree (TT); citronella (Cit); Ros + TT (50% rosemary plus 50% tea tree); Ros + Cit (50% rosemary plus 50% citronella); TT + Cit (50% rosemary plus 50% citronella); and Ros + TT + Cit (33% rosemary plus 33% tea tree plus 33% citronella). SEM = standard error of the mean; DM = dry matter; CP = crude protein; NDF = neutral detergent fiber; Lig = lignin; NFC = non-fibrous carbohydrate; Hem = hemicellulose; Cell = cellulose; IVDMD = in vitro dry matter degradability; IVNDFD = in vitro neutral detergent fiber degradability; and GE = gross energy, all expressed as g/kg, except DM, expressed as as-fed, and GE, expressed as MJ/kg DM. Means followed by the different letters in a line differ significantly according to Tukey's test ($p < 0.05$).

Table 2. Effects of the essential oils and their blends on fermentative profile of sorghum silage.

Variables	Essential Oils								SEM	*p*-Value
	CON	Ros	TT	Cit	Ros + TT	Ros + Cit	TT + Cit	Ros + TT + Cit		
T, °C after opening the silo	23.63	24.18	24.00	24.43	24.13	23.98	24.23	23.73	0.200	0.997
pH after opening the silo	4.41	4.56	4.60	4.51	4.42	4.57	4.39	4.43	0.022	0.341
NH_3-N, g/kg CP	0.65	0.89	0.68	0.58	0.80	0.59	0.67	0.74	0.028	0.251
Lactic acid, g/kg DM	35.12	65.44	40.72	21.33	61.78	37.36	31.43	42.05	4.191	0.533
Acetic acid, g/kg DM	57.97 a	55.08 a	13.24 d	28.68 c	36.27 b	37.93 b	11.88 d	37.98 b	3.446	0.031
Propionic acid, g/kg DM	0.04	0.09	0.14	0.29	0.35	0.33	0.13	0.16	0.026	0.242
Butyric acid, g/kg DM	0.04	0.09	0.07	0.19	0.15	0.08	0.12	0.08	0.014	0.610

Rosemary (Ros); tea tree (TT); citronella (Cit); Ros + TT (50% rosemary plus 50% tea tree); Ros + Cit (50% rosemary plus 50% citronella); TT + Cit (50% rosemary plus 50% citronella); and Ros + TT + Cit (33% rosemary plus 33% tea tree plus 33% citronella). SEM = standard error of the mean; T = temperature; NH_3-N = ammoniacal nitrogenMeans followed by the different letters in a line differ significantly according to Tukey's test ($p < 0.05$).

Table 3. Effects of the essential oils and their blends on losses and dry matter recovery of sorghum silage.

Variables	Essential Oils								SEM	*p*-Value
	CON	Ros	TT	Cit	Ros + TT	Ros + Cit	TT + Cit	Ros + TT + Cit		
Gas losses, g/kg DM	48.43	42.97	47.41	35.10	46.95	35.37	33.89	46.19	6.929	0.240
Effluent losses, g/kg DM	15.62	14.89	10.29	11.23	15.62	16.14	13.21	16.73	3.942	0.891
Dry matter recovery, g/kg	935.9 b	942.2 ab	942.3 ab	953.7 a	937.4 b	948.5 ab	952.9 a	952.1 b	0.764	0.012
Flieg's score	77.60	75.56	78.23	85.44	81.65	77.79	75.81	78.36	1.519	0.932

Rosemary (Ros); tea tree (TT); citronella (Cit); Ros + TT (50% rosemary plus 50% tea tree); Ros + Cit (50% rosemary plus 50% citronella); TT + Cit (50% rosemary plus 50% citronella); and Ros + TT + Cit (33% rosemary plus 33% tea tree plus 33% citronella). SEM = standard error of the mean Means followed by the different letters in a line differ significantly according to Tukey's test ($p < 0.05$).

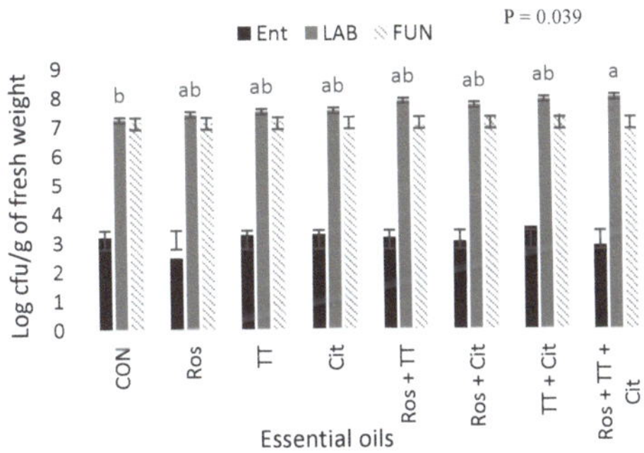

Figure 1. Microbial counts of sorghum silage with different essential oils and their blends. Control (0 mg); rosemary (Ros); tea tree (TT); citronella (Cit); Ros + TT (50% rosemary plus 50% tea tree); Ros + Cit (50% rosemary plus 50% citronella); TT + Cit (50% rosemary plus 50% citronella); and Ros + TT + Cit (33% rosemary plus 33% tea tree plus 33% citronella). Enterobacteria (Ent); lactic acid bacteria (LAB); and fungi (FUN). The dose used was 1000 mg/kg of ensiled mass (as-fed basis). Means followed by the different letters differ significantly according to Tukey's test ($p < 0.05$).

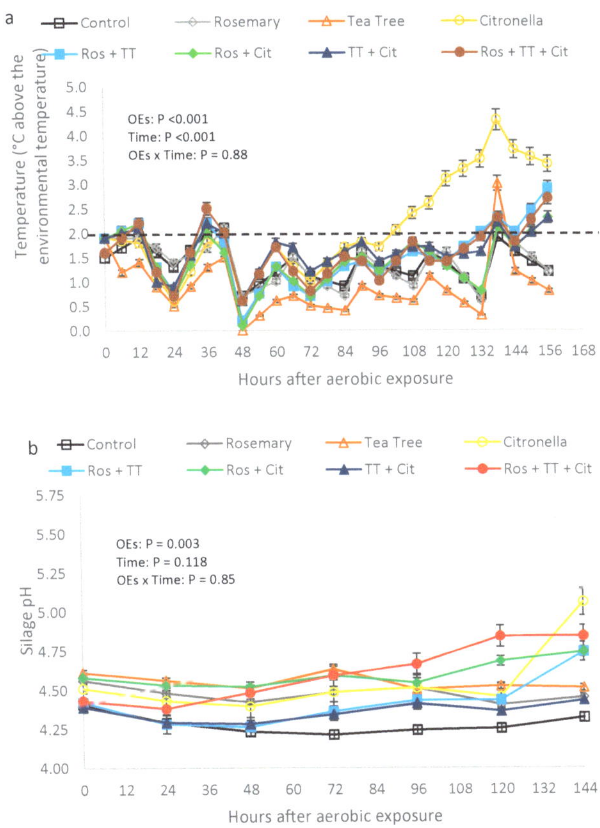

Figure 2. Temperature and pH values of sorghum silage with different essential oils and their blends for seven days. Control (0 mg); rosemary (Ros); tea tree (TT); citronella (Cit); Ros + TT (50% rosemary plus 50% tea tree); Ros + Cit (50% rosemary plus 50% citronella); TT + Cit (50% rosemary plus 50% citronella); and Ros + TT + Cit (33% rosemary plus 33% tea tree plus 33% citronella). On the panel, (**a**) temperature and (**b**) pH. The dose used was 1000 mg/kg of ensiled mass (as-fed basis).

4. Discussion

The increased CP levels in sorghum silage with the use of essential oils and their blends may be related to the antimicrobial action of the oils. Silages contain microorganisms, such as lactic acid bacteria (LAB), enterobacteria, and clostridia, with proteolytic activity [9]. According to [14], essential oils can inhibit proteolysis through antimicrobial activity. Another factor is the antioxidant action of essential oils, which can reduce protein breakdown. Antioxidants can intercept free radicals generated by cellular metabolism or exogenous sources, preventing the attack on amino acids and peptides [29]. Thus, the antioxidant properties of rosemary essential oil are attributed to 1,8-cineole, which acts as a free-radical-terminating agent and reactive oxygen species chelator [30]. According to the findings of our study, the NH_3-N remained unaffected ($p = 0.251$) despite it being an important indicator of protein degradation (Table 2). However, the antimicrobial properties of Ros and TT may likely have impacted the NDF degradability (Table 1). During the process of silage fermentation, various microorganisms, such as BAL, acetic acid bacteria, and fungi, play a crucial role in breaking down the main components of the plant cell wall, including cellulose, hemicellulose, and pectin [31]. Furthermore, the reduction may have occurred due to the antibacterial activity in the rumen fluid's Archaea, protozoa, and cellulolytic bacteria (*Fibrobacter succinogenes, Ruminococcus flavefaciens*, and *R. albus*) [32].

Regarding the fermentative profile of sorghum silage, the increased acetic acid concentration is a highly determining factor in the inhibition of yeast metabolism during fermentation (anaerobiosis), resulting in significant improvements in the aerobic stability of silages [3,6]. This fact is due to the antifungal activity of acetic acid, which can suppress the growth of lactate-assimilating fungi and yeasts that start aerobic deterioration [6,33]. Therefore, adequate concentrations of acetic acid can reduce total losses in storage and post-opening silage [33].

The production of high-quality silage faces a great challenge in reducing DM losses. These losses, combined with quality changes, occur during every step of the ensiling process, thereby reducing silage quality [34]. Although some losses are unavoidable, good practical management can help reduce or compensate for these losses to ensure a good-quality forage [34]. Essential oils and their blends were not efficient in reducing clostridial activity, as indicated by the concentration of butyric acid (Table 2), a marker of clostridial fermentation [33]; consequently, they did not reduce losses due to gases and effluents (Table 3). However, the DMR showed values higher than 93%, as oils and their blends led to improvements. For instance, Cit increased by 1.86% and TT + Cit increased by 1.78% compared to the CON (Table 3).

During the ensiling process, lactic acid (3.86 pKa) produced by lactic acid bacteria (LAB) is generally the most abundant acid found in silage. This acid contributes the most to the pH drop during fermentation, which helps to preserve the forage mass [33,35]. In the present study, we observed that the Ros + TT + Cit blend resulted in higher BAL counts than the CON (Figure 1). This may be due to the interaction of antimicrobial properties, which inhibited the growth of undesirable microorganisms, such as spoilage bacteria and fungi [9]. It creates a more favorable environment for the growth of LAB, which is beneficial for the ensiling process [9].

Both sorghum and corn have desirable characteristics for fermentation, such as epiphytic microbiota, dry matter content, water-soluble carbohydrates, and buffer capacity (BC). These characteristics are essential for the ensiling process [34]. However, sorghum silage is vulnerable to losses from aerobic deterioration due to the greater availability of substrate for spoilage microorganisms in well-fermented silages, in addition to containing a lower content of substances that inhibit such microorganisms [5]. In this study, we found that using Ros in sorghum silage increased its exposure time to air (Figure 2a). This can be attributed to two factors. The first factor is the high concentration of acetic acid in the silage (Table 2). The second factor is the antifungal properties of Ros (specifically, α-pinene and β-pinene). As per [12], essential oils can inhibit fungal growth by disrupting vital metabolic processes, inhibiting respiration, interfering with protein synthesis and DNA replication, and inducing oxidative stress through their antioxidant properties. Regarding pH, essential oils showed values between 4.3 and 4.6 during the first 48 h of air exposure (Figure 2b). Thus, certain conditions, such as temperature, absence of oxygen, and pH, are favorable for preserving forage. A lower-pH environment inhibits the growth of spoilage organisms, such as molds and yeasts, which are less active in acidic conditions. This helps in maintaining silage quality upon air exposure [3,6].

5. Conclusions

The essential oils and their blends had different effects on sorghum silage. Among the essential oils, Ros and blends containing Ros showed greater efficiency. Ros helped improve the nutritional quality and aerobic stability of sorghum silage, while the blend Ros + TT + Cit increased the lactic acid bacteria count in the silage. However, the essential oils and their blends did not reduce sorghum silage losses. More in-depth studies are needed to elucidate the action of essential oils as silage additives.

Author Contributions: E.S.A.: data curation, investigation, methodology, writing—original draft. T.S.O.: conceptualization, formal analysis, resources, project administration, writing—original draft, writing—review and editing. J.R.M.J.: investigation, methodology. I.N.S.: investigation, methodology. E.J.L.M.F.: investigation, methodology. R.S.G.: writing—original draft. J.P.A.: methodology. P.R.M.: methodology. All authors have read and agreed to the published version of the manuscript.

Funding: This research was supported by the Fundação Carlos Chagas Filho de Amparo à Pesquisa do Estado do Rio de Janeiro—FAPERJ (RJ, Brazil), grant number E-26/200.191/2023, and the Conselho Nacional de Desenvolvimento Científico e Tecnológico (CNPq). The author Raphael S. Gomes thanks the Pró-Reitoria de Pesquisa, Inovação e Pós-graduação do Instituto Federal de Rodônia for the financial support to cover the article processing charge (APC) (Process SEI—23243.002011/2024-52). The funding agencies had no role in the study design, data collection and analyses, decision to publish, or preparation of the manuscript.

Institutional Review Board Statement: All experimental procedures were approved by the Ethics Committee on the Use of Experimental Animals of the Universidade Estadual do Norte Fluminense, protocol 503/2021.

Informed Consent Statement: Not applicable.

Data Availability Statement: The data that support this study will be shared upon request to the corresponding author.

Conflicts of Interest: The authors declare no conflicts of interest.

References

1. Muck, R.E. Silage microbiology and its control through additives. *Rev. Bras. Zootec.* **2010**, *39*, 183–191. [CrossRef]
2. Pahlow, G.; Muck, R.E.; Driehuis, F.; Oude Elferink, S.J.W.H.; Spoelstra, S.F. Microbiology of ensiling. *Silage Sci. Technol.* **2003**, *42*, 31–93. [CrossRef]
3. McDonald, P.; Henderson, A.R.; Heron, S.J.E. *The Biochemistry of Silage*, 2nd ed.; Chalcomb Publications: Marlow, UK, 1991; ISBN 0948617225.
4. Camilo, M.G.; Fernandes, A.M.; Processi, E.F.; Motta, O.V.; Roseira, J.P.S.; Oliveira, T.S. Evaluation of the type of silo associated or not with additives on the nutritional value, aerobic stability, and microbiology of pearl millet silage. *An. Acad. Bras. Ciências* **2023**, *95*, e20210421. [CrossRef]
5. Wilkinson, J.M.; Davies, D.R. The aerobic stability of silage: Key findings and recent developments. *Grass Forage Sci.* **2013**, *68*, 1–19. [CrossRef]
6. Muck, R.E.; Nadeau, E.M.G.; Mcallister, T.A.; Contreras-Govea, F.E.; Santos, M.C.; Kung, L., Jr. Recent advances and future uses of silage inoculants. *J. Dairy Sci.* **2018**, *101*, 3980–4000. [CrossRef]
7. Kung, L., Jr.; Williams, P.; Schmidt, R.J.; Hu, W. A blend of essential plant oils used as an additive to alter silage fermentation or used as a feed additive for lactating dairy cows. *J. Dairy Sci.* **2008**, *91*, 4793–4800. [CrossRef]
8. Chaves, A.V.; Baah, J.; Wang, Y.; McAllister, T.A.; Benchaar, C. Effects of cinnamon leaf, oregano and sweet orange essential oils on fermentation and aerobic stability of barley silage. *J. Sci. Food Agric.* **2012**, *92*, 906–915. [CrossRef] [PubMed]
9. Foskolos, A.; Cavini, S.; Ferret, A.; Casalmiglia, S. Effects of essential oil compounds addition on ryegrass silage protein degradation. *Can. J. Anim. Sci.* **2016**, *96*, 100–103. [CrossRef]

10. Turan, A.; Soycan-Önenç, S. Effect of cumin essential oil usage on fermentation quality, aerobic stability and in vitro digestibility of alfalfa silage. *Asian-Australas. J. Anim. Sci.* **2018**, *31*, 1252–1258. [CrossRef]
11. Cantoia, R., Jr.; Capucho, E.; Garcia, T.M.; Del Valle, T.A.; Campana, M.; Zilio, E.M.C.; Azevedo, E.B.; Morais, J.P.G. Lemongrass essential oil in sugarcane silage: Fermentative profile, losses, chemical composition, and aerobic stability. *Anim. Feed Sci. Technol.* **2020**, *260*, 114371. [CrossRef]
12. Nazarro, F.; Fratianni, F.; Martino, L.; Coppola, R.; De Feo, V. Effect of essential oils on pathogenic bacteria. *Pharmaceuticals* **2013**, *6*, 1451–1474. [CrossRef] [PubMed]
13. Liang, J.; Zhang, Y.; Chi, P.; Liu, H.; Jing, Z.; Cao, H.; Du, Y.; Zhao, Y.; Qin, X.; Zhang, W.; et al. Essential oils: Chemical constituents, potential neuropharmacological effects and aromatherapy—A review. *Pharmacol. Res.—Mod. Chin. Med.* **2023**, *6*, 100210. [CrossRef]
14. Casalmiglia, S.; Busquet, M.; Cardozo, P.W.; Castillejos, L.; Ferret, A. Essential oils as modifiers of rumen microbial fermentation. *J. Dairy Sci.* **2007**, *90*, 2580–2595. [CrossRef]
15. Garcia-Garcia, R.; López-Malo, A.; Palou, E. Bactericidal action of binary and ternary mixtures of carvacrol, thymol, and eugenol against *Listeria innocua*. *J. Food Sci.* **2011**, *76*, 95–100. [CrossRef]
16. Atti-Santos, A.C.; Agostini, F.; Pansera, M.R.; Rossato, M.; Atti-Serafini, L. Estudo ds qualidade de amostras comerciais de óleos essenciais de alecrim (*Rosmarinus officinalis* L.). *Rev. Bras. Plantas Med.* **2004**, *6*, 44–47.
17. Nascimento, J.C.; Barbosa, L.C.A.; Paula, V.F.; David, J.M.; Fontana, R.; Silva, L.A.M.; França, R.S. Chemical composition and antimicrobial activity of essential oils of *Ocimum canum Sims.* and *Ocimum selloi Benth. An. Acad. Bras. Ciências* **2011**, *83*, 787–799. [CrossRef]
18. Yu, D.; Wang, J.; Shao, X.; Xu, F.; Wang, H. Antifungal modes of action of tea tree oil and its two characteristic components against Botrytis cinerea. *J. Appl. Microbiol.* **2015**, *119*, 1253–1262. [CrossRef]
19. Alvares, C.A.; Stape, J.L.; Sentelhas, P.C.; Moraes, J.L.G.; Sparovek, G. Köppen's climate classification map for Brazil. *Meteorol. Z.* **2013**, *22*, 711–728. [CrossRef] [PubMed]
20. AOAC. *Official Methods of Analysis*, 21st ed.; Association of Official Analytical Chemistry: Washington, DC, USA, 2019.
21. Detmann, E.; Souza, M.A.; Valadares Filho, S.C.; Queiroz, A.C.; Berchielli, T.T.; Saliba, E.O.E.; Cabral, L.S.; Pina, D.S.; Ladeira, M.M.; Azevedo, J.A.G. *Métodos Para Análise de Alimentos (INCT—Ciência Animal)*; Editora UFV: Viçosa, Brazil, 2012.
22. Goering, H.K.; Van Soest, P.J. *Forage Fiber Analyses (Apparatus, Reagents, Procedures, and Some Applications)*; Agric. Handbook No. 379; ARS-USDA: Washington, DC, USA, 1970.
23. McDougall, E.I. The composition and output of sheep's saliva. *Biochem. J.* **1948**, *43*, 99–109. [CrossRef]
24. Jobim, C.C.; Nussio, L.G.; Reis, R.A.; Schmidt, P. Avanços metodológicos na avaliação da qualidade da forragem conservada. *Rev. Bras. Zootec.* **2007**, *36*, 101–119. [CrossRef]
25. Kilic, A. *Silo Feed (Instruction, Education and Application Proposals)*; Bilgehan Press: Izmir, Turkey, 1986; p. 327.
26. Fenner, H. Method for determining total volatile bases in rumen fluid by steam distillation. *J. Dairy Sci.* **1965**, *48*, 249–251. [CrossRef]
27. Ranjit, N.K.; Kung, L., Jr. The effect of *Lactobacillus buchneri* and other additives on the fermentation and aerobic stability of barley silage. *J. Dairy Sci.* **2001**, *84*, 1149–1155. [CrossRef]
28. Muck, R.E.; Pitt, R.E. Aerobic deterioration in corn silage relative to the silo face. *Trans. ASAE* **1994**, *37*, 735–743. [CrossRef]
29. Lobo, V.; Patil, A.; Phatak, A.; Chandra, N. Free radicals, antioxidants and functional foods: Impact on human health. *Pharmacogn. Rev.* **2010**, *4*, 118–126. [CrossRef]
30. Rašković, A.; Milanović, I.; Pavlović, N.; Ćebović, T.; Vukmirović, S.; Mikov, M. Antioxidant activity of rosemary (*Rosmarinus officinalis* L.) essential oil and its hepatoprotective potential. *BMC Complement Altern Med.* **2014**, *14*, 225–234. [CrossRef]
31. Du, Z.; Yamasaki, S.; Oya, T.; Cai, Y. Cellulase–lactic acid bacteria synergy action regulates silage fermentation of woody plant. *Biotechnol. Biofuels Bioprod.* **2023**, *16*, 125–143. [CrossRef]
32. Patra, A.K.; Yu, Z. Effects of essential oils on methane production and fermentation by, and abundance and diversity of, rumen microbial populations. *Appl. Environ. Microbiol.* **2012**, *78*, 4271–4280. [CrossRef]
33. Kung, L., Jr.; Shaver, R.D.; Grant, R.J.; Schmidt, R.J. Interpretation of chemical, microbial, and organoleptic components of silages. *J. Dairy Sci.* **2018**, *101*, 4020–4033. [CrossRef]
34. Borreani, G.; Tabacco, E.; Schmidt, R.J.; Holmes, B.J.; Muck, R.E. Factors affecting dry matter and quality losses in silages. *J. Dairy Sci.* **2018**, *5*, 3952–3979. [CrossRef]
35. Silva, E.B.; Kung, L., Jr. A meta-analysis of the effects of a chemical additive on the fermentation and aerobic stability of whole-plant maize silage. *Agriculture* **2022**, *12*, 132. [CrossRef]

fermentation

Article

Dynamic Analysis of Fermentation Quality, Microbial Community, and Metabolome in the Whole Plant Soybean Silage

He Meng [1], Yan Jiang [1,2,*], Lin Wang [1], Yuanming Li [1,2], Sui Wang [3], Xiaohong Tong [1] and Shaodong Wang [1,*]

1 Agricultural College, Northeast Agricultural University, Harbin 150030, China; bishop92@126.com (H.M.)
2 Heilongjiang Academy Green Food Science Resarch Institute, National Soybean Engineering Technology Research Center, Harbin 150028, China
3 National Key Laboratory of Smart Farm Technologies and Systems, Harbin 150030, China
* Correspondence: yjiang@neau.edu.cn (Y.J.); wsdhlj@neau.edu.cn (S.W.)

Abstract: Soybean (*Glycine max* (L.) Merr.) is an important oilseed crop, known for its rich nutritional content and high-quality protein. To address the shortage of feed protein resources and better utilize soybeans as a raw material, this study investigated the feasibility of using whole-plant soybean (WPS) as silage. As the ensiling period is a critical fermentation parameter, identifying the optimal fermentation duration was a key objective. The research involves fermenting WPS for silage production, conducted over five fermentation durations: 7, 15, 30, 60, and 90 days. The fermentation quality, microbial community, and metabolome of WPS silage were analyzed across these different time points. WPS silage fermented for 30 days exhibited optimal fermentation characteristics, with the highest lactic acid (LA) content observed at 30 days ($p < 0.05$), while butyric acid (BA) was detected only at 60 and 90 days. At 30 days, *Enterococcus* genera reached its peak relative abundance and was identified as the dominant genus. Random forest analysis highlighted *Pantoea* genera as the most influential biomarker. Metabolomic analysis revealed that the metabolic pathways involved in the biosynthesis of essential amino acids valine, leucine, and isoleucine were significantly enhanced during the later stages of fermentation compared to the earlier stages. Under natural fermentation conditions, the optimal fermentation period for WPS silage is approximately 30 days. These findings provide a theoretical basis for the utilization of WPS and the subsequent optimization of fermentation quality.

Keywords: whole plant soybean; silage; fermentation period; fermentation quality; microbial community; metabolome

Citation: Meng, H.; Jiang, Y.; Wang, L.; Li, Y.; Wang, S.; Tong, X.; Wang, S. Dynamic Analysis of Fermentation Quality, Microbial Community, and Metabolome in the Whole Plant Soybean Silage. *Fermentation* **2024**, *10*, 535. https://doi.org/10.3390/fermentation10100535

Academic Editors: Siran Wang and Wansup Kwak

Received: 14 September 2024
Revised: 15 October 2024
Accepted: 16 October 2024
Published: 21 October 2024

1. Introduction

The rapid development of the livestock industry has led to a continuous increase in the demand for animal feed [1]. As a result, the livestock industry has been facing an increasing shortage of feed protein resources. Soybean (*Glycine max* (L.) Merr.) is one of the most important oilseed crops in the world [2]. Soybean seeds are rich in crude protein (CP), making their byproduct, soymeal, widely used in animal feed. In addition to the seeds, whole plant soybeans (WPS), including stems, leaves, and seeds, are characterized by their rich nutritional content, high biomass, and widespread distribution [3]. Soybeans are photoperiod-sensitive short-day crops. Research indicates that growing low-latitude soybean varieties in high-latitude regions significantly extends their vegetative growth period, resulting in several-fold increases in biomass [4]. However, in recent years, WPS has primarily been used inefficiently as animal feed, rural fuel, or returned to the field, and is often arbitrarily discarded [5]. Therefore, rational strategies are needed to enhance the utilization of WPS.

Ensiling is a traditional method of forage preservation that relies on anaerobic fermentation by lactic acid bacteria (LAB) [1]. The world's major ruminant feed source is silage [6]. Therefore, WPS can be better utilized as a forage resource by converting it into WPS silage through ensiling fermentation. De Morais et al. [7] demonstrated that WPS silage offers a potential high-protein roughage source for ruminant diets. During ensiling, water-soluble carbohydrates (WSC) in the green raw material are fermented into organic acids like lactic acid (LA) and acetic acid (AA), leading to a pH decline that inhibits undesirable microorganisms. Beneficial microorganisms, such as LAB, accelerate the utilization of these carbohydrates by other microorganisms, enhancing the production of organic acids and significantly contributing to the silage fermentation process, while some undesirable microorganisms, such as mold and clostridia, may cause spoilage [8]. However, as a legume, WPS may struggle to achieve satisfactory silage quality due to its low WSC content and insufficient epiphytic LAB count. For well-preserved silage, WSC content should exceed 60 g/kg dry matter, and epiphytic LAB count should be no less than 5 $\log_{10}$ CFU/g FM [9]. If these criteria are not met, the silage pH cannot decline rapidly, leading to reduced silage quality.

Silage fermentation is a process driven by microorganisms, and its quality heavily depends on the activities and diversity of the microorganisms involved [10]. Hence, revealing the bacterial community in silage is crucial for ensuring its preservation. Additionally, silage fermentation is a dynamic process. Understanding the changes in the microbial community structure during silage production helps identify key beneficial microorganisms that influence silage quality. However, due to the limitations of traditional microbial isolation and cultivation methods, advancements in next-generation sequencing (NGS) have led to an exponential increase in the discovery and characterization of microorganisms. NGS techniques do not rely on conventional cultivation methods, allowing for the detection of unculturable microorganisms [11]. Currently, many studies utilize NGS techniques to explore microbial community changes in various silage. Kharazian et al. [12] employed NGS sequencing to investigate the microbial community dynamics in sorghum silage with and without inoculation of *Lactiplantibacillus plantarum* and under different DM contents. Their findings revealed that during fermentation, the dominant microbial group shifted from *Pseudomonas congelans* to *L. buchneri*. Similarly, Liu et al. [13] used NGS sequencing to study the variations in the microbial communities of alfalfa silage with LAB additives and under varying temperatures. Their results revealed that an increase in temperature and the presence of LAB additives significantly increased the abundance of *Lactobacillus*. Silage fermentation is a complex process, and the quality of silage largely depends not only on the microbial community present but also on the metabolic products produced. Over the past few decades, research on silage metabolites has primarily focused on organic acids, ethanol, and 1,2-propanediol to evaluate the fermentation quality and aerobic stability of silage [14]. Sun et al. demonstrated that LAB generates various compounds during silage fermentation, including fatty acids, vitamins, and aromatic compounds. This indicates that the metabolites in silage have not been thoroughly identified. As one of the most popular "omics" technologies, metabolomics is considered a powerful tool due to its ability to rapidly detect, identify, and quantify numerous metabolites in biological samples [15]. It is widely used in biomedical research, environmental monitoring, and food and nutrition studies [16]. Xu et al. [17] used a multi-omics approach to study the interactions between bacterial microbiota and metabolome in whole-crop corn ensiling systems. They found that biofunctional metabolites were closely linked to the main types of lactic acid bacteria, impacting the fermentation process.

Based on previous research, there has been limited investigation into the dynamic processes of WPS silage. Therefore, this study aims to combine NGS sequencing and metabolomics techniques to examine the impact of different fermentation durations on the fermentation quality, microbial communities, and metabolites of WPS silage. By elucidating the interactions between key microorganisms and metabolites during fermentation, this research provides a foundation for producing high-quality animal feed.

2. Materials and Methods

2.1. Materials Preparation

The soybeans (variety "J8009") were planted on 20 May 2022, in the experimental fields of Northeast Agricultural University (126°3′30″ E, 45°44′34″ N, altitude 178 m) at a density of 250,000 plants per hectare. A basal fertilizer of 225 kg/ha compound fertilizer ($N{:}P_2O_5$ = 45:75:75) was applied with no additional fertilization until harvest. According to our previous research, soybeans harvested at the early podding stage are suitable for ensiling [18], so the crop was manually cut at this stage, leaving a stubble height of 10 cm. After harvesting, the soybeans were wilted on clean outdoor mats for 8 h and then chopped into 2-3 cm pieces using a forage chopper (YL100L-2, Weihai, China). The chopped material was mixed thoroughly, and 500 g samples were packed into polyethylene bags (25 cm × 30 cm, Wenzhou, China), vacuum-sealed (DZQ-420C, Ansenke, Quanzhou, China), and stored in the dark at room temperature (20 °C–25 °C). A total of 15 silage bags (5 fermentation periods × 3 replicates) were prepared. Samples were randomly taken at 7, 15, 30, 60, and 90 days of ensiling for analysis of fermentation indices, bacterial community, and metabolites.

2.2. WPS Silage Chemical Composition Analysis

At the end of the designated fermentation period, samples were taken from the silage bags. Each silage sample was weighed (100 g), dried at 75 °C for 48 h to determine the dry matter (DM) content, then ground through a 1 mm sieve and stored in a desiccator at room temperature for subsequent chemical analysis. The WSC content was determined using the anthrone-sulfuric acid method. Crude fat (CF) and CP contents were measured using the Soxhlet extraction method and the Kjeldahl nitrogen method, respectively [19]. Acid detergent fiber (ADF) and neutral detergent fiber (NDF) were analyzed using the methods of McDonald et al. [20]. Ammonia nitrogen (NH3-N) content was determined according to the method of Broderick and Kang [21].

A 10 g sample from each silage bag was mixed with 100 mL of sterile ultrapure water and homogenized for 35 s. The homogenate was stored in a refrigerator at 4 °C for 24 h. The mixture was then filtered through medical gauze, and the pH was immediately measured using a pH meter (PHS-3CW, BANTE, Shanghai, China). Organic acid contents, including lactic acid (LA), acetic acid (AA), propionic acid (PA), and butyric acid (BA), were subsequently determined according to the method of Meng et al. [22]. The aerobic stability (AS) of the silage was assessed by recording the number of hours required for the silage temperature to rise 2 °C above the ambient temperature [23].

2.3. WPS Silage Bacterial Community Sequencing Analysis

Total bacterial genomic DNA was extracted using FastDNA SPIN kits (MP Biomedicals, Santa Ana, CA, USA) according to the manufacturer's guidelines and stored at −20 °C for subsequent analyses. PCR amplification of the V3-V4 region of bacterial 16S rRNA genes was performed using primers 338F (5′-ACTCCTACGGGAGGCAGCA) and 806R (5′-GGACTACHVGGGTWTCTAAT). Sequencing was carried out on the Illumina NovaSeq platform with paired-end 2 × 300 bp reads by Biomarker Technologies Co., Ltd. Raw reads were first filtered using Trimmomatic v0.33. Primer sequences were identified and removed with Cutadapt 1.9.1, resulting in clean reads. Denoising was performed using the DADA2 method in QIIME2 2020.6 [24]. Reads were trimmed based on the error rate algorithm in DADA2, and ASVs were clustered using the DADA2 clustering algorithm [25]. Taxonomy was assigned through DADA2 utilizing the Silva database (https://www.arb-silva.de/, accessed on 1 June 2024).

2.4. WPS Silage Metabolite Analysis

Silage samples were ground post-freeze-drying and extracted with 70% aqueous methanol at 4 °C overnight. The extracts were centrifuged (12,000 rpm for 10 min) and filtered through a 0.22 μm membrane before LC-MS/MS analysis [26]. Metabolite analysis was conducted using the MetWare database by Wekemo Tech Group Co., Ltd., Shenzhen, China, with differential metabolites selected based on VIP $\geq$ 1.0 and FC $\geq$ 2.0. Metabolites were annotated using the Kyoto Encyclopedia of Genes and Genomes (KEGG) compound database and mapped to KEGG pathways (http://www.kegg.jp/kegg/pathway.html, accessed on 1 June 2024).

2.5. Statistical Analysis and Graphing

The fermentation and nutritional indicator data of the silage were analyzed using one-way ANOVA (SPSS 26.0, Chicago, IL, USA). Tukey's Honest Significant Difference (HSD) test was employed for mean comparisons among different samples, with significance set at $p < 0.05$. Sequence data analyses were primarily performed using QIIME2 2020.6 and R software (v4.4.0). PCA, PCoA, and CCA analyses were conducted using the "Vegan" package in R. Alpha diversity was calculated using the "diversity" function of the "Vegan" package, while beta diversity distances were computed using the "vegdist" function from the same package. CCA analysis was performed with the "cca" function in the "Vegan" package. Pearson correlation calculations were conducted using the "correlation_matrix" function from the "corrtable" package, and random forest analysis was conducted using the "randomForest" package. The Mantel test calculations and plotting were executed with the "linkET" package, and boxplots were drawn using the "ggpubr" package. Scatter plots with linear regression were created using the "ggscatter" function from the "ggpubr" package. PLS-DA analysis, pie chart plotting, and multiple volcano plots were generated using the OmicShare tools, a free online data analysis platform (https://www.omicshare.com/tools, accessed on 26 July 2024).

3. Results

3.1. Chemical Compositions of WPS and WPS Silage

The chemical composition of fresh WPS before ensiling is shown in Table 1. The DM content of fresh WPS is 34.26%, with nutritional indices of CP, ADF, NDF, and CF at 18.78%, 35.26%, 48.27%, and 3.77%, respectively. The WSC content of fresh WPS is 3.84%. The DM, CP, ADF, and NDF contents are higher, while the WSC content is lower than those reported by Zeng et al. [3]. This discrepancy may be due to differences in soybean varieties and sampling periods.

Table 1. Chemical composition of the WPS before ensiling.

Item	WPS
DM (%)	34.26 ± 0.28
CP (%)	18.78 ± 0.38
ADF (%)	35.26 ± 0.55
NDF (%)	48.27 ± 0.58
CF (%)	3.77 ± 0.09
WSC (%)	3.84 ± 0.95

Notes: DM, dry matter; CP, crude protein; ADF, acid detergent fiber; NDF, neutral detergent fiber; CF, crude fat; WSC, water-soluble carbohydrate.

The fermentation quality and chemical composition of different ensiling days silage treatments (7, 15, 30, 60, and 90 days) are summarized in Table 2. The DM content varied, with values ranging from 35.01% at 7 days to 37.34% at 90 days. The DM content at 15 days was significantly lower than other groups ($p < 0.05$). CP content showed significant variation, peaking at 18.35% at 15 days ($p < 0.05$) and reaching a low of 17.02% at 30 days ($p < 0.05$), with no significant difference between the 7 d and 15 d groups. CF content was highest at 7 days (3.72%) and lowest at 30 days (3.18%), with no significant difference between the 7 d and 15 d groups. NDF content was highest at 7 days (48.55%) ($p < 0.05$) and lowest at 60 days (43.11%) ($p < 0.05$), while ADF content was lowest at 7 days (31.07%) ($p < 0.05$) and highest at 90 days (34.50%) ($p < 0.05$). WSC content decreased from 2.51%

at 7 days to 1.39% at 90 days. The pH values showed significant differences, with the highest values at 7 days (5.98) and 15 days (5.96), and the lowest at 90 days (5.13); the 7 d and 15 d groups were not significantly different. The LA content in the 7 d group was significantly lower than in the other groups, while no significant differences were observed in LA content among the other groups. The AA content was highest in the 7 d group (1.70%) and the 90 d group (1.73%) ($p < 0.05$), significantly lower in the 15 d (1.50%) and 30 d (1.45%) groups ($p < 0.05$), with the lowest content observed in the 60 d group (1.27%) ($p < 0.05$). PA content was consistently low, with values ranging from 0.06% at 60 days to 0.16% at 30 days. BA was not detected at 7, 15, and 30 days of ensiling. However, BA content was 0.06% in the 60 d group and 0.12% in the 90 d group. The NH3-H content (% of TN) increased significantly over the fermentation periods, from 3.27% at 7 days to 5.17% at 90 days, with significant differences among most groups. AS improved from 54.00 h at 7 days to 112.00 h at 90 days. These results indicate significant differences in fermentation quality and chemical composition across the different fermentation periods.

Table 2. The fermentation quality and nutritional indicators of different mixed silage treatments after ensiling.

Items	7 d	15 d	30 d	60 d	90 d	*p*-Value
DM (%)	35.01 ± 0.95^{b}	37.29 ± 0.37^{a}	37.94 ± 0.38^{a}	37.16 ± 0.33^{a}	37.34 ± 0.36^{a}	<0.001
CP (%)	18.32 ± 0.46^{a}	18.35 ± 0.07^{a}	17.02 ± 0.54^{c}	17.92 ± 0.31^{ab}	17.37 ± 0.33^{bc}	<0.001
CF (%)	3.72 ± 0.07^{a}	3.63 ± 0.08^{a}	3.18 ± 0.98^{c}	3.45 ± 0.05^{b}	3.56 ± 0.03^{ab}	<0.001
NDF (%)	48.55 ± 0.12^{a}	45.87 ± 0.15^{b}	46.11 ± 0.58^{b}	43.11 ± 0.33^{c}	43.77 ± 1.60^{c}	<0.001
ADF (%)	31.07 ± 0.49^{b}	32.01 ± 0.60^{b}	31.78 ± 0.71^{b}	32.22 ± 0.82^{b}	34.50 ± 1.39^{a}	0.009
WSC (%)	2.51 ± 0.41^{a}	1.88 ± 0.02^{b}	1.63 ± 0.01^{bc}	1.65 ± 0.16^{bc}	1.39 ± 0.07^{c}	<0.001
pH	5.98 ± 0.14^{a}	5.96 ± 0.13^{a}	5.96 ± 0.34^{a}	5.42 ± 0.07^{b}	5.13 ± 0.16^{b}	<0.001
LA (%)	2.19 ± 0.04^{b}	2.68 ± 0.18^{a}	2.72 ± 0.09^{a}	2.99 ± 0.16^{a}	2.97 ± 0.11^{a}	<0.001
AA (%)	1.70 ± 0.02^{a}	1.50 ± 0.11^{b}	1.45 ± 0.04^{b}	1.73 ± 0.18^{a}	1.27 ± 0.08^{c}	<0.001
PA (%)	0.10 ± 0.02^{c}	0.08 ± 0.03^{c}	0.16 ± 0.04^{ab}	0.11 ± 0.04^{bc}	0.17 ± 0.02^{a}	0.004
BA (%)	-	-	-	0.06 ± 0.03^{b}	0.12 ± 0.01^{a}	<0.001
NH3-H (%/TN)	3.27 ± 0.13^{d}	3.85 ± 0.31^{c}	3.58 ± 0.10^{cd}	4.47 ± 0.29^{b}	5.17 ± 0.18^{a}	<0.001
AS (h)	54.00 ± 2.00^{e}	62.67 ± 2.52^{d}	100.67 ± 5.32^{c}	108.67 ± 3.38^{b}	112.00 ± 2.87^{a}	<0.001

Note: LA, lactic acid; AA, acetic acid; PA, propionic acid; BA, butyric acid; NDF, neutral detergent fiber; ADF, acid detergent fiber; DM, dry matter; NH3-N, ammonia nitrogen; TN, total nitrogen; CF, crude fat; -, not detected; WSC, water-soluble carbohydrates; CP, crude protein; AS, aerobic stability. Different superscript letters (a–e) in a row indicate significant differences ($p < 0.05$) based on Tukey's HSD test.

3.2. Microbial Community of the WPS Silage

After quality control of silage fermentation, each sample yielded an average of 76,430 reads based on Illumina sequencing. Rarefaction curves indicated that sequencing coverage was adequate (Figure S1). The alpha diversity indices, as shown in Figure 1A, include Simpson, Shannon, Chao1, and PD_whole_tree indices. There were no significant differences in the Simpson index among the groups at different ensiling times. The Shannon index for the 60 d group was significantly higher than that of the 15 d and 90 d groups ($p < 0.05$). The Chao1 index followed a similar trend to the Shannon index, with the 60 d group having a significantly higher Chao1 index than the 15 d and 30 d groups ($p < 0.05$). The PD_whole_tree index for the 60 d group was significantly higher than those of the 30 d and 90 d groups, and the 7 d group had a significantly higher PD_whole_tree index than the 15 d group. The PCoA analysis based on Bray–Curtis distances is shown in Figure 1B. The Anosim test indicates that there are significant differences in beta diversity between the samples from different groups. The first and second principal coordinates (PCoA1 and PCoA2) account for 69.2% of the total variance in the microbial composition of the silage, with PCoA1 explaining 40.41% and PCoA2 explaining 28.79%. There is a clear separation among the groups with different fermentation times along PCoA2. Specifically, the 7 d group is distinctly separated from the 15 d, 30 d, and 90 d groups, and the 30 d group is clearly separated from the 60 d and 7 d group. Additionally, the samples from the 7 d group were the most distant from those of the 30 d group, indicating the greatest difference in microbial communities between these two groups. The Venn diagram illustrates that the number of shared ASVs across the five fermentation periods is 119 (Figure 1C). The 7 d group has the highest number of unique ASVs, with 23, while the 30 d group has the fewest, with 9. The number of unique ASVs in the other groups is similar. To better explore the relationship between ensiling time and species richness, we conducted a linear regression analysis. Figure 1D illustrates the relationship between ln-transformed species richness and ln-transformed ensiling days. As presented in Figure 1E, at the phylum level, *Proteobacteria* and *Firmicutes* are the dominant bacteria across all fermentation periods. The relative abundance of *Proteobacteria* is highest in the 7 d group (69.12%), followed by the 60 d (64.08%), 15 d (58.27%), 90 d (47.98%), and 30 d (44.28%) groups. Conversely, *Firmicutes* show the highest relative abundance in the 30 d group (43.78%), followed by the 90 d (34.45%), 15 d (26.18%), 7 d (17.40%), and 60 d (15.82%) groups. As shown in Figure 1F, at the genus level, *Enterococcus* and *Lactobacillus* are prominent genera in the silage bacterial community. *Enterococcus* exhibits the highest relative abundance in the 30 d group (26.17%), followed by the 15 d (18.94%), 7 d (11.82%), 60 d (6.82%), and 90 d (6.32%) groups, indicating its dominance across all fermentation periods. *Lactobacillus*, while not consistently dominant throughout the fermentation process, shows a significant increase in abundance at 60 d (5.25%) and reaches its peak in the 90 d group (17.63%). Additionally, *Sphingomonas* genera shows considerable presence, particularly in the 15 d group (16.44%), with substantial abundance in other periods as well. *Weissella* genera peaks at 30 d (13.04%), and *Kosakonia* genera is notably abundant at 7 d (18.69%) but decreases significantly in later stages.

3.3. Biomarker Selection and Correlation with Fermentation and Nutritional Indicators

The ASVs from each group were analyzed using a random forest model (Bootstrap resampling with 1000 iterations and 1000 decision trees) to identify the most important microbial biomarkers in the silage. The top 10 important ASVs were selected based on the percentage increase in MSE% (%IncMSE). The abundance bubble chart of these biomarkers across different samples is shown in Figure 2A. *Pantoea* and *Brevundimonas* genera were significantly more abundant in the 60 d group compared to other groups, while *Enterococcus* genera was significantly more abundant in the 7 d group. A higher increase in MSE% indicates greater importance of the biomarker. The bar chart arranged by importance, as shown in Figure 2B, highlights *Pantoea* as the key biomarker, followed by *Brevundimonas*, *Paenibacillus*, *Sphingomonas*, *Enterococcus*, *Klebsiella*, *Lactobacillus*, *Bradyrhizobium*, *Pseudomonas*, and *Quadrisphaera*.

Next, a correlation analysis between the selected biomarkers and fermentation and nutritional indicators was conducted. It was observed that *Pantoea* genera was significantly negatively correlated with DM ($p < 0.05$). *Lactobacillus* genera showed a significant positive correlation with LA ($p < 0.05$), *Bradyrhizobium* genera had a highly significant positive correlation with DM ($p < 0.01$), and *Enterococcus* genera was significantly positively correlated with ADF, WSC, and LA ($p < 0.05$).

3.4. Metabolite Analysis of WPS Silage

In this study, non-targeted metabolomics analysis of WPS silage at different fermentation days was performed using LC–MS/MS (Liquid Chromatography–Mass Spectrometry). A total of 1173 metabolites were detected in the positive ion mode (ESI(+)), as shown in Figure 3A. These included 439 lipids and lipid-like molecules, 192 organic acids and derivatives, 153 organoheterocyclic compounds, 114 phenylpropanoids and polyketides, 75 benzenoids, 53 organic oxygen compounds, 41 nucleosides, nucleotides, and analogues, 24 alkaloids and derivatives, 19 organic nitrogen compounds, 9 lignans, neolignans, and related compounds, 2 homogeneous non-metal compounds, and 2 organosulfur compounds.

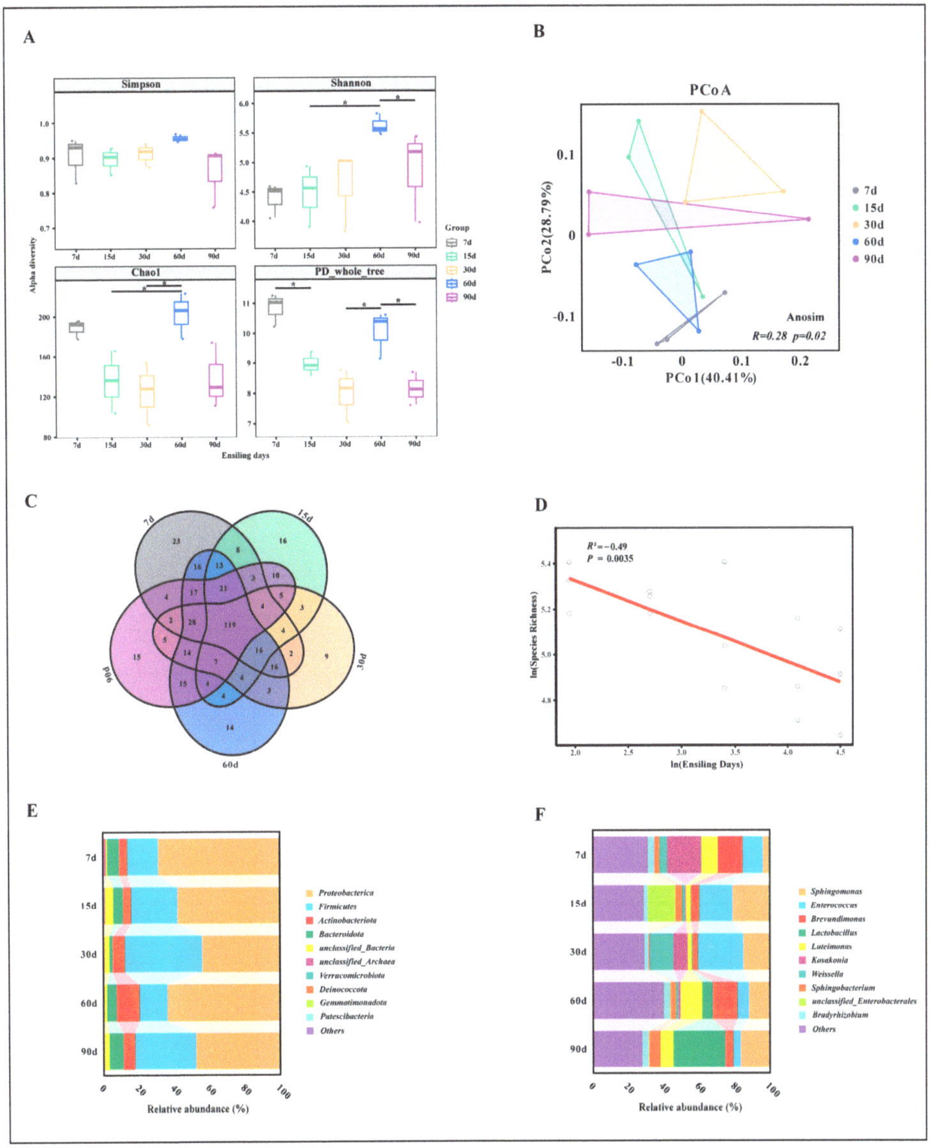

Figure 1. Bacterial diversity, community composition, and structure variations in WPS silage during the ensiling process. (**A**) Box plots of alpha diversity (Simpson, Shannon, Chao1 indices, and PD_whole_tree) with *p*-values from Kruskal–Wallis tests, where * indicates $p < 0.05$. (**B**) PCoA analysis based on Bray–Curtis distances. (**C**) Venn diagram of WPS silage at different fermentation days. (**D**) Correlation between ensiling days and species richness in WPS silage. (**E**) Stacked bar chart of bacterial community relative abundance at the phylum level. (**F**) Stacked bar chart of bacterial community relative abundance at the genus level.

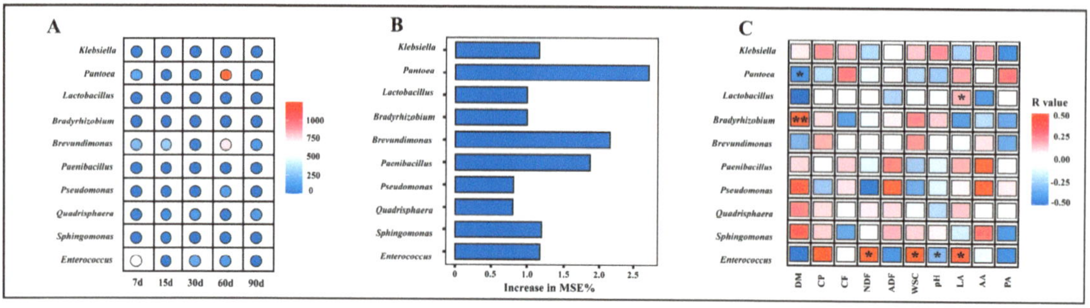

Figure 2. Biomarkers were selected based on a random forest algorithm and correlated with fermentation and nutritional indicators. (**A**) Bubble chart of biomarker abundance. (**B**) Bar chart of the top 10 most important biomarkers. (**C**) Heatmap of the correlation between biomarkers and fermentation and nutritional indicators (based on Pearson correlation), Asterisks: * denotes significant difference ($p < 0.05$); ** denotes significant difference ($p < 0.01$).

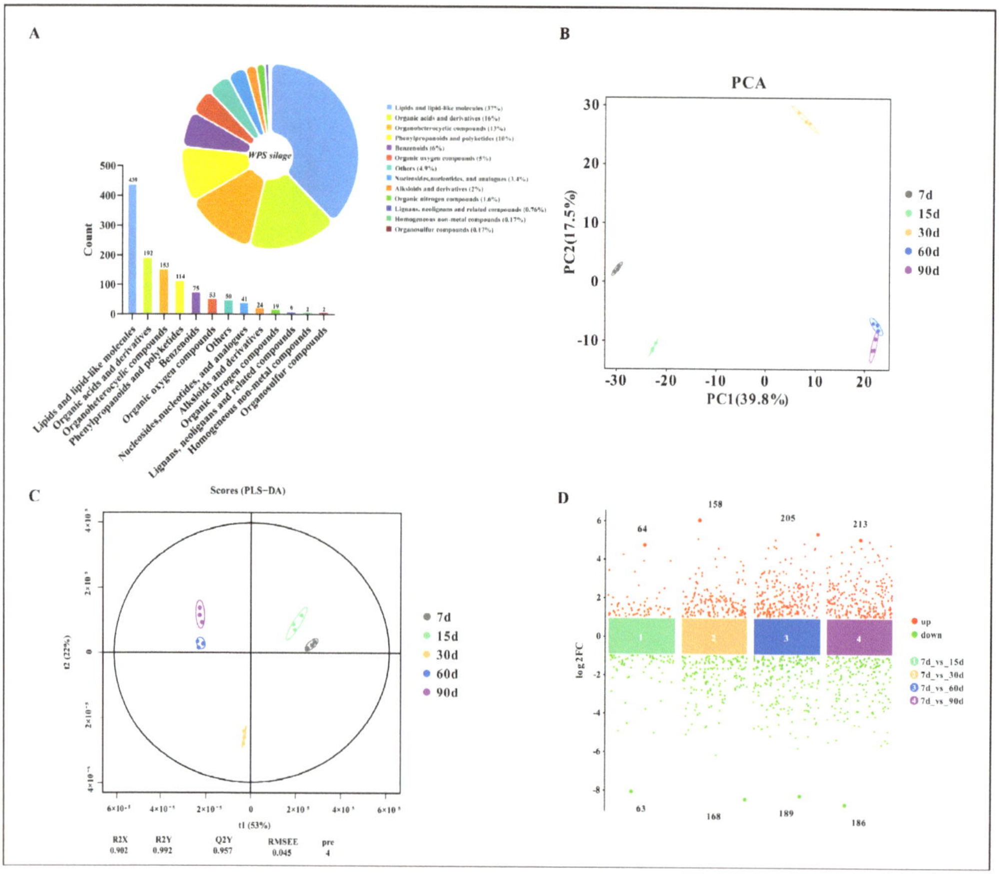

Figure 3. (**A**) Bar and pie charts of detected metabolites types. (**B**) PCA analysis of WPS silage metabolites at different fermentation days. (**C**) PLS-DA analysis of WPS silage metabolites at different fermentation days (**D**) Multiple volcano plots of differential metabolites in WPS silage.

To distinguish changes in WPS silage metabolites across different fermentation days, unsupervised PCA and supervised PLS-DA multivariate analyses were employed to investigate the relationship between metabolites and the quality changes in WPS silage. As shown in Figure 3B,C, the first two PCA components (PCA1 and PCA2) explained 57.3% of the variance. Samples from different fermentation times were dispersed between groups and clustered within groups, indicating good intra-group reproducibility and clear inter-group differences. PLS-DA was used to analyze the impact of different fermentation days on WPS silage metabolites, resulting in the identification of differential metabolites (DAMs) for WPS silage at various fermentation stages. R2X and R2Y represent the explanatory power of the PLS-DA model for the X and Y matrices, respectively, while Q2 represents the predictive ability of the model. In this study, these three indicators were close to 1, indicating that the model was stable and reliable. Given the significant differences between the fermentation quality and chemical composition of the 7 d silage group and other fermentation times, the 7 d silage group was selected as the control. DAMs with a fold change (FC) > 2 were identified through multiple volcano plots, comparing 15 d, 30 d, 60 d, and 90 d silage samples with the 7 d silage group, as shown in Figure 3D. Compared to the 7 d group, 64 DAMs were significantly upregulated, and 63 were significantly downregulated in the 15 d group; 158 DAMs were significantly upregulated, and 168 were significantly downregulated in the 30 d group; 205 DAMs were significantly upregulated, and 189 were significantly downregulated in the 60 d group; and 213 DAMs were significantly upregulated, and 186 were significantly downregulated in the 90 d group. Notably, the 30 d group exhibited the most DAMs compared to the 7 d group, while the 15 d group showed the fewest. Additionally, the 30 d group exhibited the greatest difference in microbial communities compared to the 7 d group (Figure 1B).

Metabolite enrichment analysis was conducted using differentially accumulated metabolites (DAMs) with fold change (FC) values greater than 2 and VIP scores greater than 1, based on the Kyoto Encyclopedia of Genes and Genomes (KEGG). Multiple KEGG enrichment bubble charts were generated for the top 10 metabolic pathways with the lowest p-values, comparing 15 d, 30 d, 60 d, and 90 d with 7 d, as shown in Figure 4. A total of 30 metabolic pathways were identified across the four groups, with common enrichment observed in pathways such as "Galactose metabolism", "Glutathione metabolism", "One carbon pool by folate", "Phenylalanine metabolism", "Porphyrin metabolism", "Propanoate metabolism", and "Valine, leucine and isoleucine biosynthesis". Notably, the "Valine, leucine and isoleucine biosynthesis" pathway was enriched in all four groups.

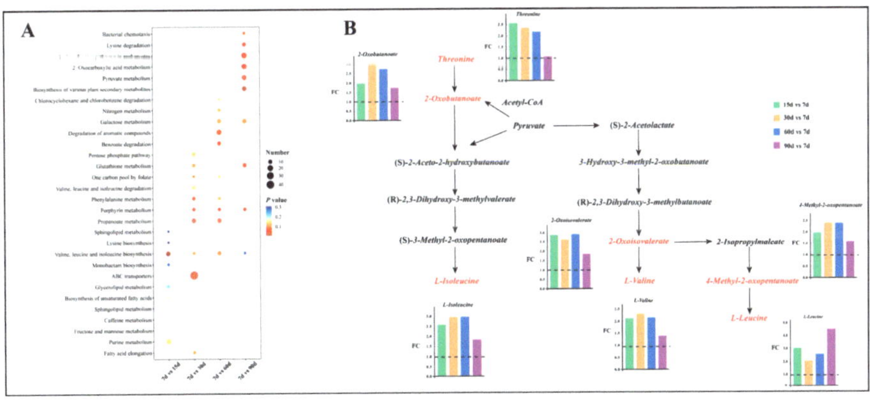

Figure 4. (**A**) Multiple KEGG enrichment bubble charts. (**B**) KEGG metabolic network analysis of the Valine, Leucine, and Isoleucine biosynthesis pathways in the WPS silage. The differentially accumulated metabolites (DAMs) are shown in red.

Subsequently, a metabolic network analysis of the "Valine, leucine and isoleucine biosynthesis" pathway was performed to observe the relative abundance changes of key metabolites across different fermentation periods (15 d, 30 d, 60 d, 90 d compared to 7 d), as depicted in Figure 4B. This pathway begins with Threonine, and the FC values represent changes in metabolite abundance between 7 days of fermentation and other time points. Most metabolites exhibited higher FC values at 30 d and 60 d, indicating a relative increase in abundance at these stages. However, at 90 d, the FC values decreased for most metabolites, except for L-Leucine, suggesting a decline in the utilization or production rate of these metabolites during the latter stages of fermentation. Specifically, 2-Oxobutanoate and L-Valine showed an overall increasing trend between 15 and 30 days, peaking at 30 days, followed by a decline at 60 and 90 days. This pattern indicates that these metabolites play a more significant role during the early to middle stages of the fermentation process.

3.5. Correlations Between Chemical Composition and Microbial Community

Based on Detrended Correspondence Analysis (DCA), the axis length value was determined to be 3.18, leading to the selection of CCA for further investigation. In this analysis, the top 10 most significant bacterial genera identified through random forest (as shown in Figure 2) were used as explanatory variables, with metabolites serving as response variables (Figure 5A). The length of the vector arrows represents the degree to which each environmental factor influences the distribution of the studied objects, with longer arrows indicating a stronger influence. From the CCA plot, it is evident that *Enterococcus* exerts the greatest impact on the metabolites, followed by *Pantoea*. Additionally, the angles between vectors less than 90 degrees indicate positive correlations, angles greater than 90 degrees indicate negative correlations and angles of 90 degrees indicate no correlation. *Enterococcus* shows a positive correlation with *Klebsiella*, *Bradyrhizobium*, and *Pseudomonas* and a negative correlation with *Sphingomonas*, *Pantoea*, *Lactobacillus*, and *Brevundimonas*. *Sphingomonas*, *Pantoea*, and *Lactobacillus* exhibit positive correlations with each other, while *Pseudomonas*, *Paenibacillus*, and *Quadrisphaera* are also positively correlated.

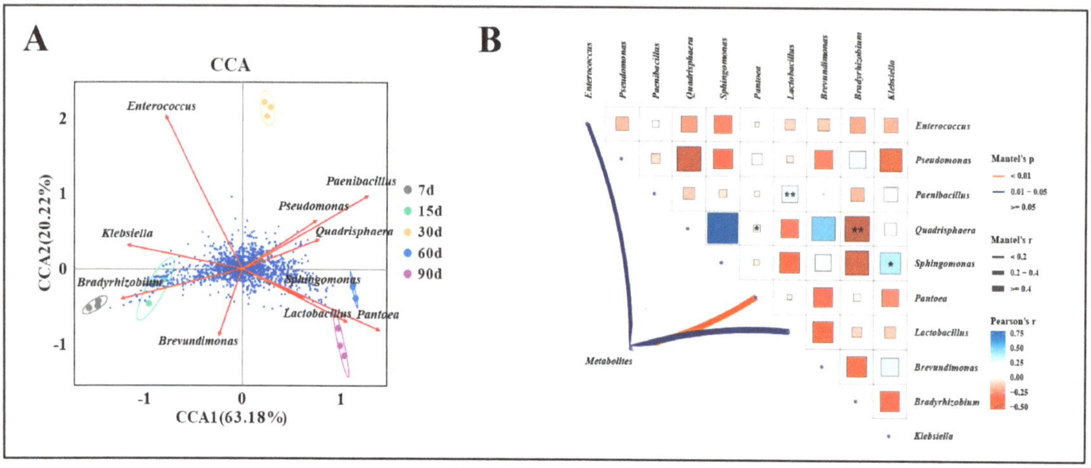

Figure 5. (**A**) CCA of bacterial and metabolites in WPS silage at different fermentation days. (**B**) The Mantel test correlation plot between bacteria and metabolites with Pearson's correlation coefficients, Asterisks: * denotes significant difference ($p < 0.05$); ** denotes significant difference ($p < 0.01$).

Subsequently, a Mantel test was conducted to analyze the relationship analyze between the relative abundances of key silage bacteria and metabolites, as shown in Figure 5B. The results demonstrated a highly significant positive correlation between *Lactobacillus* and *Paenibacillus* ($p < 0.01$) and a significant negative correlation between *Pantoea* and

Quadrisphaera ($p < 0.05$). Additionally, *Bradyrhizobium* also showed a highly significant negative correlation with *Quadrisphaera* ($p < 0.01$). The genus *Klebsiella* was significantly positively correlated with *Sphingomonas* ($p < 0.05$). Notably, only *Enterococcus*, *Pantoea*, and *Lactobacillus* had a significant impact on the metabolome, with *Pantoea* having the most substantial influence on the metabolic profile.

4. Discussion

4.1. Fermentation Quality and Nutritional Indicators of the WPS Silage

As reported by McDonald et al. [20] and Muck [27], a high moisture content in fresh material is associated with an increased risk of undesirable microbial fermentation, particularly by clostridia, which can lead to spoilage during the ensiling process. In this study, the DM content of the fresh WPS was close to the ideal range of 30–35%, which is considered suitable for producing satisfactory silage [28]. Key indicators such as CP, ADF, and NDF were analyzed. The CP content in our study was higher than that reported by Zeng and Ni et al. [29,30], while the ADF and NDF contents were lower than those reported by Ni et al. [30] but higher than those reported by Zeng et al. [29]. The WSC, which serve as fermentation substrates, are primarily utilized by bacteria to produce organic acids that lower the pH of the silage and inhibit the growth of undesirable bacteria. In our study, the WSC content in the WPS was relatively low, falling below the theoretical requirement of 60–70 g/kg DM needed for well-preserved silage [31]. Additionally, the characteristics of fresh WPS in this study differed from previous reports, likely due to variations in soybean cultivar, soil conditions, fertilization rates, and harvest timing.

The fermentation quality of silage is influenced by multiple factors, including pH, lactic acid concentration, volatile fatty acids, NH3-H levels, and others [32]. Among these, pH is a critical indicator of silage fermentation quality. In this study, a downward trend in pH was not observed until the 30th day of ensiling, with no significant differences between the pH values at 7, 15, and 30 days, which is consistent with findings by Zhang et al. [3]. Additionally, the goal of silage fermentation is to rapidly reduce the pH below 4.2 to ensure stable and high-quality silage [33]. However, in this experiment, the pH remained above 4.2 throughout all time periods, indicating a slow fermentation process and suggesting the need for silage additives to enhance fermentation. LA, a strong acid, is the primary factor contributing to the decrease in silage pH, with higher LA concentrations correlating with better silage quality [34]. In this study, the LA content at 7 days was significantly lower than in the other periods, with no significant differences observed among the other time points, indicating that LAB fermentation accelerated significantly after 7 days. Additionally, LAB also produces AA during the fermentation process, a phenomenon known as heterofermentation [35]. In this study, AA peaked at 7 days (1.70%) and 60 days (1.73%), with the lowest level observed at 90 days (1.27%). This suggests fluctuations in the abundance of LAB within the microbial community as fermentation progressed. PA levels remained consistently low throughout the study, aligning with the typical profile of LAB fermentation. BA was not detected in the early stages but emerged at 60 days (0.06%) and slightly increased at 90 days (0.12%), potentially indicating late-stage clostridial activity, which is undesirable [36]. The increase in NH3-H content indicates protein degradation, which is primarily associated with the activity of plant enzymes, enterobacteria, and clostridia [20].

4.2. Microbial Community and Key Bacteria of the WPS Silage

The ensiling period, as a crucial fermentation parameter [37], has rarely been the focus in determining the optimal fermentation duration for WPS silage. The study by Gao et al. [38] observed the impact of different time intervals and carbon sources on the organic acid content in ensiled alfalfa. However, few studies have addressed the changes in microbial communities during various fermentation periods. This study investigated the dynamic changes in microbial communities during different ensiling periods of WPS silage under natural conditions, with the goal of identifying key microbes that could optimize the fermentation process. The Simpson index showed no significant differences across all periods, indicating consistent community diversity throughout the ensiling process. However, the Shannon and Chao1 indices were significantly higher in the 60 d group compared to the 15 d and 30 d groups, suggesting that microbial richness peaked around 60 days. The Shannon index, which also accounts for evenness, highlighted a more balanced microbial community at this stage. The PD_whole_tree index further supported these findings. However, Xu et al. [17] reported a decline in alpha diversity when the dominant LAB community became relatively simplified during ensiling. The anaerobic environment of silage inhibits the growth of most aerobic microbes and reduces pH, which may contribute to decreased microbial diversity. This could suggest that the microbial community in the 60 d group may be shifting towards a less favorable fermentation profile compared to other groups. The PCoA analysis based on Bray–Curtis distances indicates significant differences in microbial community beta diversity across different fermentation periods. The clear separation between groups, particularly the marked distinction between the 7 d and 30 d groups, suggests substantial shifts in microbial composition as fermentation progresses. Furthermore, the pronounced separation along the PCoA2 axis highlights the influential role of fermentation time in driving structural differences between communities. The Venn diagram indicates a continuous decrease in unique ASVs from 7 to 30 days, reaching a minimum at 30 days, followed by a consistent increase from 60 to 90 days. The reduction in unique ASVs during the initial fermentation stages may be attributed to the anaerobic environment created by the ensiling process, which inhibits the growth of most aerobic microorganisms and allows LAB to become the dominant species [39]. This trend may indicate that fermentation quality improves from 7 to 30 days but begins to decline after 60 days. The results indicate a significant negative correlation (R^2 = 0.49, p = 0.0035), suggesting that species richness decreases significantly as ensiling time increases. The red regression line shows this trend, with the shaded area representing the 95% confidence interval. The linear regression between ln-transformed species richness and ensiling days shows a significant negative correlation from 7 to 90 days (Figure 1D), suggesting that longer ensiling may reduce microbial diversity and affect silage quality. Few studies have examined this linear relationship, and future research will include groups like WPS silage with different silage additives to further explore this trend and assess changes in the slope.

To further understand the impact of different fermentation periods on the microbial community in WPS silage, we investigated the changes in relative abundance at both the phylum and genus levels. At the phylum level, *Firmicutes* and *Proteobacteria* were identified as the dominant bacteria in the silage, consistent with findings from other studies [40]. During the initial fermentation phase from 7 to 30 days, the relative abundance of *Proteobacteria* decreased, while *Firmicutes* increased. This shift is likely due to the greater adaptability of *Firmicutes* to acidic and anaerobic conditions [41]. However, from 30 to 60 days, a decline in *Firmicutes* was observed, which is not an expected or favorable outcome. At the genus level, the top ten genera in relative abundance include *Lactobacillus*, *Enterococcus*, and *Weissella*, which are recognized as typical dominant genera in silage and are all classified as LAB [42]. *Enterococcus* is commonly used to enhance fermentation characteristics. Additionally, *Enterococcus* play a pivotal role in accelerating LA fermentation and establishing an anaerobic acidic environment conducive to the development of *Lactobacillus* [42]. Generally, *Enterococcus* are lactic acid-producing cocci that typically survive only in the early stages of fermentation due to their limited acid resistance [43]. Notably, in this study, the relative abundance of *Enterococcus* was highest at 30 days, followed by a decline. This may be attributed to the slower pH reduction in naturally fermented WPS silage, consistent with the findings of Wang et al. [44]. Additionally, *Sphingomonas* was observed in this study. Zhou et al. [45] identified *Sphingomonas* in silage, categorizing it as a harmful bacterium associated with the hydrolysis of soluble proteins. The presence of *Sphingomonas* in WPS silage could also be linked to the high NH3-H content. *Unclassified_Enterobacterales* were also detected, which are commonly regarded as harmful bacteria in silage, negatively affecting silage quality [46]. Their relative abundance peaked at 15 days, indicating bad fermentation at this stage. *Kosakonia*, a newly divided from the genus *Enterobacter* genus [47], also considered a harmful silage bacterium, showed the highest relative abundance at 7 days, suggesting that the early stages of fermentation were not ideal. Additionally, *Bradyrhizobium*, a genus of Rhizobia commonly found in leguminous plants, was observed. This genus is known for its slow growth and symbiotic relationship with alfalfa [48]. Its presence in this study may be attributed to the epiphytic bacteria on WPS.

The random forest algorithm, a non-parametric machine learning method, is particularly effective in identifying significant microbial biomarkers that differentiate between groups. In this study, we utilized the random forest algorithm to pinpoint biomarkers in WPS silage across different fermentation periods under natural conditions. The top ten genera were selected based on their importance, determined by %IncMSE. The top-ranked biomarker identified in this study was the genus *Pantoea*, a harmful bacterium that has recently been reclassified from the *Enterobacter* genus [47]. Notably, research by Sun et al. [49] found that *Pantoea* was the most dominant epiphytic bacterium in fresh whole-plant corn silage samples. Similarly, Tian et al. [50] identified *Pantoea* as the most dominant epiphytic bacterium in fresh Stylo samples. Sun et al. [51] further observed that the application of a LAB suspension as a silage additive led to a decrease in the relative abundance of *Pantoea* and a significant negative correlation between *Pantoea* and *Lactobacillus*. This suggests that *Pantoea*, being widespread as an epiphytic bacterium in fresh silage materials, plays a critical role in influencing the subsequent fermentation quality. The identification of *Pantoea* as a key biomarker in this study highlights its significant impact on the early stages of fermentation, indicating that controlling *Pantoea* abundance before the onset of fermentation could potentially steer the process in a favorable direction. However, this study did not analyze the microbial community at the start of fermentation, which is a gap that should be addressed in future research. Additionally, *Lactobacillus* and *Enterococcus* were also identified as important biomarkers, both showing a significant positive correlation with LA concentration ($p < 0.05$).

4.3. Metabolites Analysis and Correlation of Metabolites with Microbiome and Fermentation Indicators

A total of 1173 metabolites were detected in the naturally fermented WPS silage, exceeding the numbers reported in previous studies on whole-crop corn silage, paper mulberry leaf silage, and stylo silage [17,52,53]. The relatively high pH observed in this study's WPS silage likely contributed to more active microbial metabolism compared to previous studies, resulting in a greater abundance of metabolites [54]. Notably, lipids and lipid-like molecules accounted for the highest proportion at 37%, which is likely due to the lipid-rich nature of soybean grains. PCA and PLS-DA analyses revealed that the 30 day fermentation samples were the most distinct from other fermentation periods, showing the greatest metabolite variation compared to other groups. Multiple volcano plots indicated that the least number of differential metabolites was observed between the 15 days and 7 days fermentation periods, while the number of differential metabolites increased significantly in the 60 days and 90 days periods, suggesting heightened microbial activity after 30 days.

Further analysis was conducted on the DAMs identified with a FC greater than 2 and VIP scores greater than 1, leading to the generation of Multiple KEGG enrichment bubble charts for the top 10 metabolic pathways. Similar to the findings of He et al., pathways such as "Lysine degradation", "Valine, leucine and isoleucine degradation", "Lysine biosynthesis" and "Purine metabolism" were enriched. In this study, all fermentation periods compared to the 7 days period commonly enriched the "Valine, leucine and isoleucine biosynthesis" pathway. Consequently, this pathway was selected for further network analysis. Valine, leucine, and isoleucine are essential branched-chain amino acids (BCAAs) in mammals, playing critical roles in protein synthesis, muscle growth, and hormone release [55]. In mammals, these BCAAs cannot be synthesized de novo and must be obtained through diet [56]. Therefore, the production and accumulation of these amino acids during silage fermentation are crucial for enhancing the nutritional value of silage. Metabolic analysis indicates that the biosynthesis pathways for valine, leucine, and isoleucine were more active during different stages of fermentation, particularly at 30 and 60 days, where the relative abundance of precursor metabolites was higher. This suggests that the mid-stage of fermentation may be a critical period for the accumulation of these essential amino acids. This finding has significant implications for optimizing the nutritional content of silage to better meet the dietary needs of livestock. However, during the later stage of fermentation (90 days), the abundance of most of these metabolites, except for L-leucine, decreased, potentially reflecting a reduced rate of amino acid synthesis and utilization as fermentation progresses. Therefore, selecting an appropriate fermentation time, ideally not exceeding 60 days, under natural conditions for WPS silage can maximize the accumulation of these essential amino acids, thereby enhancing the quality and nutritional benefits of the silage. Building on the results from the random forest analysis, the key bacterial taxa identified were used as explanatory variables in CCA. The vector for *Enterococcus* had the greatest length, signifying that *Enterococcus*, a beneficial silage bacterium and a key taxon within the community, exerts the most significant influence on the silage metabolome. The proximity of the 30-day group samples to the *Enterococcus* vector further supports the observation that the relative abundance of *Enterococcus* was highest in this group.

To explore the relationship between the microbial community and metabolome, a Mantel test was conducted. The Mantel test was seldom used in previous silage studies to examine the relationships among multidimensional datasets, making this approach particularly noteworthy. The analysis revealed that the genera *Enterococcus*, *Pantoea*, and *Lactobacillus* had significant impacts on metabolome matrices. These findings suggest that these genera are crucial for optimizing the microbial community and metabolome in future silage production.

5. Conclusions

The study indicates that WPS silage holds potential as a feed source, but under natural fermentation conditions, it is prone to undesirable fermentation. Significant differences were observed in the microbial community and metabolome across different fermentation periods, with the optimal fermentation effect occurring at 30 days, where *Enterococcus* emerged as the dominant genus. Through random forest analysis, ten key bacterial genera were identified as biomarkers, with *Pantoea* being the most influential. Notably, compared to the early stages of fermentation, the later stages significantly enhanced the metabolic pathways involved in the biosynthesis of essential amino acids, specifically valine, leucine, and isoleucine.

Supplementary Materials: The following supporting information can be downloaded at: https://www.mdpi.com/article/10.3390/fermentation10100535/s1, Figure S1: The species dilution curve of WPS silage at different fermentation period.

Author Contributions: Conceptualization Y.J. and S.W. (Shaodong Wang); methodology, Y.J.; software, S.W (Sui Wang); validation, S.W. (Shaodong Wang), Y.J., S.W. (Sui Wang), H.M., L.W., Y.L. and X.T.; formal analysis, H.M.; investigation, H.M.; resources, Y.J. and Y.L.; data curation, H.M.; writing—original draft preparation, H.M.; writing—review and editing, H.M. and Y.J.; visualization, X.T.; supervision, S.W. (Shaodong Wang); project administration, Y.J. and S.W. (Shaodong Wang); funding acquisition, S.W. (Shaodong Wang). All authors have read and agreed to the published version of the manuscript.

Funding: This work was financially supported by the National Key Research and Development Prgram (2021YFD1201103-01-04), the Opening Project of Key Laboratory of Soybean Biology of Chinese Education Ministry, and Research Business Expense Project of Heilongjiang Provincial Research Institute (CZKYF2023-1-B017).

Institutional Review Board Statement: Not applicable.

Informed Consent Statement: Not applicable.

Data Availability Statement: The data presented in this study are available on request from the corresponding author (due to privacy).

Acknowledgments: The authors thank Biomarker Technologies, Inc. (Beijing, China) for sequencing services and thank Wekemo Tech Group Co., Ltd. (Shenzhen, China) for metabolite detection.

Conflicts of Interest: The authors declare no conflicts of interest.

References

1. You, L.; Bao, W.; Yao, C.; Zhao, F.; Jin, H.; Huang, W.; Li, B.; Kwok, L.; Liu, W. Changes in chemical composition, structural and functional microbiome during alfalfa (*Medicago sativa*) ensilage with *Lactobacillus plantarum* PS-8. *Anim. Nutr.* **2022**, *9*, 100–109. [CrossRef] [PubMed]
2. Wang, T.; He, T.; Ding, X.; Zhang, Q.; Yang, L.; Nie, Z.; Zhao, T.; Gai, J.; Yang, S. Confirmation of GmPPR576 as a fertility restorer gene of cytoplasmic male sterility in soybean. *J. Exp. Bot.* **2021**, *72*, 7729–7742. [CrossRef] [PubMed]
3. Wang, C.; Zheng, M.; Wu, S.; Zou, X.; Chen, X.; Ge, L.; Zhang, Q. Effects of gallic acid on fermentation parameters, protein fraction, and bacterial community of whole plant soybean silage. *Front. Microbiol.* **2021**, *12*, 662966. [CrossRef]
4. An, D.; Lai, X.; Han, T.; Nsigayehe, J.M.V.; Li, G.; Shen, Y. Crossing latitude introduction delayed flowering and facilitated dry matter accumulation of soybean as a forage crop. *J. Integr. Agr.* 2024, *in press*. [CrossRef]
5. Liu, Z.; Cao, Y.; Wang, Z.; Ren, H.; Amidon, T.; Lai, Y. The Utilization of Soybean Straw. II. Dissolution & Regeneration of Soybean Straw in LiCl/DMSO. *Bioresources* **2015**, *10*, 2305–2317. [CrossRef]
6. Wilkinson, J.M.; Bolsen, K.K.; Lin, C.J. History of silage. *Silage Sci. Technol.* **2003**, *42*, 1–30.
7. de Morais, J.P.G.; Cantoia Júnior, R.; Garcia, T.M.; Capucho, E.; Campana, M.; Gandra, J.R.; Ghizzi, L.G.; Del Valle, T.A. Chitosan and microbial inoculants in whole-plant soybean silage. *J. Agric. Sci.* **2021**, *159*, 227–235. [CrossRef]
8. Driehuis, F.; Elferink, S.O. The impact of the quality of silage on animal health and food safety: A review. *Vet. Quart.* **2000**, *22*, 212–216. [CrossRef]
9. Yang, J.; Tan, H.; Cai, Y. Characteristics of lactic acid bacteria isolates and their effect on silage fermentation of fruit residues. *J. Dairy Sci.* **2016**, *99*, 5325–5334. [CrossRef]
10. Peng, K.; Jin, L.; Niu, Y.D.; Huang, Q.; McAllister, T.A.; Yang, H.E.; Denise, H.; Xu, Z.; Acharya, S.; Wang, S. Condensed tannins affect bacterial and fungal microbiomes and mycotoxin production during ensiling and upon aerobic exposure. *Appl. Environ. Microb.* **2018**, *84*, e2217–e2274. [CrossRef]
11. Wensel, C.R.; Pluznick, J.L.; Salzberg, S.L.; Sears, C.L. Next-generation sequencing: Insights to advance clinical investigations of the microbiome. *J. Clin. Investig.* **2022**, *132*. [CrossRef] [PubMed]
12. Kharazian, Z.A.; Xu, D.; Su, R.; Guo, X. Effects of inoculation and dry matter content on microbiome dynamics and metabolome profiling of sorghum silage. *Appl. Microbiol. Biot.* **2024**, *108*, 257. [CrossRef] [PubMed]
13. Liu, J.; Zhao, M.; Hao, J.; Yan, X.; Fu, Z.; Zhu, N.; Jia, Y.; Wang, Z.; Ge, G. Effects of temperature and lactic acid Bacteria additives on the quality and microbial community of wilted alfalfa silage. *BMC Plant Biol.* **2024**, *24*, 844. [CrossRef]
14. Guo, X.S.; Ke, W.C.; Ding, W.R.; Ding, L.M.; Xu, D.M.; Wang, W.W.; Zhang, P.; Yang, F.Y. Profiling of metabolome and bacterial community dynamics in ensiled *Medicago sativa* inoculated without or with *Lactobacillus plantarum* or *Lactobacillus buchneri*. *Sci. Rep.* **2018**, *8*, 357. [CrossRef]
15. Saia, S.; Fragasso, M.; De Vita, P.; Beleggia, R. Metabolomics Provides Valuable Insight for the Study of Durum Wheat: A Review. *J. Agr. Food Chem.* **2019**, *67*, 3069–3085. [CrossRef]
16. Goldansaz, S.A.; Guo, A.C.; Sajed, T.; Steele, M.A.; Plastow, G.S.; Wishart, D.S. Livestock metabolomics and the livestock metabolome: A systematic review. *PLoS ONE* **2017**, *12*, e177675. [CrossRef] [PubMed]
17. Xu, D.; Wang, N.; Rinne, M.; Ke, W.; Weinberg, Z.G.; Da, M.; Bai, J.; Zhang, Y.; Li, F.; Guo, X. The bacterial community and metabolome dynamics and their interactions modulate fermentation process of whole crop corn silage prepared with or without inoculants. *Microb. Biotechnol.* **2021**, *14*, 561–576. [CrossRef]
18. Jiang, Y.; Xue, E.Y.; Lu, W.C.; Cui, G.W.; Li, Y.M.; Han, T.F.; Wang, S.D. Breeding and feeding quality analysis of a new soybean strain deficient in Kunitz trypsin inhibitor. *Acta Prataculturae Sin.* **2020**, *29*, 91. (In Chinese)
19. AOAC. *Official Methods of Analysis of the Association of Official Analytical Chemists*, 15th ed.; Association of Official Analytical Chemists: Arlington, VA, USA, 1990.
20. McDonald, P.; Henderson, A.R.; Heron, S.J.E. *The Biochemistry of Silage*; Chalcombe Publications: Aber, UK, 1991.
21. Broderick, G.A.; Kang, J.H. Automated Simultaneous Determination of Ammonia and Total Amino Acids in Ruminal Fluid and In Vitro Media1. *J. Dairy Sci.* **1980**, *63*, 64–75. [CrossRef]
22. Meng, H.; Jiang, Y.; Wang, L.; Wang, S.; Zhang, Z.; Tong, X.; Wang, S. Effects of Different Soybean and Maize Mixed Proportions in a Strip Intercropping System on Silage Fermentation Quality. *Fermentation* **2022**, *8*, 696. [CrossRef]
23. Wilkinson, J.M.; Davies, D.R. The aerobic stability of silage: Key findings and recent developments. *Grass Forage Sci.* **2013**, *68*, 1–19. [CrossRef]
24. Bolyen, E.; Rideout, J.R.; Dillon, M.R.; Bokulich, N.A.; Abnet, C.C.; Al-Ghalith, G.A.; Alexander, H.; Alm, E.J.; Arumugam, M.; Asnicar, F.; et al. Reproducible, interactive, scalable and extensible microbiome data science using QIIME 2. *Nat. Biotechnol.* **2019**, *37*, 852–857. [CrossRef]

25. Callahan, B.J.; McMurdie, P.J.; Rosen, M.J.; Han, A.W.; Johnson, A.J.A.; Holmes, S.P. DADA2: High-resolution sample inference from Illumina amplicon data. *Nat. Methods* **2016**, *13*, 581–583. [CrossRef]
26. Wang, C.; Pian, R.; Chen, X.; Zhang, Q. Effects of polyphenol oxidases on proteolysis and lipolysis during ensiling of Moringa oleifera leaves with or without pyrocatechol. *Anim. Feed Sci. Tech.* **2021**, *275*, 114870. [CrossRef]
27. Muck, R.E. Silage microbiology and its control through additives. *Rev. Bras. Zootec.* **2010**, *39*, 183–191. [CrossRef]
28. Guyader, J.; Baron, V.S.; Beauchemin, K.A. Corn Forage Yield and Quality for Silage in Short Growing Season Areas of the Canadian Prairies. *Agronomy* **2018**, *8*, 164. [CrossRef]
29. Zeng, T.; Li, X.; Guan, H.; Yang, W.; Liu, W.; Liu, J.; Du, Z.; Li, X.; Xiao, Q.; Wang, X.; et al. Dynamic microbial diversity and fermentation quality of the mixed silage of corn and soybean grown in strip intercropping system. *Bioresour. Technol.* **2020**, *313*, 123655. [CrossRef] [PubMed]
30. Ni, K.; Wang, F.; Zhu, B.; Yang, J.; Zhou, G.; Pan, Y.I.; Tao, Y.; Zhong, J. Effects of lactic acid bacteria and molasses additives on the microbial community and fermentation quality of soybean silage. *Bioresour. Technol.* **2017**, *238*, 706–715. [CrossRef]
31. Smith, L.H. Theoretical Carbohydrates Requirement for Alfalfa Silage Production. *Agron. J.* **1962**, *54*, 291–293. [CrossRef]
32. Li, M.; Zi, X.; Zhou, H.; Lv, R.; Tang, J.; Cai, Y. Silage fermentation and ruminal degradation of cassava foliage prepared with microbial additive. *Amb Express* **2019**, *9*, 180. [CrossRef]
33. Wang, S.; Yuan, X.; Dong, Z.; Li, J.; Shao, T. Effect of ensiling corn stover with legume herbages in different proportions on fermentation characteristics, nutritive quality and in vitro digestibility on the Tibetan Plateau. *Grassl. Sci.* **2017**, *63*, 236–244. [CrossRef]
34. Maneerat, W.; Prasanpanich, S.; Tumwasorn, S.; Laudadio, V.; Tufarelli, V. Evaluating agro-industrial by-products as dietary roughage source on growth performance of fattening steers. *Saudi J. Biol. Sci.* **2015**, *22*, 580–584. [CrossRef] [PubMed]
35. Wu, Y.; Hao, F.; Lv, X.; Chen, B.; Yang, Y.; Zeng, X.; Yang, F.; Wang, H.; Wang, L. Diversity of lactic acid bacteria in Moutai-flavor liquor fermentation process. *Food Biotechnol.* **2020**, *34*, 212–227. [CrossRef]
36. Li, R.; Jiang, D.; Zheng, M.; Tian, P.; Zheng, M.; Xu, C. Microbial community dynamics during alfalfa silage with or without clostridial fermentation. *Sci. Rep.* **2020**, *10*, 17782. [CrossRef]
37. Mohd-Setapar, S.H.; Abd-Talib, N.; Aziz, R. Review on Crucial Parameters of Silage Quality. *Apcbee Procedia* **2012**, *3*, 99–103. [CrossRef]
38. Gao, R.; Wang, B.; Jia, T.; Luo, Y.; Yu, Z. Effects of different carbohydrate sources on alfalfa silage quality at different ensiling days. *Agriculture* **2021**, *11*, 58. [CrossRef]
39. Muck, R.E.; Nadeau, E.M.G.; McAllister, T.A.; Contreras-Govea, F.E.; Santos, M.C.; Kung, L. Silage review: Recent advances and future uses of silage additives. *J. Dairy Sci.* **2018**, *101*, 3980–4000. [CrossRef]
40. Zhang, Q.; Guo, X.; Zheng, M.; Chen, D.; Chen, X. Altering microbial communities: A possible way of lactic acid bacteria inoculants changing smell of silage. *Anim. Feed Sci. Tech.* **2021**, *279*, 114998. [CrossRef]
41. Keshri, J.; Chen, Y.; Pinto, R.; Kroupitski, Y.; Weinberg, Z.G.; Sela, S. Microbiome dynamics during ensiling of corn with and without *Lactobacillus plantarum* inoculant. *Appl. Microbiol. Biot.* **2018**, *102*, 4025–4037. [CrossRef]
42. Cai, Y.; Benno, Y.; Ogawa, M.; Ohmomo, S.; Kumai, S.; Nakase, T. Influence of *Lactobacillus* spp. from an Inoculant and of *Weissella* and *Leuconostoc* spp. from Forage Crops on Silage Fermentation. *Appl. Environ. Microb.* **1998**, *64*, 2982–2987. [CrossRef]
43. Wang, S.; Zhao, J.; Dong, Z.; Li, J.; Kaka, N.A.; Shao, T. Sequencing and microbiota transplantation to determine the role of microbiota on the fermentation type of oat silage. *Bioresour. Technol.* **2020**, *309*, 123371. [CrossRef] [PubMed]
44. Wang, Y.; Chen, X.; Wang, C.; He, L.; Zhou, W.; Yang, F.; Zhang, Q. The bacterial community and fermentation quality of mulberry (Morus alba) leaf silage with or without Lactobacillus casei and sucrose. *Bioresour. Technol.* **2019**, *293*, 122059. [CrossRef] [PubMed]
45. Zhou, Y.; Chen, Y.; Guo, J.; Shen, Y.; Yang, J. The correlations and spatial characteristics of microbiome and silage quality by reusing of citrus waste in a family-scale bunker silo. *J. Clean. Prod.* **2019**, *226*, 407–418. [CrossRef]
46. Östling, C.; Lindgren, S. Influences of enterobacteria on the fermentation and aerobic stability of grass silages. *Grass Forage Sci.* **1995**, *50*, 41–47. [CrossRef]
47. Dong, C.; Liu, P.; Wang, X.; Zhang, W.; He, L. Effects of Phenyllactic Acid on Fermentation Parameters, Nitrogen Fractions and Bacterial Community of High-Moisture Stylo Silage. *Fermentation* **2023**, *9*, 572. [CrossRef]
48. Baldwin, I.L.; MacCoy, E. *Root Nodule Bacteria and Leguminous Plants*; University of Wisconsin: Madison, WI, USA, 1932.
49. Sun, R.; Yuan, X.; Li, J.; Tao, X.; Dong, Z.; Shao, T. Contributions of epiphytic microbiota on the fermentation characteristics and microbial composition of ensiled six whole crop corn varieties. *J. Appl. Microbiol. Appl. Microbiol.* **2021**, *131*, 1683–1694. [CrossRef]
50. Tian, J.; Huang, L.; Tian, R.; Wu, J.; Tang, R.; Zhang, J. Fermentation quality and bacterial community of delayed filling stylo silage in response to inoculating lactic acid bacteria strains and inoculating time. *Chem. Biol. Technol. Agric.* **2023**, *10*, 44. [CrossRef]
51. Sun, L.; Bai, C.; Xu, H.; Na, N.; Jiang, Y.; Yin, G.; Liu, S.; Xue, Y. Succession of bacterial community during the initial aerobic, intense fermentation, and stable phases of whole-plant corn silages treated with lactic acid bacteria suspensions prepared from other silages. *Front. Microbiol.* **2021**, *12*, 655095. [CrossRef] [PubMed]
52. Li, M.; Lv, R.; Zhang, L.; Zi, X.; Zhou, H.; Tang, J. Melatonin is a promising silage additive: Evidence from microbiota and metabolites. *Front. Microbiol.* **2021**, *12*, 670764. [CrossRef]
53. He, Q.; Zhou, W.; Chen, X.; Zhang, Q. Chemical and bacterial composition of Broussonetia papyrifera leaves ensiled at two ensiling densities with or without *Lactobacillus plantarum*. *J. Clean. Prod.* **2021**, *329*, 129792. [CrossRef]

54. Xia, G.; Wu, C.; Zhang, M.; Yang, F.; Chen, C.; Hao, J. The metabolome and bacterial composition of high-moisture Italian ryegrass silage inoculated with lactic acid bacteria during ensiling. *Biotechnol. Biofuels Bioprod.* **2023**, *16*, 91. [CrossRef] [PubMed]
55. Xiao, Z.; Zhang, Z.; Huang, S.; Lon, J.R.; Xie, S. Metabolic Profiling of Serum for Osteoarthritis Biomarkers. *Dis. Markers* **2022**, *2022*, 1800812. [CrossRef] [PubMed]
56. Burlikowska, K.; Stryjak, I.; Bogusiewicz, J.; Kupcewicz, B.; Jaroch, K.; Bojko, B. Comparison of Metabolomic Profiles of Organs in Mice of Different Strains Based on SPME-LC-HRMS. *Metabolites* **2020**, *10*, 255. [CrossRef] [PubMed]

fermentation

MDPI

Article

Effects of *Lactobacillus plantarum* and Cellulase on Mixed Silages of *Amaranthus hypochondriacus* and Cornmeal: Fermentation Characteristics, Nutritional Value, and Aerobic Stability

Xinxin Li †, Yitong Jin †, Fuhou Li , Meng Yu, Jiarui Du, Qixuan Yi , Tianyue Zhao, Bao Yuan * and Peng Wang *

College of Animal Sciences, Jilin University, Changchun 130062, China; jlulxx@163.com (X.L.); jinyitong0102@163.com (Y.J.); lifh@jlu.edu.cn (F.L.); yumeng10165865@163.com (M.Y.); dujiarui626@163.com (J.D.); 13540649853@163.com (Q.Y.); zty521630@163.com (T.Z.)

* Correspondence: yuan_bao@jlu.edu.cn (B.Y.); pengwang@jlu.edu.cn (P.W.)

† These authors contributed equally to this work.

Abstract: In order to develop new feed resources, the aim of this study was to investigate the effects of moisture content, additives, and their interactions on the fermentation quality, aerobic stability, and in vitro digestibility of mixed silage of amaranth and cornmeal. The mass ratios of amaranth and cornmeal were 69:31, 76:24, and 84:16 for adjusting the moisture content of silage to 60% (W1), 65% (W2), and 70% (W3), respectively. The silage treatments included no additives (U), the addition of *Lactobacillus plantarum* (L), the addition of cellulase (E), and the addition of *Lactobacillus plantarum* + cellulase (M) mixed reagents. The results revealed that the pH and ammonia nitrogen (NH_3-N/TN) ratios were significantly lower in W1 than in W2 and W3 (3.66,19.3 g kg^{-1} TN vs. 3.70, 3.70, 20.0 kg^{-1} TN, 25.1 kg^{-1} TN, $p < 0.05$). Moreover, dry matter (DM), organic matter (OM), in vitro dry matter digestibility (*iv*DMD), in vitro organic matter digestibility (*iv*OMD), and in vitro crude protein digestibility (*iv*CPD) significantly increased ($p < 0.05$). Meanwhile, the aerobic stability of mixed silage containing amaranth and cornmeal decreased with increasing water content. The aerobic stability of the L, E, and M treatment groups was improved by 15, 105, and 111 h, respectively, compared with that of the control group at W1. The pH and NH_3-N/TN ratios were lower with the addition of E (E and M) than with the absence of E (U and L) (3.73, 20.1 g kg^{-1} DM vs. 3.64, 22.9 g kg^{-1} DM, $p < 0.05$). NDF and ADF were significantly lower with the addition of E than without the addition of E (598 g kg^{-1} DM, 145 g kg^{-1} DM vs. 632 g kg^{-1} DM, 160 g kg^{-1} DM, $p < 0.05$). However, CP, *iv*DMD, *iv*OMD, and *iv*CPD were significantly higher ($p < 0.05$). AA and NH_3-N/TN were significantly lower ($p < 0.05$) with the addition of L (L and M) than without the addition of L (U and E). In conclusion, the best fermentation quality, in vitro digestibility, and aerobic stability of amaranth and cornmeal mixed silage treated with *Lactobacillus plantarum* + cellulase (M) were achieved at 60% water content. The present study confirmed the potential of amaranth as silage and its potential application for improving feed quality and animal performance.

Keywords: amaranth; fermentation quality; nutritional value; in vitro digestibility; aerobic stability

Citation: Li, X.; Jin, Y.; Li, F.; Yu, M.; Du, J.; Yi, Q.; Zhao, T.; Yuan, B.; Wang, P. Effects of *Lactobacillus plantarum* and Cellulase on Mixed Silages of *Amaranthus hypochondriacus* and Cornmeal: Fermentation Characteristics, Nutritional Value, and Aerobic Stability. *Fermentation* **2024**, *10*, 378. https://doi.org/10.3390/fermentation10080378

Academic Editors: Siran Wang and Wansup Kwak

Received: 9 June 2024
Revised: 18 July 2024
Accepted: 23 July 2024
Published: 24 July 2024

1. Introduction

With the development of livestock and poultry farming and the increased demand for feed resources, traditional silage cannot fully meet the needs of the farming industry. However, some of the other proteins required by ruminants, such as dairy cows, usually come from feed [1]. Therefore, the pursuit of reasonably priced, highly productive, and efficient novel protein feeds to completely or partially replace traditional protein feeds is highly important for improving the quality of animal products [2]. In this context, amaranth, as a plant-based protein feed rich in nutrients with unique properties, is considered a feed resource with great potential. Amaranth can be grown in many areas of China, and the yield is generally 75,000–150,000 kg per hectare of fresh weight, with some varieties yielding as much as 225,000 kg, which is considered a promising feed resource [3].

Amaranth belongs to the genus Amaranthus of the family Amaranthaceae and is used as both human food and animal feed. It is rich in vitamins and minerals, with high protein content, high resistance, and high yield, making it a high-quality feed resource for ruminants [4,5]. In addition, the dry matter of amaranth has good degradability and fermentation properties, which can add value to ruminant feed. The addition of amaranth silage has been reported to increase body weight gain and reduce rumen methane emissions in male lambs [6–9]. However, amaranth cannot be preserved for a long period via conventional methods because of its high protein content, high water content, and thick stalks that cannot be easily dried into hay [4,5]. Therefore, silage is a good way to improve the utilization of amaranth, which can not only prolong the preservation time but also improve the palatability of the feed. Amaranth is characterized by high moisture and low soluble carbohydrate content. Thus, preserving amaranth directly via conventional silage methods is considered difficult. The soluble carbohydrate content of amaranth was found to be 50.46 g kg^{-1} dry matter (DM), meeting just the minimum threshold of 50 g kg^{-1} soluble carbohydrate content recommended for producing high-quality silage [10]. The addition of 10% cornmeal has been reported to improve the fermentation quality and apparent digestibility of silage [11]. Moreover, cornmeal, which is characterized by high contents of soluble carbohydrates (WSC) and dry matter, is a good mixed silage auxiliary that can directly increase the fermentation substrate, compensate for the lack of fermentable carbohydrates in amaranth, and reduce the water content to improve the fermentation success of silage. However, to our knowledge, few studies have investigated the effects of mixing amaranth silage with cornmeal during ensiling.

Silage success also depends on appropriate biological and chemical conditions that allow a rapid and sufficient decrease in the pH of the silage. Therefore, silage additives are recommended to manipulate fermentation and prolong aerobic stability [4]. Microbial additives such as lactic acid bacteria and cellulase can lead to a rapid drop in pH, facilitating the silage process and improving the fermentation quality [12].

The effects of water content, lactic acid bacteria, and cellulase on fermentation quality, in vitro digestibility, and aerobic stability of amaranth silage have been shown in previous studies, but their interactions have not been explored [13]. Therefore, the aim of this study was to explore the effects of moisture and additives on the fermentation quality, aerobic stability, and in vitro digestibility of mixed silage of amaranth and cornmeal by adding lactobacilli and cellulase individually or in a mixture. This study can provide more feed choices and ways to utilize the resources in the farming industry and promote the sustainable development of the livestock and poultry farming industry.

2. Materials and Methods

2.1. Experimental Materials and Design

The cultivation experiment was conducted in the experimental field of Jilin University (123.3° E, 44.1° N), China. Planting was in June 2020, and the plants were harvested in September. Amaranth was in full maturity at the time of harvesting.

The amaranth was chopped into lengths of approximately 1 to 2 cm before ensiling. The mixing ratios (w/w) of amaranth and cornmeal ingredients were 69:31, 76:24, and 84:16 in order to regulate the water content of the silage ingredients to 60%, 65%, and 70%, respectively. For each silage moisture level, the silage treatment was designed as follows: no additive (U), lactic acid bacteria inoculant (L), cellulase (E), and a mixed preparation of lactic acid bacteria and cellulase (M). For the lactic acid bacterial inoculant, Chikusou-1 (*Lactobacillus plantarum*) was obtained from Snow Brand Seed Co., Ltd. (Tokyo, Japan). Acremonium cellulase was obtained from Meiji Seika Pharma Co., Ltd., Tokyo, Japan (Lot No.: ACCF-6940). The lactic acid bacteria were dissolved in distilled water according to the manufacturer's instructions, after which the samples were sprayed evenly with a micro sprayer and mixed thoroughly. The dosage of lactic acid bacteria was 4.7 × 10^6 colony-forming units (cfu) per gram of fresh weight (FW). Meanwhile, cellulase was applied at 50 mg kg^{-1} FW. It was dissolved in distilled water according to the manufacturer's instructions, after which the samples were sprayed evenly with a micro sprayer and mixed well. The actual enzyme activity was 4.2 × 10^3 U g^{-1} FW. The mixed silage was subsequently loaded into a 5-liter plastic silo. The silage density was 550.1 ± 20.0 kg m^{-3} FW, and the silo was kept at room temperature (21–25 °C) for anaerobic fermentation. After 60 days of fermentation, three replicate silos were opened for the determination of the chemical composition, fermentation quality, and in vitro digestibility of the silage. The remaining silage was repeatedly mixed for the aerobic stability tests.

2.2. Fermentation Quality Analysis

Once opened, silage samples were taken via the "tetrad" method. Subsequently, 20 g of silage was thoroughly mixed with 180 mL of distilled water and homogenized in polyethylene vacuum bags for 1 min. The sample was then extracted in a refrigerator at a constant temperature of 4 °C for 24 h and filtered through 4 layers of gauze and qualitative filter paper [14]. A portion of the resulting extract was used to measure the pH via a pH meter (PHSJ-4F, Yidian Scientific Instruments Co., Ltd., Shanghai, China). The other part was frozen and stored at −20 °C for the determination of organic acids and ammonia nitrogen (NH_3-N) contents. The NH_3-N content was determined by the Robinson method [15]. Lactic acid (LA), acetic acid (AA), propionic acid (PA), and butyric acid (BA) contents were determined by using high-performance liquid chromatography (column: ShodexRspak KC-811s-DVB gel column, Japan; detector: SPD-M10AVP; mobile phase: 3 mmol L^{-1} perchloric acid; flow rate: 1 mL min^{-1}; column temperature: 50 °C; detection wavelength: 210 nm; injection volume: 5 μL).

2.3. Chemical Composition Analysis, Energy, and In Vitro Degradability Analysis

The dry matter (DM) contents of fresh samples and silages were determined in a 65 °C oven for 48 h. The dried samples were ground and passed through a 1.0 mm sieve for chemical analysis. The contents of organic matter (OM) and crude protein (CP) were determined via the methods of the Official Association of Analytical Chemists [16]. The neutral detergent fiber (NDF), acid detergent fiber (ADF), and acid detergent lignin (ADL) contents were determined according to the methods reported by Van et al. [17]. The water-soluble carbohydrate (WSC) content was measured via the anthrone sulfate colorimetric method [18]. The buffering capacity (BC) was measured using the method of Playne and McDonald [19]. The gross energy (GE) content was determined by an oxygen cartridge calorimeter (SDAC1000, Sundy, Changsha, China).

In vitro degradability experiments were conducted according to the principles of the Laboratory Animal Guidelines for the Ethical Review of Animal Welfare. The study protocol was reviewed and approved by the Animal Ethics and Welfare Committee of Jilin University (Jilin, China; Approval Number: SY202009600). The dried silage sample (0.5 g) was placed in a filter bag (ANKOM F57; diameter of hole 25 μm; Ankang Technology; Macedon, NY, USA) and sealed with a hand pressure sealing machine (PFS-400; Zhejiang Dongfeng Packing Machine Co., Ltd. Wenzhou, China) for subsequent in vitro incubation. Before the samples were placed in them, the fiber bags were rinsed with acetone and then thoroughly air-dried for 5 h at 105 °C in a forced convection drying oven (VL 115, VWR, Shanghai, China). A total of 196 fiber bags were prepared (48 silage silos × 4 parallel samples + 4 blank controls). The filter bag was processed and then loaded into a 130 mL serum bottle. Four small-tailed billy goats were fed a mixture consisting of 40% corn silage, 20% alfalfa hay, and 40% concentrate (DM-based) twice daily. Rumen fluid was collected from these animals. The rumen fluid was maintained at a temperature of 39 °C under an atmosphere of carbon dioxide. The medium was filtered through four layers of coarse cotton cloth, and then the filtrate was mixed with McDougall artificial saliva at a 1:4 (*v*/*v*) ratio. Each serum vial was supplemented with a 60 mL mixture followed by incubation in a CO_2 atmosphere at 39 °C. The incubation was performed in a water bath. After 72 h of incubation, the filter bag was removed from the serum vial and gently rinsed with cold distilled water until the water become clear. The fiber bags were gently squeezed to remove excess moisture and then dried in a forced convection oven at 100 °C for 24 h. The residue was then weighed and measured for in vitro dry matter digestibility (*iv*DMD) [20]. In vitro neutral deterrent fiber digestibility (*iv*NDFD) was determined by analyzing the residual NDF [21]. The formulas for *iv*DMD, *iv*NDFD, in vitro crude protein digestibility (*iv*CPD), and in vitro organic matter digestibility (*iv*OMD) are as follows: (respective weights of DM, NDF, CP, and OM before digestion-respective residual weights of DM, NDF, CP, and OM)/(respective weights of DM, NDF, CP, and OM before digestion).

2.4. Microbiological Analysis

Fresh samples (20 g) were mixed uniformly with 180 mL of sterile saline solution (0.85% NaCl) and shaken on a shaker for 30 min. Then, 1 mL of the homogenate was subjected to 10 × serial dilutions. Each gradient was prepared as 3 parallel replicates and poured into dishes. Finally, 100 μL of the dilutions at various concentrations were evenly applied to agar media as described below with coated rods. *Lactobacillus bacteria* were cultivated on De Man, Rogosa, and Sharp agar media (Budweiser Technology Co., Ltd., Shanghai, China) via incubation for 48 h under anaerobic conditions at 37 °C. Aerobic bacteria were cultivated on nutrient agar media via incubation at 37 °C (Hope Biotechnology Co., Ltd., Qingdao, China). Yeast and molds were grown on potato glucose agar media at 28 °C for 48 h (Budweiser Technology Co., Ltd., Shanghai, China). The numbers of microorganisms were counted on plates of 20–200 cfu. All microbial data were converted to log10 cfu g^{-1}. The results are reported on a fresh weight basis.

2.5. Aerobic Stability Analysis

Amaranth silage from each treatment was placed in a clean 1 L plastic bucket. A thermocouple wire was placed in the center of the amaranth silage, and the ambient temperature was recorded by the thermocouple line in the empty bucket using a data recorder (OHR-G100T; Hongrun Company, Ltd., Fuzhou, China). The silage temperature was recorded at 1 h intervals. The ambient temperature was also recorded every hour as a blank. Aerobic stability is the time taken for silage to reach a temperature 2 °C higher than the ambient temperature.

2.6. Statistical Analysis

The data were analyzed via the GLM program of SPSS statistical software (version 26; International Business Machine Corporation; Armonk, NY, USA) for each indicator according to the following model:

$$Y_{ijk} = a + W_i + E_j + L_k + (W \times E)_{ij} + (W \times L)_{ik} + (L \times E)_{jk} + (W \times E \times L)_{ijk} + b_{ijk}$$

In the above model, Y_{ijk} is the response variable, a is the overall mean, W_i is the fixed effect of the moisture content of silage material i (i = 1, 2, 3), E_j is the fixed effect of cellulase j (j = 1, 2), and L_k is the fixed effect of lactic acid bacteria k (k = 1, 2). $(W \times E)_{ij}$ is the interaction of silage feedstock moisture content i and cellulase j. $(W \times L)_{ik}$ is the interaction of silage feedstock moisture content i and lactic acid bacteria k. $(L \times E)_{jk}$ is the interaction of cellulase j and lactic acid bacteria k. $(W \times E \times L)_{ijk}$ is the interaction of moisture i, cellulase j, and lactic acid bacteria k. b_{ijk} is the residual error.

Multiple comparisons were made via Tukey's test on the basis of the results of significance tests for water content, enzyme treatment, bacterial treatment, and interaction [22].

3. Results

3.1. Chemical Composition and Microbial Counts of Fresh Materials

The chemical composition, gross energy, buffering capacity, and microbial counts of amaranth and cornmeal mixed silage are shown in Table 1. The DM, CP, and WSC contents of amaranth were 185 g kg^{-1}, 124 g kg^{-1} DM, and 50.46 g kg^{-1} DM, respectively. The buffering capacity value of amaranth was 340 mEq kg^{-1} DM, which was 4.0 times higher than that of cornmeal (85.9 mEq kg^{-1} DM). The numbers of lactic acid bacteria, yeast, and mold adhering to the surface of amaranth were 2.42 $\log_{10}$ cfu^{-1}, 2.00 $\log_{10}$ cfu^{-1}, and 0.41 $\log_{10}$ cfu^{-1}, respectively.

Table 1. Characteristics of amaranth and cornmeal.

Item	Amaranth	Cornmeal
Chemical composition, energy, and buffering capacity		
Dry matter (g kg^{-1} FW)	185	873
Organic matter (g kg^{-1} DM)	876	981
Crude protein (g kg^{-1} DM)	124	91.4
Neutral detergent fiber (g kg^{-1} DM)	651	303
Acid detergent fiber (g kg^{-1} DM)	377	85.0
Acid detergent lignin (g kg^{-1} DM)	111	16.7
Water-soluble carbohydrate (g kg^{-1} DM)	50.46	103.32
Gross energy (MJ kg^{-1} DM)	18.0	19.5
Buffering capacity (mEq kg^{-1} DM)	340	85.9
Microbial counts		
Lactic acid bacteria ($\log_{10}$ cfu^{-1} FW)	2.42	ND
Yeast ($\log_{10}$ cfu^{-1} FW)	2.00	ND
Mold ($\log_{10}$ cfu^{-1} FW)	0.41	ND

DM, dry matter; FW, fresh weight; cfu, colony-forming units; ND, not detected.

3.2. Fermentation Quality of Amaranth and Cornmeal Mixed Silage

Table 2 showed the fermentation quality of the amaranth and cornmeal mixed silage. All the treatment groups had pH values less than 4.0 after 60 days of silage. The addition of L (L and M groups) or E (E and M groups) significantly ($p < 0.05$) decreased the pH of the silages. In addition, there was an interaction effect of W × E on pH ($p < 0.001$). Without the addition of E (U and L groups) significantly increased pH compared to with the addition of E (3.74, 3.72 vs. 3.64, 3.63, $p < 0.05$). However, this effect was greater in W2 than in W1 and W3. In contrast, the addition of L led to a decrease in pH (3.72, 3.63 vs. 3.74, 3.64), but L and W did not interact.

The water content significantly affected the LA and PA contents, with W2 having significantly higher LA content than W1 and W3 (20.3 g kg^{-1} DM vs. 16.9 g kg^{-1} DM, 14.6 g kg^{-1} DM, $p < 0.05$) and a significantly lower PA content (0.00 g kg^{-1} DM vs. 0.03 g kg^{-1} DM, 7.34 g kg^{-1} DM, $p < 0.05$). The AA content of amaranth and cornmeal mixed silage was significantly lower (16.3 g kg^{-1} DM, vs. 17.8 g kg^{-1} DM) in group L than in group U. The AA content of silage was significantly lower in group L than in group U. The AA content of silage was significantly lower in group L than in group U.

There was an interaction effect between W × E and W × L on the BA content ($p < 0.05$). The BA content resulting from the addition of E was significantly higher than that resulting from the addition of BA without E (1.26 g kg^{-1} DM, 1.81 g kg^{-1} DM vs. 0.649 g kg^{-1} DM, 0.246 g kg^{-1} DM). With the addition of E, W3 had the lowest BA content. However, without the addition of E, W1 had the lowest BA content. The BA content with the addition of L was significantly lower than that without the addition of L (0.246 g kg^{-1} DM, 1.26 g kg^{-1} DM vs. 0.649 g kg^{-1} DM, 1.81 g kg^{-1} DM). Without the addition of L, the BA content decreased with increasing water content. However, with the addition of L, the change in BA content with water content was not significant ($p > 0.05$).

Table 2. Fermentation quality of amaranth and cornmeal mixed silage prepared with lactic acid bacteria and cellulose.

Item ‡	Moisture	Average	Additives †				SEM	Significance of Main Effects and Interactions (p-Value)						
			U	E	L	M		W	E	L	W × E	W × L	L × E	W × L × E
pH value	W1	3.66	3.69 ^{Ab}	3.63 ^{a}	3.69 ^{Ab}	3.62 ^{a}	0.002	<0.001	<0.001	0.001	<0.001	0.179	0.217	0.093
	W2	3.70	3.77 ^{Bb}	3.64 ^{a}	3.73 ^{Bb}	3.64 ^{a}								
	W3	3.70	3.77 ^{Bc}	3.64 ^{a}	3.72 ^{Bb}	3.62 ^{a}								
	Average	3.68	3.74	3.64	3.72	3.63								
LA (g kg^{-1} DM)	W1	16.9	16.6 ^{Aab}	20.6 ^{b}	12.2 ^{a}	18.1 ^{b}	0.078	0.020	0.811	0.338	0.116	0.599	0.953	0.455
	W2	20.3	22.4 ^{B}	17.7	21.1	19.8								
	W3	14.6	14.3 ^{A}	16.3	15.7	12.0								
	Average	17.2	17.8	18.2	16.3	16.6								
AA (g kg^{-1} DM)	W1	17.1	13.7	18.3	9.10 ^{A}	10.1	0.090	0.041	0.941	0.047	0.090	0.199	0.514	0.126
	W2	18.3	22.0	13.7	20.4 ^{B}	17.2								
	W3	17.5	16.5	24.2	15.0 ^{AB}	14.1								
	Average	16.2	17.4	18.7	14.8	13.8								
PA (g kg^{-1} DM)	W1	0.03	0.11	ND	ND	ND ^{A}	0.131	0.002	0.082	0.681	0.053	0.836	0.535	0.671
	W2	ND	ND	ND	ND	ND ^{A}								
	W3	7.34	ND	19.3	8.30	1.76 ^{B}								
	Average	2.46	0.37	6.43	2.77	0.59								
BA (g kg^{-1} DM)	W1	1.67	0.902 ^{a}	3.85 ^{Cb}	0.152 ^{a}	1.78 ^{a}	0.009	<0.001	<0.001	0.006	<0.001	0.008	0.433	0.139
	W2	1.06	0.721	1.59 ^{B}	0.330	1.58								
	W3	0.198	0.323	ND ^{A}	0.255	0.017								
	Average	0.990	0.649	1.81	0.246	1.26								
NH_3-N (g kg^{-1} TN)	W1	19.3	21.0 ^{Ab}	19.9 ^{Aab}	18.1 ^{Aa}	18.3 ^{Aa}	0.016	<0.001	<0.001	<0.001	<0.001	0.036	<0.001	<0.001
	W2	20.0	21.4 ^{Ab}	19.2 ^{Aab}	21.1 ^{Bb}	18.4 ^{Aa}								
	W3	25.1	30.7 ^{Bc}	21.9 ^{Ba}	24.7 ^{Cb}	22.9 ^{Bab}								
	Average	21.5	24.4	20.3	21.3	19.9								

$^{A–C}$ Means of water contents within a column with different superscripts differ in the same additive treatment ($p < 0.05$); $^{a–c}$ Means of additives treatments within a row with different superscripts differ on the same water content ($p < 0.05$); SEM, standard error of the mean; W, moisture; U, no additive. † L, lactic acid bacteria; E, cellulase; M, the mixture of lactic acid bacteria and cellulase. ‡ DM, dry matter; LA, lactic acid; AA, acetic acid; PA, propionic acid; BA, butyric acid; NH_3-N, ammonia nitrogen; TN, total nitrogen; ND, not detected.

There was an interaction effect between W × E, W × L, and L × E on NH_3-N/TN ($p < 0.05$). NH_3-N/TN decreased with decreasing water content (25.1 g kg^{-1} DM, 20.0 g kg^{-1} DM, and 19.3 g kg^{-1} DM, $p < 0.05$). However, without the addition of E, the mean NH_3-N/TN ratio was higher than that of the addition of the E group (22.9 g kg^{-1} DM vs. 20.1 g kg^{-1} DM, $p < 0.05$). Moreover, the NH_3-N/TN ratio with the addition of L was significantly lower than that without the addition of L (20.6 g kg^{-1} DM vs. 22.4 g kg^{-1} DM, $p < 0.05$). With the addition of L, W1 had the lowest NH_3-N/TN. Without the addition of L, W2 had the lowest NH_3-N/TN. The addition of L significantly reduced NH_3-N/TN, but with the addition of E, the mean NH_3-N/TN ratio was lower than that without the addition of E (19.9 g kg^{-1} DM vs. 21.3 g kg^{-1} DM, $p < 0.05$).

3.3. *Chemical Composition of Amaranth and Cornmeal Mixed Silage*

Table 3 showed the chemical composition of the amaranth and cornmeal mixed silage. The effects of water content and additives on the in vitro digestibility of silage are shown in Table 4. The DM content increased with decreasing water content (287 g kg^{-1}, 330 g kg^{-1}, 389 g kg^{-1}, $p < 0.05$). There was an interaction effect of L × E on the DM content ($p = 0.010$). The addition of L reduced the DM content of the silage. However, the effect of the E addition was greater than that without the addition of E (329 g kg^{-1} vs. 340 g kg^{-1}, $p < 0.05$). There was an interaction effect of W × E on the OM content ($p < 0.001$). With the addition of E, the OM content was significantly lower than that without the addition of E (940 g kg^{-1} DM, 941 g kg^{-1} DM vs. 942 g kg^{-1} DM, 942 g kg^{-1} DM, $p < 0.05$). However, this effect was lower for W2 than for W1 and W3.

There was an interaction effect between W × E and W × L on the CP content ($p < 0.05$). With the addition of E, the CP content was higher than that of the group without E (121 g kg^{-1} DM vs. 118 g kg^{-1} DM, $p < 0.05$). However, this effect is smaller for W1 compared to W2 and W3. With the addition of L, CP content increased ($p < 0.05$) with increasing water content. However, it remained unchanged without the addition of L.

There was an interaction effect of W × L on NDF content ($p = 0.024$). The NDF content in the addition of L treatment was significantly lower than that in the treatment without the L addition (631 gkg^{-1} DM, 588 g kg^{-1} DM vs. 632 g kg^{-1} DM, 607 g kg^{-1} DM, $p < 0.05$). With the addition of L, W1 had the highest NDF content, whereas without the addition of L, W3 had the highest NDF content.

There was an interaction effect of W × E on the ADF content ($p = 0.033$). The ADF content decreased with decreasing water content (190 g kg^{-1} DM, 153 g kg^{-1} DM, 115 g kg^{-1} DM, $p < 0.05$). Without the addition of E, the ADF content was significantly higher than that with the addition of E (160 g kg^{-1} DM, 160 g kg^{-1} DM vs. 146 g kg^{-1} DM, 143 g kg^{-1} DM, $p < 0.05$). However, this effect was smaller for W1 compared to W2 and W3.

There was an interaction effect between W × E and W × L on GE ($p < 0.05$). The addition of E increased the GE content under W2 (18.9 MJ kg^{-1} DM, 19.0 MJ kg^{-1} DM vs. 18.7 MJ kg^{-1} DM, 18.8 MJ kg^{-1} DM, $p < 0.05$), but there was no significant difference between W1 and W3 conditions with the addition of E. The effect of W × E on the GE content of silage with the addition of L was not significant ($p < 0.05$), and the effect of W × L on the GE content of silage with the addition of L was not significant ($p < 0.05$). In silage with the L addition, GE decreased ($p < 0.05$) with increasing water content but was unaffected ($p > 0.05$) in silage without the L addition.

Table 3. Chemical composition and energy of amaranth and cornmeal mixed silage prepared with lactic acid bacteria and cellulose.

Item [‡]	Moisture	Average	Additives [†]				SEM	Significance of Main Effects and Interactions (p-Value)						
			U	E	L	M		W	E	L	W × E	W × L	L × E	W × L × E
DM	W1	389	400 ^{Cb}	382 ^{Ca}	394 ^{Cb}	381 ^{Ca}								
(g kg^{-1})	W2	330	338 ^{Bb}	323 ^{Ba}	335 ^{Bb}	325 ^{Ba}	0.050	<0.001	<0.001	0.074	0.492	0.468	0.010	0.971
	W3	287	296 ^{Ab}	280 ^{Aa}	292 ^{Ab}	280 ^{Aa}								
	Average	336	345	328	340	329								
OM	W1	952	954 ^{Cb}	951 ^{Ca}	953 ^{Cb}	951 ^{Ca}								
(g kg^{-1} DM)	W2	941	940 ^{B}	941 ^{B}	941 ^{B}	942 ^{B}	0.015	<0.001	<0.001	0.832	<0.001	0.108	0.924	0.957
	W3	931	933 ^{Ab}	929 ^{Aa}	933 ^{Ab}	929 ^{Aa}								
	Average	941	942	940	942	941								
CP	W1	118	119	118 ^{A}	116 ^{A}	118 ^{A}								
(g kg^{-1} DM)	W2	120	117 ^{a}	122 ^{Bb}	118 ^{ABa}	121 ^{Bb}	0.022	<0.001	<0.001	0.766	0.002	0.017	0.773	0.035
	W3	122	119 ^{a}	122 ^{Bb}	121 ^{Bab}	124 ^{Cb}								
	Average	120	118	121	118	121								
NDF	W1	627	643 ^{B}	608	644	613 ^{B}								
(g kg^{-1} DM)	W2	595	598 ^{A}	584	623	575 ^{A}	0.368	0.004	<0.001	0.190	0.924	0.024	0.248	0.553
	W3	621	654 ^{Bb}	628 ^{ab}	625 ^{ab}	576 ^{Aa}								
	Average	615	632	607	631	588								
ADF	W1	115	114 ^{A}	115 ^{A}	121 ^{A}	110 ^{A}								
(g kg^{-1} DM)	W2	153	162 ^{B}	145 ^{B}	159 ^{B}	144 ^{B}	0.151	<0.001	<0.001	0.828	0.033	0.900	0.655	0.508
	W3	190	204 ^{Cb}	177 ^{Ca}	201 ^{Cb}	176 ^{Ca}								
	Average	152	160	146	160	143								
ADL	W1	19.9	19.1 ^{A}	21.1 ^{A}	19.4 ^{A}	19.8 ^{A}								
(g kg^{-1} DM)	W2	25.5	25.5 ^{B}	25.9 ^{BC}	24.3 ^{B}	26.1 ^{B}	0.033	<0.001	0.478	0.713	0.393	0.854	0.193	0.110
	W3	34.0	35.7 ^{C}	32.1 ^{C}	33.2 ^{C}	35.2 ^{C}								
	Average	26.5	26.8	26.4	25.6	27.0								
GE	W1	18.7	18.7 ^{B}	18.7 ^{B}	18.7 ^{B}	18.6 ^{B}								
(MJ kg^{-1} DM)	W2	18.9	18.7 ^{B}	18.9 ^{C}	18.8 ^{B}	19.0 ^{C}	0.009	<0.001	0.274	0.467	<0.001	0.017	0.734	0.920
	W3	17.9	18.0 ^{Ac}	17.9 ^{Aab}	18.0 ^{Abc}	17.8 ^{Aa}								
	SEM	18.5	18.5	18.5	18.5	18.5								

$^{A–C}$ Means of water contents within a column with different superscripts differ with the same additive treatment ($p < 0.05$); $^{a–c}$ Means of additives treatments within a row with different superscripts differ with the same water content ($p < 0.05$); SEM, standard error of the mean; W, moisture; U, no additive. [†] L, lactic acid bacteria; E, cellulase; M, the mixture of lactic acid bacteria and cellulase. [‡] DM, dry matter; OM, organic matter; CP, crude protein; NDF, neutral detergent fiber; ADF, acid detergent fiber; ADL, acid detergent lignin; GE, gross energy.

Table 4. In vitro digestibility of amaranth and cornmeal mixed silage prepared with lactic acid bacteria and cellulase.

Item ‡	Moisture	Average	Additives †				SEM	Significance of Main Effects and Interactions (*p*-Value)						
			U	E	L	M		W	E	L	W × E	W × L	L × E	W × L × E
*iv*DMD ($g\ kg^{-1}$)	W1	737	738^{C}	737^{C}	732^{C}	741^{C}	0.117	<0.001	<0.001	0.828	0.033	0.900	0.655	0.508
	W2	709	701^{B}	715^{B}	703^{B}	715^{B}								
	W3	680	669^{Aa}	690^{Ab}	671^{Aa}	690^{Ab}								
	Average	709	703	714	702	715								
*iv*OMD ($g\ kg^{-1}$ DM)	W1	777	778^{C}	777^{C}	772^{C}	781^{C}	0.118	<0.001	<0.001	0.828	0.033	0.900	0.655	0.508
	W2	748	740^{B}	754^{B}	742^{B}	754^{B}								
	W3	719	708^{Aa}	729^{Ab}	710^{Aa}	729^{Ab}								
	Average	748	742	753	741	756								
*iv*CPD ($g\ kg^{-1}$ DM)	W1	589	590^{Cb}	589^{Bab}	586^{Ca}	591^{Bb}	0.040	<0.001	<0.001	0.733	0.001	0.204	0.585	0.078
	W2	582	577^{Ba}	586^{Bb}	578^{Ba}	585^{ABb}								
	W3	574	568^{Aa}	578^{Ab}	571^{Aa}	580^{Ab}								
	Average	581	578	584	578	585								
*iv*NDFD ($g\ kg^{-1}$ DM)	W1	579	594^{B}	560	595	565^{B}	0.355	0.004	<0.001	0.190	0.924	0.024	0.248	0.553
	W2	548	550^{A}	537	575	528^{A}								
	W3	572	605^{Bb}	579^{ab}	576^{ab}	529^{Aa}								
	Average	567	583	559	583	541								

$^{A–C}$ Means of water contents within a column with different superscripts differ with the same additive treatment ($p < 0.05$). $^{a–b}$ Means of additives treatments within a row with different superscripts differ with the same water content ($p < 0.05$). SEM, standard error of the mean; W, moisture; U, no, additive. † L, lactic acid bacteria; E, cellulase; M, the mixture of lactic acid bacteria and cellulase. ‡ *iv*DMD, in vitro dry matter digestibility; *iv*OMD, in vitro organic matter digestibility; *iv*CPD, in vitro crude protein digestibility; *iv*NDFD, in vitro neutral detergent fiber digestibility.

3.4. In Vitro Digestibility of Amaranth and Cornmeal Mixed Silage

There was an interaction effect of W × E on *iv*DMD, *iv*OMD, and *iv*CPD ($p < 0.05$). *iv*DMD, *iv*OMD, and *iv*CPD were significantly lower without the addition of E than with the addition of E (715 g kg^{-1} DM, 755 g kg^{-1} DM, 585 g kg^{-1} DM vs. 703 g kg^{-1} DM, 742 g kg^{-1} DM, 578 g kg^{-1} DM, $p < 0.05$). However, this effect was smaller in W1 compared to W2 and W3. There was an interaction effect of W × L on the *iv*NDFD ($p = 0.024$). The addition of L increased the *iv*NDFD content under W1 (595 g kg^{-1} DM, 565 g kg^{-1} DM vs. 594 g kg^{-1} DM, 560 g kg^{-1} DM, $p < 0.05$), but there was no significant difference between W2 and W3 conditions with the addition of E.

3.5. Aerobic Stability of Amaranth and Cornmeal Mixed Silage

The effects of moisture content and additives on temperature changes in mixed amaranth and cornmeal silage under aerobic conditions are shown in Figures 1–4. At 60% moisture content, the temperatures increased from 0 to 69 h in all the treatment groups, with greater increases in the U and L treatment groups. At 65% moisture content, the temperature of all the treatment groups increased from 0 to 78 h. At 70% moisture content, the temperature of all the treatment groups increased from 0 to 69 h, with greater increases in the U and L treatment groups. At 70% moisture content, the temperature increased significantly after 51 h in the L treatment group and after 69 h in the E treatment group. The aerobic stability decreased significantly ($p < 0.05$) with increasing water content. The aerobic stability of the E and M treatments was significantly higher than that of the U treatment at all water contents.

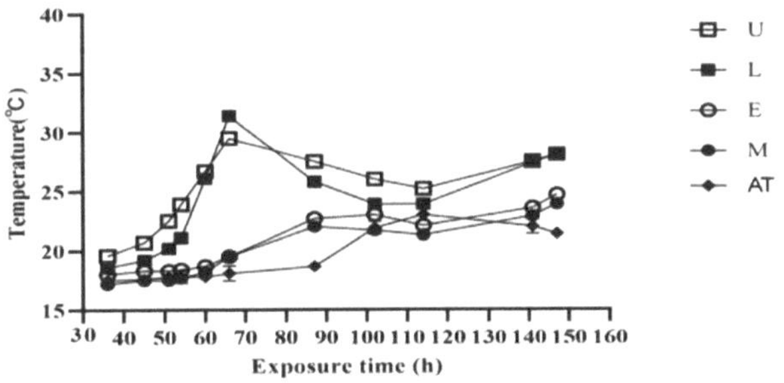

Figure 1. Effect of additive application on the temperature change in amaranth and corn powder mixed silage after exposure to aerobic conditions at 60% moisture content. U, control; L, lactic acid bacteria; E, cellulase; M, mixture of lactic acid bacteria and cellulase; AT, ambient temperature.

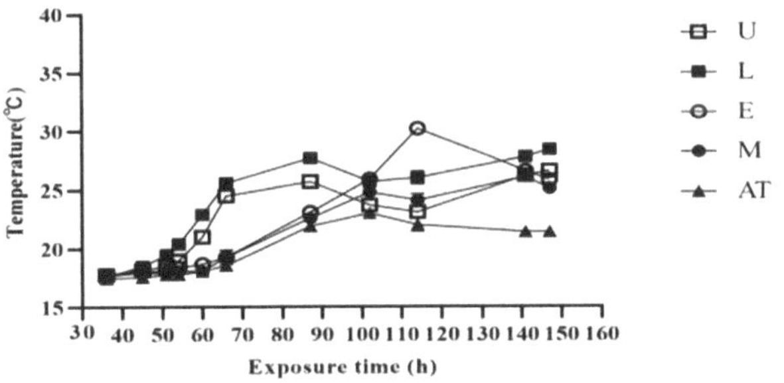

Figure 2. Effect of additive application on the temperature change in amaranth and corn powder mixed silage after exposure to aerobic conditions at 65% moisture content. U, control; L, lactic acid bacteria; E, cellulase; M, mixture of lactic acid bacteria and cellulase; AT, ambient temperature.

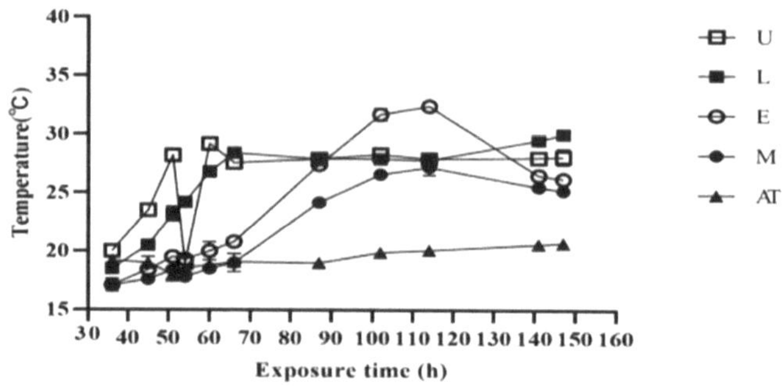

Figure 3. Effect of additive application on the temperature change in amaranth and corn powder mixed silage after exposure to aerobic conditions at 70% moisture content. U, control; L, lactic acid bacteria; E, cellulase; M, mixture of lactic acid bacteria and cellulase; AT, ambient temperature.

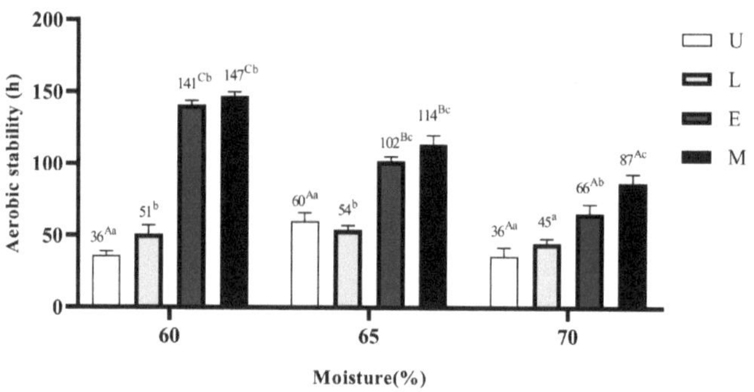

Figure 4. Time required for the temperature of amaranth and corn powder mixed silage to exceed room temperature by 2 °C after exposure to air. U, control; L, lactic acid bacteria; E, cellulase; M, mixture of lactic acid bacteria and cellulase. $^{A–C}$ Means of moisture contents with different superscripts differ under the same additive treatment ($p < 0.05$). $^{a–c}$ Means of additives treatments with different superscripts differ under the same moisture content.

4. Discussion

In our experimental hypotheses, we speculated that the addition of lactic acid bacteria and cellulase at W1 could induce earlier lactic acid fermentation in the mixed silage and improve fermentation quality, in vitro digestibility, and aerobic stability. As shown in Table 2, the M treatment had the best quality. According to Table 4, *iv*DMD and *iv*OMD were the highest in the M treatment at all water contents. As shown in Figure 1, the M treatment had the highest aerobic stability, which is consistent with our previous speculation.

4.1. Effects of Moisture and Additives on the Fermentation Quality of Amaranth and Cornmeal Mixed Silage

Water content is the main factor affecting silage quality. When the silage water content is too high, it can lead to a negative silage quality [23]. However, when the water content is too low, more pores are present in silage silos than in silages with higher water contents. Moreover, the low content of organic acids with antifungal activity (acetic acid) is not sufficient to inhibit the growth of yeasts, which can deteriorate quickly after opening [24]. Muck et al. reported that fermentation quality improved and nutrient losses decreased at a silage moisture of approximately 65% [25]. Therefore, three moisture levels of 60%, 65%, and 70% were used in the current study. The decrease in pH was more pronounced at lower water contents. This may be due to the higher DM content at low water content, which provides more fermentation substrate. It enables lactic acid bacteria to produce large amounts of lactic acid while inhibiting the respiration of plant cells and reducing glycogen consumption and protein degradation. Yahaya et al. reported that high-moisture silage with a high pH value was not as effective in fermentation as low-moisture fermentation was, which is consistent with the results of the present study [26]. In the present study, W $\times$ E had an interactive effect, and the E addition treatment further reduced the pH of the silage. This is due to the addition of cellulase, which breaks down the plant cell wall during ensiling and provides soluble sugars to lactic acid bacteria. The increased sugar content during the early stages of ensiling promoted lactic acid bacteria colonization. This leads to a rapid increase in lactic acid and a decrease in pH, which in turn inhibits the protein hydrolyzing activity of harmful microorganisms and plant enzymes [27,28]. Generally, a low pH indicates a high lactic acid concentration, and the typical concentration of lactic acid in silage ranges from 2% to 4% DM. Interestingly, although the pH in this experiment was less than 4.0, the lactic acid content was not high. This may be because Enterobacteriaceae can convert nitrate to nitrite, which is then converted to NO and NO_3 in a 2:1 ratio under acidic conditions, resulting in a lower pH [24,29]. The BA content of the E-added or L-added treatments ranged from 0.00 to 1.78 g kg^{-1} DM. The low BA content indicated that lactic acid bacteria and cellulase preparations can reduce clostridial fermentation [30].

The NH_3-N/TN ratio is an indicator of protein hydrolysis activity, amino acid deamination, and decarboxylation. This is mainly because protein hydrolysis by *Clostridium perfringens* ferments amino acids through valine and leucine deamination and redox reactions between alanine and glycine. This usually indicates the degradation of nutrients in mixed silage [31]. The NH_3-N/TN ratios of all the silages in this study were within satisfactory limits (< 10% TN), indicating that extensive protein hydrolysis did not occur [2]. Li et al. reported that the addition of cellulase to cassava leaf silage significantly reduced NH_3-N, supporting the results of the present treatment. In addition, the addition of lactic acid bacteria can reduce the microbial diversity of clover, annual ryegrass, and their mixed silage and improve silage quality [32]. This may be due to the addition of exogenous lactic acid bacteria, shifting fermentation towards lactic acid with homofermentative lactic acid bacteria or towards acetic acid with fermentative lactic acid bacteria. It also reduces the growth of clostridia and molds in silage, which reduces the degradation of proteins via the silage process and results in the retention of more nutrients in the silage, which is consistent with the results of this experiment [33]. The combined action of lactic acid bacteria and cellulase improved fermentation quality, reduced the plant cell wall fraction and protein loss, provided more digestible substrates for rumen microbial fermentation, and promoted rumen digestion. The combined treatment of lactic acid bacteria and cellulase may have beneficial synergistic effects on the fermentation quality of amaranth and cornmeal mixed silage [28].

4.2. Effects of Moisture and Additives on the Chemical Composition and In Vitro Digestibility of Amaranth and Cornmeal Mixed Silage

The DM, GE, and in vitro digestibility of silage tended to decrease with increasing moisture. These results indicated that high-moisture mixed silage had high losses of WSC and hemicellulose and low digestibility. This is in line with the results of Yahaya et al.'s study on orchard grass [34]. At the same time, we found an interesting phenomenon: There was the highest DM in all treatment groups at W1, but the CP content was the lowest. This may be because the addition of cornmeal regulates moisture. However, the protein content of cornmeal was 32.6 g kg^{-1} DM lower than that of amaranth, thereby resulting in the lowest CP content in all the treatment groups at W1. This finding is similar to that of Mehrangiz et al., who reported that the addition of molasses could lead to amaranth fermentation, increasing DM concentration [35]. Mehrangiz et al. reported that the soluble and degradable CP fractions of amaranth, as well as effective CP degradability, were not affected by wilting or the addition of any additives to silage [35]. However, compared with the no E treatment, the addition of the E treatment significantly increased the CP content in the mixed silage. This may be because cellulase disrupts the plant cell wall and releases more plant proteins. The plant proteins continue to synthesize new bacterial proteins that are more easily digested and absorbed by the animals, which in turn promotes digestion and degradation and improves the *iv*CPD. This was also indicated by the results of a previous study on the mixed silage of soybean residue and corn stover by Zhao et al. [36]. Compared with W2 and W3, W1 reduced the levels of ADF and ADL. This is due to the increase in raw material, which leads to a higher WSC content and lower NDF and ADF levels in silage [37]. High-moisture silage tends to have relatively high cellulose digestibility. Morrison reported a similar increase in cellulose digestibility due to the action of extracellular cellulase, which leads to the shortening of the cellulose chain length and makes it more susceptible to enzymatic attack [38]. Compared with the no E treatment, the addition of the E treatment significantly reduced the NDF and ADF contents, which is similar to the findings of Lynch et al. on corn silage [39]. This may be because the added fibrocystic enzymes increased the hydrolysis of cell wall carbohydrates, decreased their fiber content, and increased the WSC content. This result is in agreement with the findings of Foster et al., who reported that the addition of cellulase to warm-season legumes and Bahia grass silage increased the WSC content [37,40].

The *iv*DMD and *iv*OMD of the mixed silage in the M treatment were greater than those in the other groups, which may have been due to the reduction in DM loss from the silage with the addition of the L and E treatments. As a result, the levels of *iv*DMD and *iv*OMD in the rumen were elevated. A low *iv*NDFD was observed in mixed silage under the M treatment. This result may be due to two reasons. One is related to the hydrolysis of hemicellulose due to silage fermentation. Hemicellulose is acid-unstable under strongly acidic conditions, and silage fermentation leads to hydrolysis of the most readily available structural carbohydrates in feed [41]. Secondly, the addition of lactic acid bacteria enhanced NDF fermentation and increased hydrolysis. Moreover, cellulase treatment reduced the amount of available NDF degraded by rumen microorganisms in mixed silage [42]. At W3, in vitro digestibility was significantly increased in mixed silage under E and M treatments than under the U and L treatments. Therefore, we can infer that in vitro digestibility and NDF and ADF contents were negatively correlated, and our conclusions were the same as those of Bao et al. [37].

4.3. *Effects of Moisture Content and Additives on the Aerobic Stability of Amaranth and Cornmeal Mixed Silage*

Aerobic instability is the underlying cause of the loss of nutrients and DM, and mycotoxins produced from undesirable microorganisms also lead to health risks in human beings and animals. Therefore, aerobic stability is an important factor affecting the nutritional quality and subsequent feeding value of silage in ruminants [32]. Aerobic microorganisms metabolize and consume nutrients, and a change in silage temperature is usually used as an important parameter to evaluate the aerobic stability of silage [43].

AA is one of the most effective substances for inhibiting spoilage microorganisms to improve aerobic stability [44]. Interestingly, in this experiment, the aerobic stability of the mixed silage of amaranth and cornmeal decreased with increasing water content (Figure 4). However, the AA content did not decrease with increasing water content. This may be because a moist environment is more favorable for the growth of microorganisms such as yeasts and acetic acid bacteria, and acid-tolerant yeasts can survive in silage [45]. Increased yeast growth rate in high-moisture treatments was also demonstrated in a study of total mixed rations by Hao et al. [46].

The aerobic stability of the L, E, and M treatments improved at all water contents in this experiment. This occurred because the inoculated lactic acid bacteria have an anisolactic acid metabolic pathway that is capable of producing acetic acid during fermentation after the silos are opened. Thus, effectively controlling the yeast and filamentous fungi could improve aerobic stability [47]. In addition, according to Kaewpila et al., the addition of cellulase can improve the aerobic stability of Napier Pakchong grass, which was consistent with our experimental results [20]. The exposure time of all the M treatment groups was longer than that of the other groups, which may be due to the synergistic effect of lactic acid bacteria and cellulase when used together. Many studies have shown that lactic acid bacteria or cellulase can improve the aerobic stability of mixed silage by lowering the pH and NH_3-N contents and reducing the abundance of yeasts and clostridia [27]. As a result, the M treatment group was more stable during aerobic exposure and presented reduced spoilage losses during silage fermentation.

4.4. Discussion of the Effects of Mixing Seed Amaranth and Maize Meal on Actual Production

The competence of veterinarians in the field of animal nutrition is essential for the promotion and maintenance of good health in livestock [48,49]. Moreover, the general public is becoming increasingly concerned with the process of food production and animal welfare [50]. However, in the veterinary training system, most graduating veterinarians lack knowledge in the field of animal nutrition [51]. This may be due to insufficient teaching time regarding animal nutrition in veterinary schools. Therefore, practitioners should increase their reading of the literature in the area of animal nutrition. This study may provide new guidance programs and solutions for veterinary feeding in terms of additive application and water content control to provide up-to-date best practice insights for preparing future professionals to meet the challenges discussed in this paper.

5. Conclusions

In summary, the silage water content, lactic acid bacteria, and cellulase affect the fermentation quality, nutrient content, in vitro digestibility and aerobic stability of mixed amaranth and cornmeal silage. In this study, the simultaneous addition of Lactobacillus and cellulase at 60% water content (the mass ratio of amaranth and cornmeal was 69:31) resulted in the lowest pH, PA, AA, and NH_3-N/TN and therefore the best fermentation quality. In addition, mixed silage under the above conditions presented the lowest content of ADF, the highest contents of *iv*DMD, *iv*OMD, and *iv*CPD, and higher contents of DM and OM, thus providing higher nutritional value and digestibility. However, further in suit experiments are needed in this experiment to evaluate the effects of amaranth and cornmeal silage mixtures on rumen growth performance.

Author Contributions: Conceptualization, X.L., B.Y. and F.L.; methodology, Y.J., J.D., Q.Y., M.Y. and F.L.; software, X.L., T.Z., M.Y., J.D. and Y.J.; validation, Y.J., B.Y., P.W., J.D., M.Y. and T.Z.; formal analysis, X.L., P.W. and J.D.; investigation, F.L.; resources, B.Y. and F.L.; data analysis, Y.J., P.W. and M.Y; writing—original draft preparation, Y.J. and P.W.; writing—review and editing, X.L., P.W. and T.Z; visualization, B.Y. and P.W.; supervision, B.Y. and P.W.; project administration, B.Y. and P.W.; funding acquisition, P.W. and Y.J. All authors have read and agreed to the published version of the manuscript.

Funding: This study was supported by the Jilin Province Agricultural Key Core Technology Demonstration and Promotion (Industrial Technology System) Project (JARS-2024-0703) and funds from the China Agriculture Research System (CARS-37).

Institutional Review Board Statement: This study strictly followed the Chinese Laboratory Animal Welfare Ethical Review Guidelines and was approved by the Laboratory Animal Welfare Ethics Committee of Jilin University (permit number: SY202009600).

Informed Consent Statement: Not applicable.

Data Availability Statement: The author is solely responsible for the completeness and accuracy of all data in the article. Further inquiries can be directed to the corresponding authors.

Conflicts of Interest: The authors declare no conflicts of interest.

References

1. Cavallini, D.; Mammi, L.M.E.; Biagi, G.; Fusaro, I.; Giammarco, M.; Formigoni, A.; Palmonari, A. Effects of 00-rapeseed meal inclusion in Parmigiano Reggiano hay-based ration on dairy cows' production, reticular pH and fibre digestibility. *Ital. J. Anim. Sci.* **2021**, *20*, 295–303. [CrossRef]
2. Dong, Z.; Wang, S.; Zhao, J.; Li, J.; Shao, T. Effects of additives on the fermentation quality, in vitro digestibility and aerobic stability of mulberry (*Morus alba* L.) leaves silage. *Asian Australas. J. Anim. Sci.* **2020**, *33*, 1292–1300. [CrossRef] [PubMed]
3. Cui, D.W. High-yield cultivation techniques of *Amaranthus hypochondriacus*. *Mod. Anim. Husb. Technol.* **2020**, *3*, 28–29. [CrossRef]
4. Nogoy, K.M.C.; Yu, J.; Song, Y.G.; Li, S.; Chung, J.-W.; Choi, S.H. Evaluation of the Nutrient Composition, In Vitro Fermentation Characteristics, and In Situ Degradability of *Amaranthus caudatus*, *Amaranthus cruentus*, and *Amaranthus hypochondriacus* in Cattle. *Animals* **2020**, *11*, 18. [CrossRef]
5. Zhao, S.; Zhou, S.; Zhao, Y.; Yang, J.; Lv, L.; Zheng, Z.; Lu, H.; Ren, Y. Comparative Study of the Nutritional Value and Degradation Characteristics of Amaranth Hay in the Rumen of Goats at Different Growth Stages. *Animals.* **2022**, *13*, 25. [CrossRef] [PubMed]
6. Pond, W.G.; Lehmann, J.W. Nutritive value of a vegetable amaranth cultivar for growing lambs. *J. Anim. Sci.* **1989**, *67*, 3036–3039. [CrossRef]
7. Calabro, S.; Oteri, M.; Vastolo, A.; Cutrignelli, M.I.; Todaro, M.; Chiofalo, B.; Gresta, F. Amaranthus grain as a new ingredient in diets for dairy cows: Productive, qualitative, and in vitro fermentation traits. *J. Sci. Food Agric.* **2022**, *102*, 4121–4130. [CrossRef] [PubMed]
8. Rezaei, J.; Rouzbehan, Y.; Fazaeli, H.; Zahedifar, M. Carcass characteristics, non-carcass components and blood parameters of fattening lambs fed on diets containing amaranth silage substituted for corn silage. *Small Ruminant Res.* **2013**, *114*, 225–232. [CrossRef]
9. Shadi, H.; Rouzbehan, Y.; Rezaei, J.; Fazaeli, H. Yield, chemical composition, fermentation characteristics, in vitro ruminal variables, and degradability of ensiled amaranth (*Amaranthus hypochondriacus*) cultivars compared with corn (*Zea mays*) silage. *Transl. Anim. Sci.* **2020**, *4*, txaa180. [CrossRef]
10. Cajarville, C.; Britos, A.; Garciarena, D.; Luis Repetto, J. Temperate forages ensiled with molasses or fresh cheese whey: Effects on conservation quality, effluent losses and ruminal degradation. *Anim. Feed Sci. Technol.* **2012**, *171*, 14–19. [CrossRef]
11. Zhang, H.; Cheng, X.; Elsabagh, M.; Lin, B.; Wang, H.-r. Effects of formic acid and corn flour supplementation of banana pseudostem silages on nutritional quality of silage, growth, digestion, rumen fermentation and cellulolytic bacterial community of Nubian black goats. *J. Integr. Agric.* **2021**, *20*, 2214–2226. [CrossRef]
12. de Assis Pires, F.P.A.; Tomich, T.R.; Pereira, L.G.R.; Machado, F.S.; Campos, M.M.; de Oliveira, A.F.; Menezes, G.L.; de Menezes, R.A.; de Sousa, P.G.; Jayme, D.G.; et al. Effect of the *Lactiplantibacillus plantarum* and *Lentilactobacillus buchneri* on corn and sorghum silage quality and sheep energy partition under tropical conditions. *Grass Forage Sci.* **2023**, *78*, 224–235. [CrossRef]
13. Yi, Q.; Wang, P.; Yu, M.; Zhao, T.; Li, X.; Tang, H. Effects of Additives on the Fermentation Quality, In Vitro Digestibility, and Aerobic Stability of Amaranth (*Amaranthus hypochondriacus*) and Wheat Bran Mixed Silage. *Fermentation* **2023**, *9*, 711. [CrossRef]
14. Owens, V.N.; Albrecht, K.A.; Muck, R.E. Protein degradation and ensiling characteristics of red clover and alfalfa wilted under varying levels of shade. *Can. J. Plant Sci.* **1999**, *79*, 209–222. [CrossRef]
15. Robinson, D. Compensatory changes in the partitioning of dry matter in relation to nitrogen uptake and optimal variations in growth. *Ann. Botany* **1986**, *58*, 841–848. [CrossRef]
16. AOAC. *Official Methods of Analysis*, 18th ed.; AOAC Int.: Gaithersburg, MD, USA, 2005.
17. Van Soest, P.J.; Robertson, J.B.; Lewis, B.A. Methods for dietary fiber, neutral detergent fiber, and nonstarch polysaccharides in relation to animal nutrition. *J. Dairy Sci.* **1991**, *74*, 3583–3597. [CrossRef] [PubMed]
18. Wylam, C.B. Analytical studies on the carbohydrates of grasses and clovers. IV.-further developments in the methods of estimation of mono-, di- and oligo-saccharides and fructosan. *J. Sci. Food Agric.* **1954**, *5*, 167–172. [CrossRef]
19. Playne, M.J.; McDonald, P. The buffering constituents of herbage and of silage. *J. Sci. Food Agric.* **1966**, *17*, 264–268. [CrossRef]
20. Kaewpila, C.; Khota, W.; Gunun, P.; Kesorn, P.; Cherdthong, A. Strategic addition of different additives to improve silage fermentation, aerobic stability and in vitro digestibility of napier grasses at late maturity stage. *Agriculture* **2020**, *10*, 262. [CrossRef]
21. He, L.; Zhou, W.; Wang, Y.; Wang, C.; Chen, X.; Zhang, Q. Effect of applying lactic acid bacteria and cellulase on the fermentation quality, nutritive value, tannins profile and in vitro digestibility of *Neolamarckia cadamba* leaves silage. *J. Anim. Physiol. Anim. Nutr.* **2018**, *102*, 1429–1436. [CrossRef]
22. Bordin, C.; Raspa, F.; Greppi, M.; Harris, P.; Ellis, A.D.; Roggero, A.; Palestrini, C.; Cavallini, D.; Bergero, D.; Valle, E. Pony feeding management: The role of morphology and hay feeding methods on intake rate, ingestive behaviors and mouth shaping. *Front. Vet. Sci.* **2024**, *11*, 1332207. [CrossRef] [PubMed]
23. Coblentz, W.K.; Fritz, J.O.; Bolsen, K.K.; King, C.W.; Cochran, R.C. The effects of moisture concentration and type on quality characteristics of alfalfa hay baled under two density regimes in a model system. *Anim. Feed Sci. Technol.* **1998**, *72*, 53–69. [CrossRef]
24. Kung, L.; Shaver, R.D.; Grant, R.J.; Schmidt, R.J. Silage review: Interpretation of chemical, microbial, and organoleptic components of silages. *J. Dairy Sci.* **2018**, *101*, 4020–4033. [CrossRef] [PubMed]
25. Muck, R.E. Dry matter level effects on alfalfa silage quality. II. Fermentation products and starch hydrolysis. *Trans. ASAE.* **1990**, *33*, 373–381. [CrossRef]

26. Yahaya, M.S.; Kawai, M.; Takahashi, J.; Matsuoka, S. The Effects of different moisture content and ensiling time on silo degradation of structural carbohydrate of orchardgrass. *Asian Australas. J. Anim. Sci.* **2002**, *15*, 213–217. [CrossRef]
27. Muck, R.E.; Nadeau, E.M.G.; McAllister, T.A.; Contreras-Govea, F.E.; Santos, M.C.; Kung, L. Silage review: Recent advances and future uses of silage additives. *J. Dairy Sci.* **2018**, *101*, 3980–4000. [CrossRef]
28. Li, M.; Zhou, H.; Zi, X.; Cai, Y. Silage fermentation and ruminal degradation of stylo prepared with lactic acid bacteria and cellulase. *Anim. Sci. J.* **2017**, *88*, 1531–1537. [CrossRef]
29. Pahlow, G.; Muck, R.E.; Driehuis, F.; Oude Elferink, S.J.W.H.; Spoelstra, S.F. Microbiology of ensiling. *Silage Sci. Technol.* **2015**, *42*, 31–93.
30. Yang, W.; Yang, F.; Feng, C.; Zhao, S.; Zhang, X.; Wang, Y. Fermentation Properties and Bacterial Community Composition of Mixed Silage of Mulberry Leaves and Smooth Bromegrass with and without *Lactobacillus plantarum* Inoculation. *Fermentation* **2023**, *9*, 279. [CrossRef]
31. Zanine, A.d.M.; Bonelli, E.A.; Souza, A.L.d.; Ferreira, D.d.J.; Santos, E.M.; Ribeiro, M.D.; Geron, L.J.V.; Pinho, R.M.A. Effects of *Streptococcus bovis* Isolated from Bovine Rumen on the Fermentation Characteristics and Nutritive Value of Tanzania Grass Silage. *Sci. World J.* **2016**, *2016*, 8517698. [CrossRef]
32. Li, P.; Zhang, Y.; Gou, W.; Cheng, Q.; Bai, S.; Cai, Y. Silage fermentation and bacterial community of bur clover, annual ryegrass and their mixtures prepared with microbial inoculant and chemical additive. *Anim. Feed. Sci. Technol.* **2019**, *247*, 285–293. [CrossRef]
33. Oliveira, A.S.; Weinberg, Z.G.; Ogunade, I.M.; Cervantes, A.A.P.; Arriola, K.G.; Jiang, Y.; Kim, D.; Li, X.; Gonçalves, M.C.M.; Vyas, D.; et al. Meta-analysis of effects of inoculation with homofermentative and facultative heterofermentative lactic acid bacteria on silage fermentation, aerobic stability, and the performance of dairy cows. *J. Dairy. Sci.* **2017**, *100*, 4587–4603. [CrossRef] [PubMed]
34. Yahaya, M.S.; Kawai, M.; Takahashi, J.; Matsuoka, S. The effect of different moisture contents at ensiling on silo degradation and digestibility of structural carbohydrates of orchardgrass. *Anim. Feed. Sci. Technol.* **2002**, *101*, 127–133. [CrossRef]
35. Abbasi, M.; Rouzbehan, Y.; Rezaei, J.; Jacobsen, S.E. The effect of lactic acid bacteria inoculation, molasses, or wilting on the fermentation quality and nutritive value of amaranth (*Amaranthus hypochondriaus*) silage. *J. Anim. Sci.* **2018**, *96*, 3983–3992. [CrossRef] [PubMed]
36. Zhao, C.; Wang, L.; Ma, G.; Jiang, X.; Yang, J.; Lv, J.; Zhang, Y. Cellulase interacts with lactic acid bacteria to affect fermentation quality, microbial community, and ruminal degradability in mixed silage of soybean residue and corn stover. *Animals.* **2021**, *11*, 334. [CrossRef] [PubMed]
37. Bao, J.; Wang, L.; Yu, Z. Effects of different moisture levels and additives on the ensiling characteristics and in vitro digestibility of *Stylosanthes* silage. *Animals* **2022**, *12*, 1555. [CrossRef] [PubMed]
38. Morrison, I.M. Changes in the cell wall components of laboratory silages and the effect of various additives on these changes. *J. Agric. Sci.* **1979**, *93*, 581–586. [CrossRef]
39. Lynch, J.P.; Baah, J.; Beauchemin, K.A. Conservation, fiber digestibility, and nutritive value of corn harvested at 2 cutting heights and ensiled with fibrolytic enzymes, either alone or with a ferulic acid esterase-producing inoculant. *J. Dairy Sci.* **2015**, *98*, 1214–1224. [CrossRef] [PubMed]
40. Foster, J.L.; Carter, J.N.; Sollenberger, L.E.; Blount, A.R.; Myer, R.O.; Maddox, M.K.; Phatak, S.C.; Adesogan, A.T. Nutritive value, fermentation characteristics, and in situ disappearance kinetics of ensiled warm-season legumes and bahiagrass. *J. Dairy Sci.* **2011**, *94*, 2042–2050. [CrossRef]
41. Zhao, J.; Dong, Z.; Li, J.; Chen, L.; Bai, Y.; Jia, Y.; Shao, T. Effects of lactic acid bacteria and molasses on fermentation dynamics, structural and nonstructural carbohydrate composition and in vitro ruminal fermentation of rice straw silage. *Asian Australas. J. Anim. Sci.* **2019**, *32*, 783–791. [CrossRef]
42. Chen, L.; Guo, G.; Yuan, X.; Zhang, J.; Li, J.; Shao, T. Effects of applying molasses, lactic acid bacteria and propionic acid on fermentation quality, aerobic stability and in vitro gas production of total mixed ration silage prepared with oat-common vetch intercrop on the Tibetan Plateau. *J. Sci. Food Agric.* **2016**, *96*, 1678–1685. [CrossRef]
43. Ma, J.; Fan, X.; Ma, Z.; Huang, X.; Tang, M.; Yin, F.; Zhao, Z.; Gan, S. Silage additives improve fermentation quality, aerobic stability and rumen degradation in mixed silage composed of amaranth and corn straw. *Front. Plant Sci.* **2023**, *14*, 1189747. [CrossRef]
44. Qiu, X.; Guo, G.; Yuan, X.; Shao, T. Effects of adding acetic acid and molasses on fermentation quality and aerobic stability of total mixed ration silage prepared with hulless barley straw in Tibet. *Grassl. Sci.* **2014**, *60*, 206–213. [CrossRef]
45. Cai, Y.; Du, Z.; Yamasaki, S.; Nguluve, D.; Tinga, B.; Macome, F.; Oya, T. Influence of microbial additive on microbial populations, ensiling characteristics, and spoilage loss of delayed sealing silage of Napier grass. *Anim. Biosci.* **2020**, *33*, 1103–1112. [CrossRef]
46. Hao, W.; Wang, H.L.; Ning, T.T.; Yang, F.Y.; Xu, C.C. Aerobic stability and effects of yeasts during deterioration of non-fermented and fermented total mixed ration with different moisture levels. *Asian Australas. J. Anim. Sci.* **2015**, *28*, 816–826. [CrossRef]
47. Wambacq, E.; Latré, J.P.; Haesaert, G. The effect of *Lactobacillus buchneri* inoculation on the aerobic stability and fermentation characteristics of alfalfa-ryegrass, red clover and maize silage. *Agr. Food Sci.* **2013**, *22*, 127–136. [CrossRef]
48. Buonaiuto, G.; Lopez-Villalobos, N.; Costa, A.; Niero, G.; Degano, L.; Mammi, L.M.E.; Cavallini, D.; Palmonari, A.; Formigoni, A.; Visentin, G. Stayability in Simmental cattle as affected by muscularity and body condition score between calvings. *Front. Vet. Sci.* **2023**, *10*, 1141286. [CrossRef] [PubMed]

49. Roberts, J.L.; Murray, J.-A. Survey of Equine Nutrition: Perceptions and Practices of Veterinarians in Georgia, USA. *J. Equine Vet. Sci.* **2013**, *33*, 454–459. [CrossRef]
50. Muca, E.; Cavallini, D.; Raspa, F.; Bordin, C.; Bergero, D.; Valle, E. Integrating New Learning Methods into Equine Nutrition Classrooms: The Importance of Students' Perceptions. *J. Equine Vet. Sci.* **2023**, *126*, 104537. [CrossRef]
51. Becvarova, I.; Prochazka, D.; Chandler, M.L.; Meyer, H. Nutrition Education in European Veterinary Schools: Are European Veterinary Graduates Competent in Nutrition? *J. Vet. Med. Educ.* **2016**, *43*, 349–358. [CrossRef]

fermentation

Article

The Impact of *Lactobacillus delbrueckii* Hepatic Metabolism in Post-Weaning Piglets

Xiaolong Wang [1], Longteng Ma [1], Zhuying Liu [2,*] and Xinguo Huang [1,*]

1 College of Animal Science and Technology, Hunan Agricultural University, Changsha 410128, China; wangxiaolong1982@gmail.com (X.W.); malongteng2021@163.com (L.M.)
2 College of Animal Science and Technology, Hunan Biological and Electromechanical Polytechnic, Changsha 410128, China
* Correspondence: liuzhuying@hnbemc.edu.cn (Z.L.); hxg68989@hunau.edu.cn (X.H.)

Abstract: *Lactobacillus delbrueckii* garners interest for its contributions to gut microecological balance, diarrheal prevention and treatment, immune modulation, growth promotion, and meat quality enhancement in livestock. However, its impact on the gut microbiota and liver metabolism in weaned piglets is less documented. This study involved 80 Duroc-Landrace-Yorkshire weaned piglets aged 28 days, randomized into two groups with four replicates each and ten piglets per replicate. Over a 28-day period, the piglets were fed either a basal diet (control group) or the same diet supplemented with 0.1% *Lactobacillus delbrueckii* microcapsules ($\geq 1.0 \times 10^{10}$ CFU/g) (*Lactobacillus delbrueckii* group). The principal findings are as follows: During the initial phase of the experiment, supplementation with *Lactobacillus delbrueckii* increased the levels of L-phenylalanine and L-lysine in the liver while reducing the L-alanine levels, thereby enhancing the aminoacyl–tRNA synthesis pathway in weaned piglets. In the later phase, *Lactobacillus delbrueckii* supplementation boosted the liver arachidonic acid content, strengthening the arachidonic acid metabolic pathway in the piglets. The gut microbiota and their metabolites likely play a role in regulating these processes. These results indicate that, compared to the control group, *Lactobacillus delbrueckii* reduced weaning stress-induced liver damage and metabolic disorders, increased liver glycogen content, and enhanced liver antioxidant function by improving the metabolism of lipids and carbohydrates. Consequently, the liver functioned more healthily.

Keywords: liver glycogen; aminoacyl–tRNA; arachidonic acid; metabolic disorders; weaning stress

Citation: Wang, X.; Ma, L.; Liu, Z.; Huang, X. The Impact of *Lactobacillus delbrueckii* Hepatic Metabolism in Post-Weaning Piglets. *Fermentation* **2024**, *10*, 286. https://doi.org/10.3390/fermentation10060286

Academic Editors: Siran Wang and Wansup Kwak

Received: 30 April 2024
Revised: 22 May 2024
Accepted: 22 May 2024
Published: 28 May 2024

1. Introduction

Lactobacillus delbrueckii was first discovered and studied in 1901 and was subsequently named after the renowned bacteriologist M. Delbruck [1]. *Lactobacillus delbrueckii*, one of the most extensively utilized lactobacilli, encompasses the following three subspecies: *Lactobacillus delbrueckii* subsp. *bulgaricus*, *Lactobacillus delbrueckii* subsp. *lactis*, and *Lactobacillus delbrueckii* subsp. *delbrueckii*; it belongs to the facultative anaerobic Gram-positive bacteria. It can be sourced from a diverse range of origins, including human and animal oral cavities and intestines, as well as dairy products. Numerous scholars have demonstrated that *Lactobacillus delbrueckii* plays a pivotal role in regulating intestinal microecological balance [2–4], immune response modulation [5], and antioxidant activity enhancement [6], thereby facilitating healthy intestinal development while preventing diarrhea occurrence [7,8]. Moreover, it also contributes to growth promotion and improvements in meat quality. Consequently, this probiotic exhibits immense potential with significant advantages over antibiotics [2,4,6–12].

Weaning is a critical event in pig production, involving significant changes in feeding, management, and the environment. The abrupt transition from predominantly breast milk to solid feed often leads to reduced daily energy and nutrient intake by piglets [13]. Moreover, piglets experience stress due to separation from sows and alterations in the piggery environment, which can further elevate their stress levels. Dietary modifications may disrupt the balance of the gastrointestinal microbiota and increase susceptibility to intestinal pathogen colonization [14]. Additionally, weaning induces substantial morphological and functional changes in the small intestine, such as villi atrophy and crypt proliferation, that result in decreased digestion and absorption capacity [15–18]. Probiotics offer a potential alternative for enhancing the health and productivity of weaned piglets.

Lactobacillus delbrueckii is a probiotic strain with superior characteristics. Our group's prior research indicates its ability to colonize the animal gut, modulate microecological balance, enhance gut immunity, facilitate nutrient absorption, regulate metabolism, and counteract oxidative stress, thus improving animal productivity. However, the mechanisms by which it affects the gut microbiota to regulate liver metabolism and enhance productivity are not yet clear.

The weaning stress experienced by piglets results in a sudden decline in feed intake, necessitating the increased consumption of fat, sugar, and protein to meet the body's metabolic demands [19]. The liver, as a vital organ involved in energy supply and nutrient metabolism regulation, plays a crucial role in animal growth, metabolism, anti-stress capacity, and overall health. Weaning stress can reduce the glycogen content in piglet livers; however, the liver can compensate for this by enhancing gluconeogenesis and inhibiting glycolysis function to maintain glucose metabolic balance [20,21]. *Lactobacillus delbrueckii* has been found to modulate intestinal bile acid absorption and liver cholesterol metabolic enzyme activity while also influencing the gut microbiota composition. Consequently, it can regulate liver health and metabolism in weaned piglets [22]. In our preliminary experiment with *Lactobacillus delbrueckii* supplementation on weaned piglets' serum indexes related to liver function—alanine aminotransferase and aspartate aminotransferase—we observed certain effects. To further investigate the mechanism underlying these effects on the nutritional metabolism of weaned piglets supplemented with *Lactobacillus delbrueckii*, we conducted a PAS staining analysis to observe the changes in liver morphology and glycogen content, along with a non-targeted full spectrum analysis of liver metabolites using liquid chromatogram–mass spectrometry (LC-MS).

This experiment employs metabolomics to study the impact of *Lactobacillus delbrueckii* on the liver metabolism of weaned piglets. The objective is to elucidate how *Lactobacillus delbrueckii* influences liver metabolism changes in weaned piglets, explore the mechanisms by which it affects their productive performance, and provide crucial reference data and theoretical support for the application of *Lactobacillus delbrueckii* formulations in antibiotic-free diets and production practices for weaned piglets.

2. Materials and Methods

2.1. Testing Materials and Sources

The protocol of this study was approved by the Institutional Animal Care and Use Committee of the College of Animal Science and Technology, Hunan Agricultural University (Changsha, China), and was conducted in accordance with the National Institutes of Health (Changsha, China) guidelines for the care and use of experimental animals (No. 2017-09). The strain utilized in the experiment was *Lactobacillus delbrueckii*, which was acquired from the Laboratory of Animal Science at Hunan Agricultural University and authenticated and preserved by the Strain Preservation Center at Wuhan University (storage number M207096). Hunan Perfly Biotechnology Co., Ltd. (Changsha, China) was entrusted with preparing a microencapsulated granule formulation of *Lactobacillus delbrueckii* (vital cell counting $\geq 1.0 \times 10^{10}$ CFU/g).

2.2. Design of Experiments and Formulation of Dietary Composition

The experiment was approved by the Experimental Animal Welfare Ethics Committee of Hunan Agricultural University (Changsha, China). A single-factor design was employed in this study. Eighty 28-day-old Duroc × Landrace × Yorkshire trihybrid weaned piglets (half male and half female) with similar parity, body weight (approximately 7.5 kg), and good health were selected and randomly divided into two groups with four replicates per group and ten piglets per replicate. The duration of the experiment was 28 days. The control group received a basal diet, while the *Lactobacillus delbrueckii* group received a basal diet supplemented with 0.1% *Lactobacillus delbrueckii* (vital cell counting $\geq 1.0 \times 10^7$ CFU/g feed). The basic diet consisted of powder without any antibiotics, and its nutritional level followed the NRC (2012) standard for nutritional requirements (Table 1).

Table 1. Composition and nutrition levels of basal diets.

Diet Ingredient	Content (%)	Nutrition Levels [2]	Content
Corn	62.00	DE (MJ/Kg)	14.11
Extruded soybean	10.00	CP (%)	18.47
Soybean meal	14.00	EE (%)	4.40
Whey powder	5.00	Lys (%)	1.30
Fermented soybean meal	3.00	Met (%)	0.39
Fish meal	2.50	Thr (%)	0.80
Calcium hydrophosphate	1.40	Met + Cys (%)	0.70
Calcium carbonate	0.40	Ca (%)	0.70
Choline	0.10	P (%)	0.63
Salt	0.20	AP (%)	0.44
Lysine	0.42		
Methionine	0.08		
Threonine	0.10		
Premix [1]	0.80		
Total	100.00		

Note: [1] Each kilogram of the diet contains 90 mg of iron (Fe), 80 mg of zinc (Zn), 100 mg of copper (Cu), 40 mg of manganese (Mn), 0.3 mg of selenium (Se), 0.6 mg of iodine (I), 9000 IU of vitamin A (VA), 2800 IU of vitamin D (VD), 22 IU of vitamin E (VE), 3 mg of vitamin K3, 3 mg of vitamin B1, 7 mg of vitamin B2, 4 mg of vitamin B6, 0.03 mg of vitamin B12, 30 mg of niacin, 10 mg of pantothenic acid, 0.32 mg folate, and 0.20 mg of biotin. [2] Crude protein, calcium, and total phosphorus in nutrients are measured values, while other indexes are calculated values.

2.3. *Management of Animal Experiments*

The experiment was conducted at Taiping Pig farm in Changsha County, Hunan Province, under the supervision of designated personnel. All pigs were provided with unrestricted access to food and water and immunized following the conventional protocols of the pig farm. Feeding occurred three times daily (at 8:00, 12:00, and 18:00) using a dry mix diet that ensured no residual material remained in the troughs. Huts were cleaned twice daily to maintain cleanliness standards. Natural ventilation was employed throughout the entire hut. Daily records were maintained for pig feeding patterns, the occurrence of diarrhea, disease incidence, and medication administration.

2.4. *Detection Indices and Methodologies*

2.4.1. The Liver Tissue Was Examined Using Hematoxylin and Eosin (HE) Staining

On the 14 d and 28 d of the experiment, piglets were slaughtered under anesthesia after venous blood collection, and part of the liver was fixed in 4% paraformaldehyde with the size of about 1 cm^3 and divided into two pieces for section preparation. The steps of hematoxylin–eosin (HE) staining after a paraffin section of liver tissue are as follows: bake the slices at 60 °C for 1–2 h; after dewaxing to water, dye with hematoxylin for 5–10 min, rinse with distilled water, return to blue with PBS, slice into eosin dye solution for 3–5 min, and rinse with distilled water; and dehydrate the slices successively with gradient alcohol (95–100%) for 5 min per stage. After removal, the liver was placed in xylene for 10 min, sealed with neutral gum twice, examined by a microscope, and photographed to observe the histological changes.

2.4.2. Revealing Hepatic Glycogen Accumulation through PAS Staining in Liver Tissue

To guarantee the integrity of the data and the reliability of the experimental outcomes, adherence to the established protocols for sample collection and processing was of paramount importance. As outlined in Section 2.4.1, these procedures were meticulously followed to the letter. This section provides a detailed explanation of the steps for sample collection, required materials, environmental conditions, and precautions to maintain the quality of the samples. Following paraffin sectioning, the liver tissue was subjected to periodic acid-Schiff (PAS) staining using the following protocol: The sections were baked at 60 °C for 30–60 min. Dewaxing of the slices was performed by immersing them in xylene twice for 10 min each time, followed by sequential immersion in 100%, 95%, 85%, and 75% ethanol for 5 min at each stage. Subsequently, the sections were soaked in distilled water for an additional 5 min. To initiate the PAS reaction, a quick application of 50 μL periodic acid onto the tissue was carried out, allowing it to stand for a duration of 10 min. After rinsing with tap water for another period of ten minutes, Schiff's solution was applied to facilitate dye binding during a subsequent incubation step lasting ten minutes. The sections were then thoroughly rinsed with water until no further color change occurred. Hematoxylin staining (for nuclear counterstaining) was performed by immersing the sections in a hematoxylin solution for a duration ranging from five to ten minutes, subsequently returning them to blue using phosphate-buffered saline (PBS), followed by drying with a hair dryer. Gradual dehydration through graded alcohols (from low concentration to high) ensued before two rounds of immersion in xylene for ten-minute intervals each time took place prior to sealing with neutral gum and microscopic examination. The evaluation revealed that PAS-positive structures exhibited purplish-red coloration accompanied by blue nuclei.

2.4.3. Comprehensive Analysis of Hepatic Metabolites across the Entire Spectrum

After administering the anesthesia and conducting the slaughter procedure, venous blood samples were collected from piglets on days 14 and 28 of the trial. Subsequently, a portion of the liver was swiftly excised using surgical scissors and transferred into a diethyl pyrocarbonate (DEPC)-treated Eppendorf tube. To ensure sample integrity, the liver specimen was pre-cooled with liquid nitrogen and stored at −80 °C for comprehensive spectrum analysis aimed at investigating liver metabolites.

For metabolomics determination, the following procedure was employed:

(a) Sample pretreatment: The liver sample was slowly thawed on ice, and then 50 mg of freeze-dried sample was placed in a 1.5 mL centrifuge tube containing 800 μL of 80% methanol. The mixture was ground for 90 s at 60 Hz and thoroughly mixed using vortex oscillations. Subsequently, ultrasound treatment was performed at 4 °C for 30 min followed by standing at −40 °C for 1 h, vortexing for an additional 30 s, and standing again at 4 °C for another half an hour. Next, centrifugation was carried out at 4 °C and 12,000 rpm for 15 min. All the supernatant in the centrifuge tube was collected and allowed to stand at −40 °C for 1 hour before undergoing another round of centrifugation (4 °C and 12,000 rpm; 15 min). Finally, 200 μL of supernatant was withdrawn from the resulting solution and combined with the dichlorophenylalanine internal standard (140 μg/mL; 5 μL) in a vial.

(b) Instrumental analysis conditions: Chromatographic conditions: The chromatographic column used was a Hyper gold C18 liquid chromatography column, maintained at a temperature of 40 °C, with a sample size of 4 μL. For the positive ion mode test, the mobile phase consisted of water +5% acetonitrile +0.1% formic acid (A) and acetonitrile +0.1% formic acid (B). For the negative ion mode tests, water +5% acetonitrile +0.05% acetic acid (A) and acetonitrile +0.05% acetic acid (B) were employed as the mobile phases. The elution gradient is presented in Table 2.

Table 2. The gradient of the mobile phase.

Time (min)	Flow Rate (mL/min)	A (%)	B (%)
0	0.3	100	0
0	0.3	100	0
1.5	0.3	80	20
9.5	0.3	0	100
14.5	0.3	0	100
14.6	0.3	100	0
18	0.3	100	0

Mass spectrum detection parameters for positive mode were set as follows: heater temperature at 300 °C; sheath gas flow rate at 45 arb; auxiliary gas flow rate at 15 arb; tail gas flow rate at 1 arb; electric spray voltage at 3.0 KV; capillary temperature at 350 °C; and S-Lens RF Level set to 30%.

For the negative mode, mass spectrum detection parameters included heater temperature set to 300 °C; sheath gas flow rate adjusted to 45 arb; auxiliary gas flow rate maintained at 15 arb; tail gas flow rate kept constant at 1 arb; electric spray voltage set to 3.2 KV; capillary temperature held steady at 350 °C; and S-Lens RF Level adjusted to 60%.

Scanning was performed using a primary full scan from m/z 70 to 1050 and a data-dependent secondary mass spectrometry scan (dd-MS2, TopN = 10); resolution settings were configured as 70,000 (primary mass spectrometry) and 17,500 (secondary mass spectrometry). Collision mode utilized high-energy collision dissociation (HCD).

(c) Data extraction: The Compound Discoverer software 3.0 (Thermo Company, Waltham, MA, USA) was utilized for the extraction and pre-processing of LC/MS detection data, which were subsequently organized into a two-dimensional data matrix format. This matrix encompasses essential information such as retention time (RT), molecular weight, observed quantity (sample name), and peak intensity, among others.

2.5. *Revealing the Identification of Metabolic Pathways*

Enrichment analysis of metabolite differences was performed using the KEGG function and Metaboanalyst (http://www.metaboanalyst.ca/ (accessed on 29 April 2024)), based on the Bos Taurus pathway enrichment library. The significance of enrichment pathways was determined by evaluating the *P*-value and influence factor, with pathways having a $P < 0.05$ or influence factor ≥ 1.0 considered key metabolic pathways for differential metabolite enrichment.

2.6. *Data Processing and Analysis Procedures*

The principal component analysis (PCA) and partial least squares discriminant analysis (PLS-DA) were conducted using SIMCA-P+ software 14.1.0. The univariate statistical analysis (UVA) model was employed to screen the variables, where a Student's *t*-test with a *p*-value less than 0.05 and a variable importance in the projection (VIP) greater than 1 for the first principal component of the PLS-DA model were considered significant screening conditions.

3. Results and Analysis

3.1. *The Impact of Lactobacillus delbrueckii on Hepatic Morphology in Post-Weaning Piglets*

The liver histopathology observed through HE staining is presented in Figure 1. On day 14 of the experiment, the control group exhibited slight hepatic damage, characterized by well-organized hepatocyte cords with a few nuclei displaying dissolution and shrinkage, the absence of noticeable inflammatory cell infiltration, and an increased presence of lipid vacuoles. In contrast, the *Lactobacillus delbrueckii* group displayed clear liver architecture with a radial arrangement of hepatocytes around the central vein and distinct hepatocyte cords. A minimal number of liver nuclei showed signs of dissolution and contraction without any evident inflammatory cell infiltration or lipid vacuole accumulation. By day 28 of the experiment, there was significant improvement in the liver morphology for the control group compared to day 14, as evidenced by a well-defined arrangement of hepatocyte cords, with only a small fraction showing dissolution and shrinkage along with reduced inflammatory cell infiltration and diminished lipid vacuoles. The liver structure in the *Lactobacillus delbrueckii* group remained clear, resembling that observed on day 14.

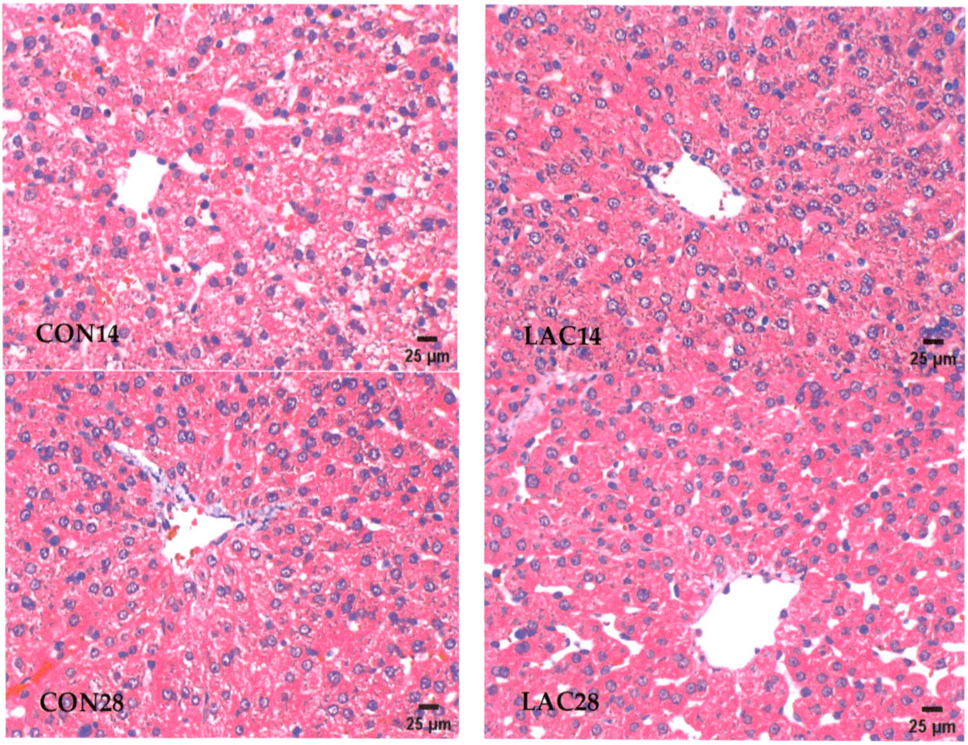

Figure 1. Effects of LAC on liver morphology of weaned piglets (original magnification 400×, hematoxylin–eosin staining). Note: CON: control group; LAC: basal diet group with 0.1% *Lactobacillus delbrueckii*.

3.2. *Impact of Lactobacillus delbrueckii on Hepatic Glycogen Content in Post-Weaning Piglets*

The morphological observation of PAS glycogen staining in the liver is presented in Figure 2.

On the 14th day of the experiment, the control group exhibited a higher abundance of liver glycogen, characterized by glycogen aggregation and fat vacuole accumulation. In contrast, the *Lactobacillus delbrueckii* group displayed a sporadic distribution of liver glycogen with reduced levels.

By day 28 of the experiment, the liver glycogen content decreased in the control group, along with a significant reduction in fat vacuoles. Conversely, there was a substantial increase in the liver glycogen content within the *Lactobacillus delbrueckii* group, distributed as granules and focal points.

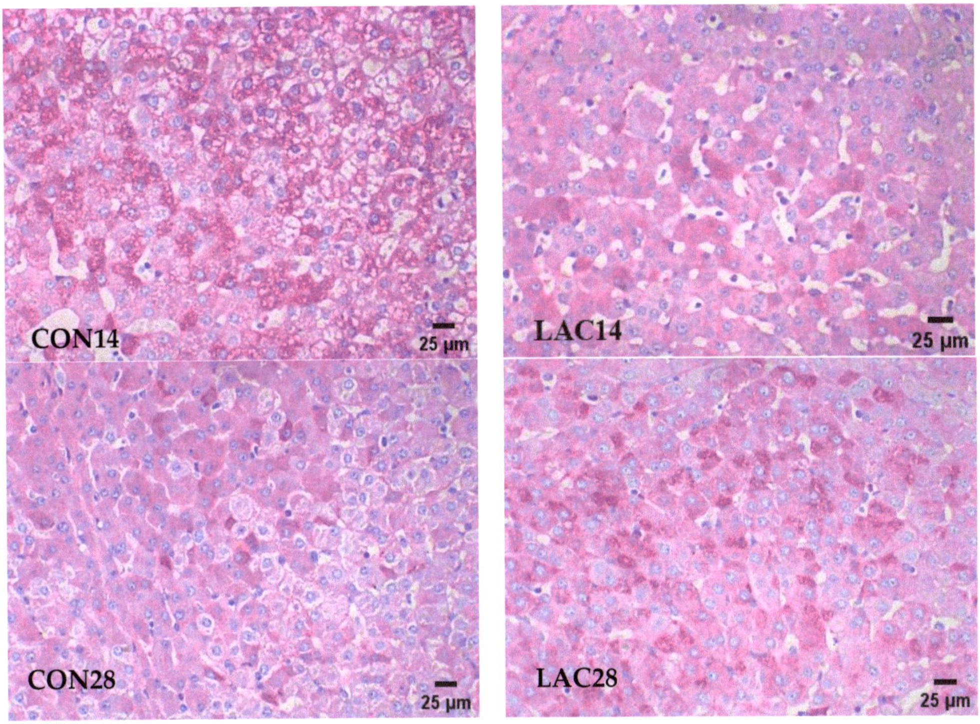

Figure 2. Effects of LAC on the liver glycogen content of weaned piglets (original magnification 400×, periodic acid Schiff staining). Note: CON: control group; LAC: basal diet group with 0.1% *Lactobacillus delbrueckii*.

3.3. The Impact of *Lactobacillus delbrueckii* on Hepatic Metabolism in Post-Weaning Piglets

Observations from the principal component analysis (PCA) graph in Figure 3 reveal an interesting phenomenon. On days 14 and 28 of the experiment, the liver metabolite profiles within the control group appeared consistent, without significant differentiation. In contrast, the *Lactobacillus de brueckii* group displayed a distinct trend, with clear separation of the liver metabolites on days 14 and 28, suggesting that the addition of *Lactobacillus de brueckii* may have modulated the liver metabolic activity. This variance indicates that *Lactobacillus de brueckii* could have a positive effect on liver function, particularly in terms of metabolic regulation.

According to the volcano plot in Figure 4, under a positive ion mode, we observed the differential metabolites identified in the liver of weaned piglets. By day 14 of the experiment, compared to the control group, the *Lactobacillus delbrueckii* group identified 90 known differential metabolites. Specifically, 44 of these metabolites showed an increase in the relative concentration, while the relative concentration of the other 46 metabolites decreased. This distribution indicates that the addition of *Lactobacillus delbrueckii* significantly impacted the metabolites in the piglets' liver, altering not only the types of metabolites but also their concentration levels, which may have significant implications for the health and growth development of the piglets.

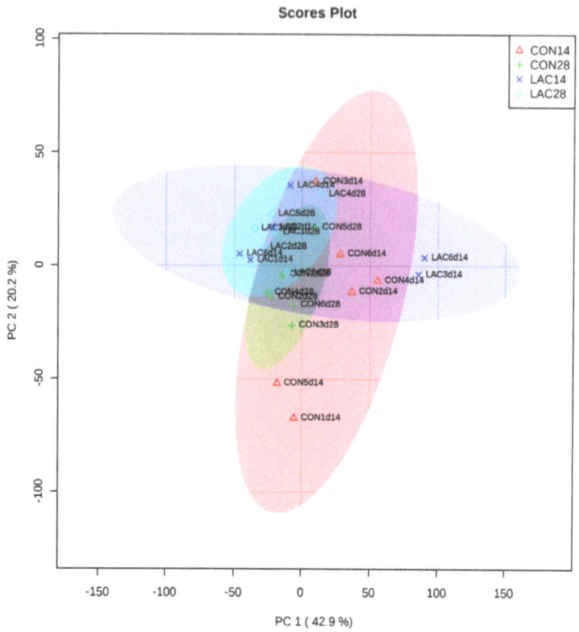

Figure 3. PCA analysis of weaned piglets liver metabolites. Note: CON: control group; LAC: basal diet group with 0.1% *Lactobacillus de brueckii.*

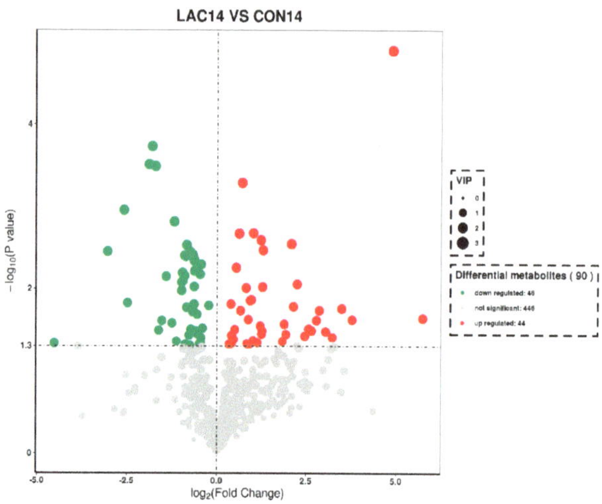

Figure 4. Volcano plot of liver differences in metabolites on the 14th day of the trial period. Note: CON: control group; LAC: basal diet group with 0.1% *Lactobacillus delbrueckii.* The horizontal axis of the volcano plot represents the fold change in metabolite levels (as log2 multiples), while the vertical axis indicates the level of statistical significance. Each point on the plot corresponds to an individual metabolite, with its position determined by the magnitude of the metabolite's effect and its significance. Points to the right signify an increase in metabolite levels (upregulation), whereas points to the left indicate a decrease (downregulation). The higher the position of a point, the greater its statistical significance, which means a smaller p-value. Typically, points that are situated far from the center line and towards the top of the volcano plot represent metabolites that are statistically significant and exhibit notable changes.

3.4. Impact of Lactobacillus delbrueckii on Hepatic Glycogen Content in Post-Weaning Piglets

According to the fold change ranking, the top 20 differential metabolites that are upregulated and downregulated are shown in Figure 5. Compared with the control group, Guanosine 5′-diphospho-L-fucose, F420-0, Phosphoric acid, Dibutyl malate, and guanosine 5′ -diphospho-l-fucose in the liver of the Lactobacillus deli group, Uridine monophosphate (UMP), 2-deoxyguanosine 5′-monophosphate (dGMP), Uridine 5′-monophosphate, Phosphatidylcholine lyso 17:0, 2-Aminoadipic acid (2-Aminoadipic acid), S-Lactoylglutathione (S-Lactoylglutathione), Saccharin, aspartate leucine (asp-leu), leu-leu, Alanine, Inosinic acid 1-stearoyl-sn-glycero-3-phosphoethanolamine, D-glutamine (D-(-)-Glutamine), and Phosphatidylethanolamine lyso 20:4. The relative concentrations of Phosphatidylcholine lyso 20:4 and Azelaate were significantly decreased (VIP $\geq$ 1.0 and $P < 0.05$).

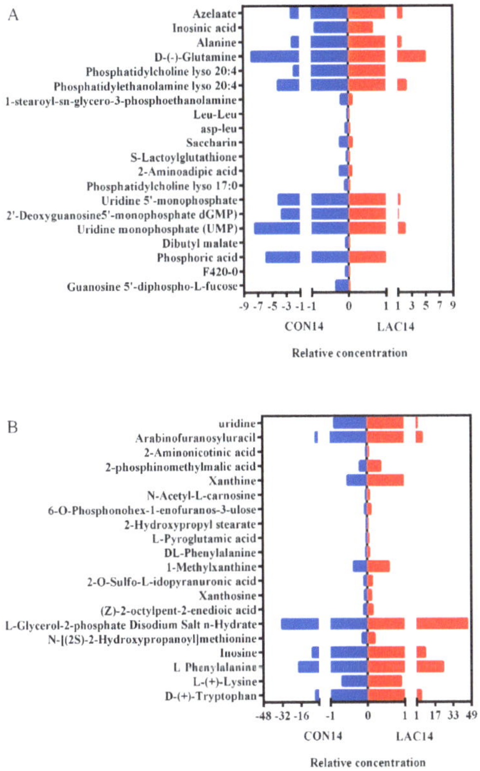

Figure 5. On day 14 of the trial period, through metabolite analysis of the weaned piglets' livers, we successfully identified 20 downregulated (**A**) and 20 upregulated (**B**) metabolites. These changes reveal significant differences in liver function and metabolic status. Note: CON: control group; LAC: basal diet group with 0.1% *Lactobacillus delbrueckii.* The identification of these differential metabolites is crucial, as it provides a snapshot of the dynamic metabolic adjustments within the liver. Group A metabolites showcase a marked decrease in concentration, suggesting a diminished role or a lower demand for these compounds within the physiological processes of the liver. Conversely, Group B metabolites have shown a significant increase in levels, indicating an elevated significance or enhanced utilization in the liver's metabolic activities. This categorization helps in pinpointing specific pathways that may be responsible for critical liver functions and health, offering valuable insights into how the liver adapts during the trial period's advanced stage.

Compared to the control group, the liver of the Lactobacillus delsoni group exhibited detectable levels of L-tryptophan (D-(+)Tryptophan), L-lysine (L-(+) Lysine), L-phenylalanine, Inosine, and Xanthosine. Additionally, N-[(2S)-2-hydroxypropanoyl]methionine, DL-Phenylalanine, L-Glycerol-2-phosphate Disodium Salt n-Hydrate, (Z)-2-octylpent-2-enedioic acid, (Z)-2-octylp-2-enendioic acid, Sulfo-l-idopyranuronic acid, and 2-O-Sulfo-L-idopyranuronic acid were detected. Furthermore, an increase in the relative concentrations of 1-Methylxanthine, L-Pyroglutamic acid, 6-O-phosphonohexa-1-enofuranos-3-ulose, N-Acetyl-L-carnosine(N-acetyl-L-Carnosine), Xanthine, and 6-O-phosphonohexa -1-enofuranos-3-ulose was observed, along with significantly elevated levels of 2-Aminonicotinic acid and Arabinofuranosyluracil (VIP $\geq$ 1.0 and $P < 0.05$).

The differential metabolite enrichment analysis of KEGG results is visualized using a bubble plot. As depicted in Figure 6, on day 14 of the experiment, the key metabolic pathways were enriched with differential metabolites, including Aminoacyl tRNA biosynthesis, purine metabolism, histidine metabolism, phenylalanine, tyrosine and tryptophan biosynthesis, lysine degradation, glutathione metabolism, and other metabolic pathways. Additionally, the identified pathways are alanine, aspartate, and glutamate metabolism; nitrogen metabolism; D-glutamine and D-glutamate metabolism; vitamin B6 metabolism; biotin metabolism; caffeine metabolism; phenylalanine metabolism; arginine biosynthesis; butyrate metabolism; selenium compound metabolism; pentose phosphate pathway; pyruvate metabolism; porphyrin and chlorophyll metabolism; glyoxylic acid and dicarboxylic acid metabolism; arginine and proline metabolism; and pyrimidine metabolism. The supplementation of *Lactobacillus delbrueckii* significantly enhanced the Aminoacyl tRNA biosynthesis pathway on day 14 of weaning piglets, with significant increases observed in intermediate products such as L-phenylalanine and L-lysine ($P < 0.05$).

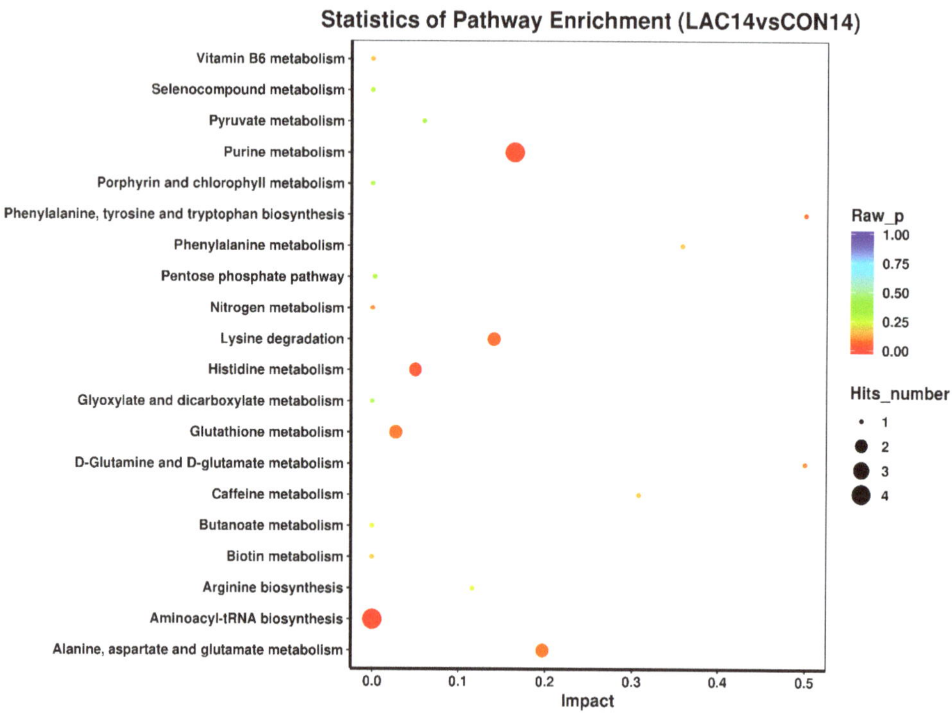

Figure 6. On the 14th day of the trial, an analysis was conducted on the changes in metabolites within the weaned piglets' livers, and their enrichment in KEGG pathways was investigated. Note: CON: control group; LAC: basal diet group with 0.1% *Lactobacillus delbrueckii*. This enrichment highlights the crucial

biochemical routes that are either upregulated or downregulated in response to the physiological changes occurring in the piglets post-weaning. The pathways identified provide a comprehensive view of the metabolic adjustments and adaptations that the liver undergoes during this critical developmental phase. Enrichment in these pathways might indicate shifts in energy production, detoxification processes, or the biosynthesis of essential biomolecules, all of which are vital for the growth, health, and overall wellbeing of the piglets. Understanding these enriched pathways offers invaluable insights into the metabolic strategies employed by the liver to cope with the nutritional and environmental challenges encountered post-weaning.

The volcanic map in Figure 7 clearly shows the metabolite differences between the *Lactobacillus delbrueckii* group and the control group on day 28 of the trial. The data showed that a total of 121 known differential metabolites were identified in the *Lactobacillus delbrueckii* group. Among these different metabolites, the relative concentration of 73 showed an increasing trend, while the relative concentration of the other 48 metabolites showed a decreasing trend. This result further confirms the significant influence of *Lactobacillus delbrueckii* on the liver metabolic activity of piglets, suggesting that this strain may influence the physiological status and health of piglets by regulating the concentration of specific metabolites. This difference in metabolic regulation may have important biological implications for the growth and development of piglets.

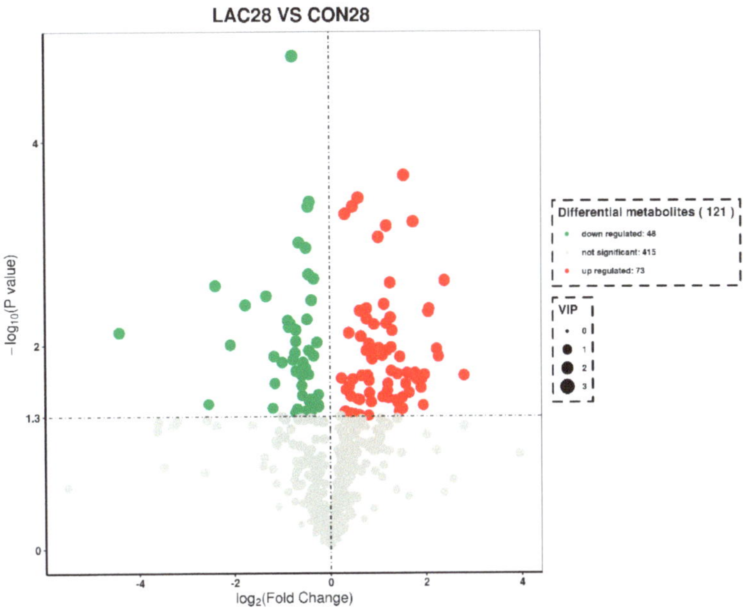

Figure 7. On the 28th day of the trial period, a volcano plot was meticulously crafted to showcase the differential metabolites present in the liver samples. Note: CON: control group; LAC: basal diet group with 0.1% *Lactobacillus delbrueckii*.

The top 20 differential metabolites upregulated and downregulated are shown in Figure 8. Compared with the control group, the liver Phosphatidylethanolamine lyso 20:4, Phosphatidylcholine lyso 16:1, Cyclohexanecarboxylate, Myristoleic acid, Phosphatidylcholine lyso, L-Aspartic acid, 4-Methylene L-glutamic acid, L-(+)-lactic acid, 1-stearoyl-SN-glycero-3-phosphoethanolamine, D-Ribulose-5-phosphate sodium salt, 2-Acetamido-2-deoxy-D-glucono-1,5-lactone, phosphoenol (1-O-Phosphonopentitol), glycine phenylalanine (Gly-Phe) and the relative concentrations of Leucylproline and 2-Acetamido-

2-deoxy-D-glucono-1,5-lactone were significantly decreased (VIP ≥ 1.0 and $P < 0.05$). Compared with the control group, the *Lactobacillus delbrueckii* group had higher levels of Oleic acid, Ethyl myristate, Lauric acid, Arachidonic acid, and ethyl myristate in the liver. gamma-Glutamylleucine, Reduced Glutathione, L-Proline, DL-TYROSINE, DL-4-Hydroxyphenyllactic acid, Docosapentaenoic acid, Taurine, N-Isovalerylglycine, Xanthine, 2-O-Sulfo-L-idopyranuronic acid, (10S)-Juvenile hormone III diol, GLUTATHIONE, 3-Hydroxybutyric acid, and N-arachidonoylglycine significantly increased the relative concentrations of S-(Formylmethyl) glutathione and 11,12-Epoxy-(5Z,8Z,11Z)-icosatrienoic acid (VIP ≥ 1.0 and $P < 0.05$).

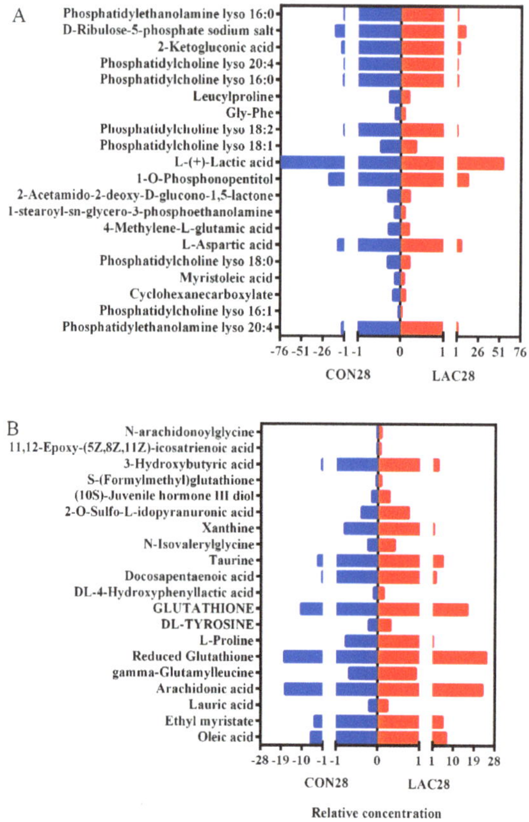

Figure 8. On the 28th day of the trial period, comprehensive data analysis revealed a list of the top 20 downregulated (Group (**A**)) and upregulated (Group (**B**)) liver metabolites. Note: CON: control group; LAC: basal diet group with 0.1% *Lactobacillus delbrueckii.* The identification of these differential metabolites is crucial, as it provides a snapshot of the dynamic metabolic adjustments within the liver. Group A metabolites showcase a marked decrease in concentration, suggesting a diminished role or a lower demand for these compounds within the physiological processes of the liver. Conversely, Group B metabolites have shown a significant increase in levels, indicating an elevated significance or enhanced utilization in the liver's metabolic activities. This categorization helps in pinpointing specific pathways that may be responsible for critical liver functions and health, offering valuable insights into how the liver adapts during the trial period's advanced stage.

The enrichment analysis of differential metabolites in the KEGG results is visualized using a bubble plot. As depicted in Figure 9, on the 28th day of the experiment, the significant metabolic pathways associated with differential metabolite enrichment include arachidonic acid metabolism, ketone body synthesis and degradation, taurine and hypotaurine metabolism, unsaturated fatty acid biosynthesis, primary bile acid biosynthesis, metabolic pathways, Aminoacyl tRNA biosynthesis, arginine biosynthesis, butyrate metabolism, niacin and niacinamide metabolism, histidine metabolism, pantothenic acid and Coenzyme A biosynthesis, fructose and mannose metabolism, beta-alanine metabolism, pyruvate metabolism, propionic acid metabolism, alanine, aspartic acid and glutamate metabolism, arginine and proline metabolism, fatty acid biosynthesis, and purine metabolism. The supplementation of *Lactobacillus delbrueckii* significantly enhanced the arachidonic acid metabolic pathway on the 28th day of the weaning experiment by significantly increasing the levels of intermediate products such as arachidonic acid and eicosatrienoic acid within this pathway ($P < 0.05$).

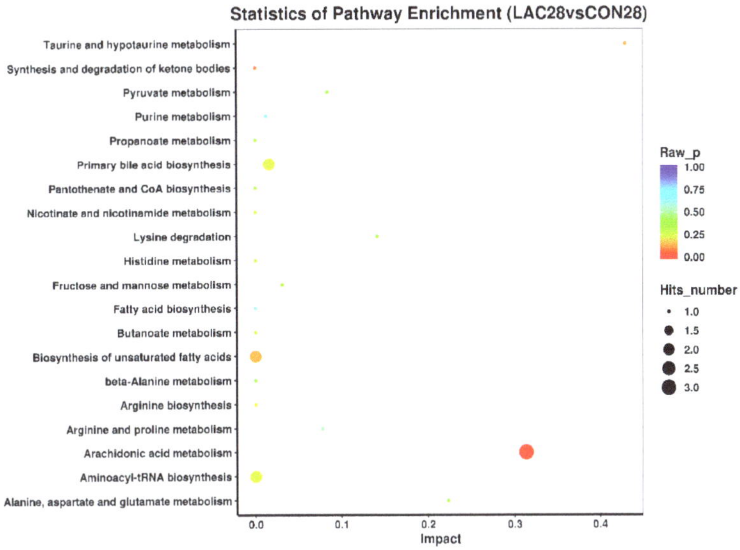

Figure 9. By the 28th day of the trial period, an extensive analysis revealed that the liver metabolites exhibiting differential expression in weaned piglets were significantly enriched in specific KEGG (Kyoto Encyclopedia of Genes and Genomes) pathways. Note: CON: control group; LAC: basal diet group with 0.1% *Lactobacillus delbrueckii.*

4. Discussion

Lipopolysaccharide (LPS) and enterotoxic Escherichia coli are commonly employed to establish immune or enterotoxic stress models in piglets. Previous studies have demonstrated that stimulation by LPS or enterotoxic Escherichia coli can induce intestinal immune stress and enteritis, subsequently impacting liver energy supply deficiency, liver damage, and metabolic dysfunction through the "enterohepatic axis" [23,24]. Numerous investigations have indicated that lipopolysaccharide-induced immune stress can alter the hepatic morphological structure of piglets [25,26]. Oxidative stress can trigger inflammatory responses and apoptosis, accelerate cellular oxidative damage, and enhance triglyceride (TG) accumulation within the liver, leading to lipid deposition [27,28]. Intestinal probiotics such as lactic acid bacteria and their metabolites modulate host the NF-κB signaling pathway or the Nrf2/ARE signaling pathway through various mechanisms to alleviate host immune stress and oxidative stress [29–31]. Studies suggest that *Lactobacillus delbrueckii* may initially ameliorate immunity and oxidative stress in weaned piglets by regulating immune antibodies, enzymes (GSH-Px, HO-1), or cytokines (8-OHdG) in intestinal tissues, thereby mitigating liver damage [6,32].

This study revealed that piglets in the control group exhibited mild hepatic morphological damage and increased liver fat content on day 14, indicating impaired hepatic lipid metabolism, possibly due to weaning stress. However, by day 28 of the experiment, hepatic morphology had recovered and the effects of weaning stress had been mitigated. Furthermore, intervention with *Lactobacillus delbrueckii* alleviated weaning stress-induced immune and oxidative stress in weaned piglets 14 days prior to the trial period, thereby safeguarding liver morphology and function.

Impaired liver function in animals can adversely affect their production performance and disease resistance. Previous studies have demonstrated that weaning stress in piglets leads to the increased consumption of liver glycogen and decreased glycogen content [21]. Chen Xiaole et al. [33] observed inhibitory phosphorylation of GSK-3 in rats following LPS pretreatment, which promotes liver glycogen synthesis and storage. Additionally, oxidative stress induced by diquat injection has been found to reduce the expression of the GLUT2 transporter, resulting in slowed glycogen decomposition and increased liver glycogen content in weaned piglets [34]. Our study revealed higher levels of liver glycogen and fat content on the 14th day among the control group weaned piglets, indicating impaired metabolism of hepatic sugar and fat affecting their growth. On the 28th day, there was a sharp decrease in the liver glycogen levels, suggesting the alleviation of stress but excessive consumption of glycogen. However, intervention with *Lactobacillus delbrueckii* prior to the experiment mitigated weaning stress and metabolic disorders among piglets, as evidenced by the increased glycogen content on the 28th day along with the enhanced ability to maintain energy metabolism balance.

Lactobacillus delbrueckii exerts a significant impact on liver metabolites. On day 14 of the experiment, the differentially enriched metabolites were primarily involved in the aminoacyl tRNA biosynthesis pathway, including L-phenylalanine, L-alanine, and L-lysine. The supplementation of *Lactobacillus delbrueckii* increased the levels of hepatic L-phenylalanine and L-lysine while decreasing the content of L-alanine, thereby enhancing the synthesis pathway of aminoacyl tRNA. Correlation analysis revealed a significant positive association between bifidobacterium and lactobacillus in the intestinal tract with phenylalanine and uridine content in the liver. Phenylalanine is an aromatic amino acid that undergoes hydroxylase-catalyzed oxidation to tyrosine within the body, playing a crucial role in sugar metabolism and fat metabolism [35]. The liver plays a vital role in amino acid metabolism as it contains most of the reaction enzymes involved in phenylalanine metabolism. Additionally, glutamic acid biosynthesis, along with tyrosine and tryptophan synthesis pathways, glutathione metabolism, and the pentose phosphate pathway, are major routes for differential metabolite enrichment, encompassing compounds such as erythritos-4-phosphate, D-glutamine ester derivative (4-pyridoxinol), uridine, etc. Supplementation with *Lactobacillus delbrueckii* resulted in decreased levels of D-glutamine and γ-glutamate while increasing uridine content within the liver metabolite pool. D-Erythrose 4-phosphate serves as an intermediate compound within the pentose phosphate pathway responsible for generating five-carbon sugar phosphates, whereas uridine contributes to the nucleic acid composition within animal cells, ultimately improving antibody levels. Simultaneously, on day 14 of the experiment, supplementation with Lactobacillus delsoni resulted in a reduction in guanosine diphosphate, mannose, and fucose levels in liver metabolites. Guanosine diphosphate mannose serves as a donor for mannose and is involved in the biosynthesis of mannosides, mannose-containing lipids, and glycoproteins in vivo. Fucose is predominantly found at the non-reducing end of complex hepatic sugar chains, such as glycoproteins, sugars, and lipids, playing a crucial role in physiological processes like cell transformation through sugar binding interactions. Fucosylation can reflect certain aspects of disease states within the body; moreover, L-fucosylation may facilitate pathogen transmission under specific circumstances [36]. Consequently, patients with cholangiocarcinoma, liver cancer, cirrhosis, or pancreatic cancer exhibit significantly higher levels of L-fucose compared to healthy individuals [37]. In this study involving weaned piglet livers, supplementation with *Lactobacillus delbrueckii* led to reduced guanosine diphosphate, mannose, and fucose contents. This suggests an improvement in liver damage caused by weaning stress and overall liver health enhancement. The observed differences in early-stage liver metabolism among weaned piglets might be associated with alterations in aminoacyl tRNA biosynthesis pathways. Supplementation with *Lactobacillus delbrueckii* promoted aminoacyl tRNA biosynthesis while improving liver amino acid metabolism among weaned piglets.

On the 28th day of the experiment, the main pathways enriched with differential metabolites were arachidonic acid metabolism, unsaturated fatty acid biosynthesis, and primary bile acid biosynthesis. The key differential metabolites identified were arachidonic acid, docosapentenoic acid, and eicosahexaenoic acid. Arachidonic acid serves as a precursor for various bioactive substances such as prostaglandins and leukotrienes, thereby exhibiting diverse biological functions, including lipid-lowering effects, anti-inflammatory properties, and the inhibition of lipid peroxidation [38]. Dietary supplementation with arachidonic acid has been shown to enhance liver function and lipid metabolism in eels [39]. Moreover, this study observed an increase in the levels of arachidonic acid, docosapentaenoic acid, and eicosahexaenoic acid, along with the improved biosynthesis of unsaturated fatty acids and saturated fatty acids in the liver. Additionally, there was a significant elevation in the relative contents of glutathione (GSH), reduced glutathione (GSH-R), and S-(formyl methyl) glutathione (FM-GSH) within the Lactobacillus deli group's liver samples. Glutathione is abundantly present in animal livers and red blood cells, where it plays crucial roles in detoxification processes as well as antioxidant defense mechanisms. Reduced glutathione contributes to hepatic protection by participating in carboxymethyl reactions and transpropylamino reactions [40]. In conclusion, differences observed during early-stage liver metabolism among weaned piglets may be associated with the aminoacyl tRNA biosynthesis pathway, while variations detected during late-stage liver metabolism could be linked to the arachidonic acid metabolism pathway specific to weaned piglets. Intestinal microorganisms, along with their metabolites, are likely involved in regulating these metabolic processes.

5. Conclusions

Lactobacillus delbrueckii effectively preserves the hepatic morphology in weaned piglets and alleviates the liver damage and metabolic disorders caused by weaning stress.

During the initial phase of the trial, *Lactobacillus delbrueckii* enhanced the aminoacyl–tRNA synthesis in weaned piglets, and in the later phase, it boosted the metabolism of arachidonic acid in the piglets.

Author Contributions: Conceptualization, Z.L. and X.H.; Investigation, Z.L., X.W. and L.M.; Methodology, Z.L. and X.W.; Writing—original draft, Z.L. and X.W.; Writing—review and editing, Z.L. and L.M. All authors have read and agreed to the published version of the manuscript.

Funding: This study was supported by the Hunan Province Natural Science Foundation key project (2024JJ3020), National Natural Science Foundation of China (U20A2055), and Scientific Research Project of Education Department of Hunan Province (22C1048).

Institutional Review Board Statement: Not applicable.

Informed Consent Statement: Not applicable.

Data Availability Statement: The original contributions presented in the study are included in the article, further inquiries can be directed to the corresponding author.

Conflicts of Interest: The authors declare no conflicts of interest.

References

1. Rogosa, M.; Hansen, P.A. Nomenclatural Considerations of Certain Species of Lactobacillus Beijerinck. *Int. J. Syst. Evol. Microbiol.* **1971**, *21*, 177–186. [CrossRef]
2. Wen, W.; Yang, T.; Wei, L.; Huang, X. Biological Function of *Lactobacillus delbrueckii* and Its Application in Animal Production. *Chin. J. Anim. Nutr.* **2019**, *31*, 1533–1539. [CrossRef]

3. Toba, T.; Yoshioka, E.; Itoh, T. Lacticin, a bacteri-ocin produced by *Lactobacillus delbrueckii* subsp. Lactis. *Lett. Appl. Microbiol.* **1991**, *12*, 43–45. [CrossRef]
4. Dai, Y.; Tian, Z.; Nan, X. Progress of the Lactobacillus and Its Physiological Function. *Agric. Prod. Process. (J. Agric. Sci.)* **2009**, *7*, 24–26+29.
5. Hansen, P.A. *Type Strains of Lactobacillus Species. A Report by the Taxonomic Subcommittee on Lactobacilli and Closely Related Organisms. R. A Subcommittee of the International Committee on Nomenclature of Bacteria of the International Association of Microbiological Societies*; American Culture Collection: Rockville, MD, USA, 1968; Corpus ID: 89670364.
6. Li, Y.; Hou, S.; Peng, W.; Lin, Q.; Chen, F.; Yang, L.; Li, F.; Huang, X. Oral Administration of *Lactobacillus delbrueckii* during the Suckling Phase Improves Antioxidant Activities and Immune Responses after the Weaning Event in a Piglet Model. *Oxidative Med. Cell Longev.* **2019**, *2019*, 6919803. [CrossRef] [PubMed]
7. Guo, S.; Tian, S.; Bai, H. Lactic acid bacteria and lactic acid bacteria fermented food. *J. Food Food Ind.* **2005**, *12*, 3. [CrossRef]
8. Giang, H.H.; Viet, T.Q.; Ogle, B.; Lindberg, J.E. Effects of Supplementation of Probiotics on the Performance, Nutrient Digestibility and Faecal Microflora in Growing-finishing Pigs. *Asian-Australas. J. Anim. Sci.* **2011**, *24*, 655–661. [CrossRef]
9. Liu, H.; Zhang, J.; Zhang, S.; Yang, F.; Thacker, P.A.; Zhang, G.; Qiao, S.; Ma, X. Oral Administration of *Lactobacillus fermentum* I5007 Favors Intestinal Development and Alters the Intestinal Microbiota in Formula-Fed Piglets. *J. Agric. Food Chem.* **2014**, *62*, 860–866. [CrossRef] [PubMed]
10. Giang, H.H.; Viet, T.Q.; Ogle, B.; Lindberg, J.E. Growth performance, digestibility, gut environment and health status in weaned piglets fed a diet supplemented with potentially probiotic complexes of lactic acid bacteria. *Livest. Sci.* **2010**, *129*, 95–103. [CrossRef]
11. He, T.; Zhu, Y.-H.; Yu, J.; Xia, B.; Liu, X.; Yang, G.-Y.; Su, J.-H.; Guo, L.; Wang, M.-L.; Wang, J.-F. *Lactobacillus johnsonii* L531 reduces pathogen load and helps maintain short-chain fatty acid levels in the intestines of pigs challenged with *Salmonella enterica* Infantis. *Veter Microbiol.* **2019**, *230*, 187–194. [CrossRef]
12. Li, Y.; Hou, S.; Chen, J.; Peng, W.; Wen, W.; Chen, F.; Huang, X. Oral administration of Lactobacillus delbrueckii during the suckling period improves intestinal integrity after weaning in piglets. *J. Funct. Foods* **2019**, *63*, 103591. [CrossRef]
13. Pluske, J.R.; Thompson, M.J.; Atwood, C.S.; Bird, P.H.; Williams, I.H.; Hartmann, P.E. Maintenance of villus height and crypt depth, and enhancement of disaccharide digestion and monosaccharide absorption, in piglets fed on cows' whole milk after weaning. *Br. J. Nutr.* **1996**, *76*, 409–422. [CrossRef] [PubMed]
14. Fuller, R. Probiotics in man and animals. *J. Appl. Bacteriol.* **1989**, *66*, 365–378. [PubMed]
15. Meale, S.J.; Li, S.C.; Azevedo, P.; Derakhshani, H.; DeVries, T.J.; Plaizier, J.C.; Steele, M.A.; Khafipour, E. Weaning age influences the severity of gastrointestinal microbiome shifts in dairy calves. *Sci. Rep.* **2017**, *7*, 198. [CrossRef] [PubMed]
16. Stokes, C.R. The development and role of microbial-host interactions in gut mucosal immune development. *J. Anim. Sci. Biotechnol.* **2017**, *8*, 307–316. [CrossRef] [PubMed]
17. Yang, H.; Xiong, X.; Wang, X.; Tan, B.; Li, T.; Yin, Y. Effects of Weaning on Intestinal Upper Villus Epithelial Cells of Piglets. *PLoS ONE* **2016**, *11*, e0150216. [CrossRef] [PubMed]
18. Gresse, R.; Chaucheyras-Durand, F.; Fleury, M.A.; Van de Wiele, T.; Forano, E.; Blanquet-Diot, S. Gut Microbiota Dysbiosis in Postweaning Piglets: Understanding the Keys to Health. *Trends Microbiol.* **2017**, *25*, 851–873. [CrossRef] [PubMed]
19. Zhang, W.; Liu, P.; Guan, W.; Chen, F.; Deng, Y. Effects of Dietary Supplementation of Immuno-globulin of Yolk on Growth Performance, Serum Biochemical Parameters, Intestinal Morphology and Microbial Flora of Weaned Piglets. *Chin. J. Anim. Nutr.* **2017**, *29*, 271–279. [CrossRef]
20. Nafikov, R.A.; Beitz, D.C. Carbohydrate and Lipid Metabolism in Farm Animals. *J. Nutr.* **2007**, *137*, 702–705. [CrossRef] [PubMed]
21. Yang, H.; Yi, S.; Xu, E.; Chen, X.; Dai, B.; Xiao, Y. Effects of Early Weaning on Glycogen Concent and Expression of Hepatic Genes Associated with Gluconeogenesis and Glycolysis in Piglets. *Chin. J. Anim. Nutr.* **2018**, *30*, 1019–1026.
22. Li, R. *Effects of Lactobacillus delbrueckii on Cholesterol Metabolism in Fattening Pigs*; Hunan Agricultural University: Changsha, China, 2014.
23. Nielsen, S.S.; Grøfte, T.; Tygstrup, N.; Vilstrup, H. Synthesis of acute phase proteins in rats with cirrhosis exposed to lipopolysaccharide. *Comp. Hepatol.* **2006**, *5*, 3. [CrossRef] [PubMed]
24. Wang, L.M.; Wang, X.Y.; Wu, H.T.; Wang, H.B.; Chen, S.K.; Zhu, H.L.; Liu, Y.L. Effects of glutamate on energy metabolism and mRNA expression of key enzymes and related regulatory factors in liver of piglets. *Chin. J. Anim. Sci.* **2016**, *52*, 38–43.
25. Li, Q.; Liu, Y.; Che, Z.; Zhu, H.; Meng, G.; Hou, Y.; Ding, B.; Yin, Y.; Chen, F. Dietary L-arginine supplementation alleviates liver injury caused by *Escherichia coli* LPS in weaned pigs. *J. Endotoxin Res.* **2012**, *18*, 804–814. [CrossRef] [PubMed]
26. Wang, H.B. *Regulative Effect of Flaxseed Oil on Intestinal and Liver Injury of Piglets after Lipopolysaccharide Challenge*; Wuhan University of Light Industry: Wuhan, China, 2016.
27. Baiceanu, A.; Mesdom, P.; Lagouge, M.; Foufelle, F. *Endoplasmic reticulum* proteostasis in hepatic steatosis. *Nat. Rev. Endocrinol.* **2016**, *12*, 710–722. [CrossRef] [PubMed]
28. He, Q.; Shao, Y.; Yao, W.; Wang, T.; Xia, J.; Huang, F. Effects of Garcinol Supplementation on Growth Performance, Antioxidant Function and Liver Lipid Synthesis of Piglets under Oxidative Stress. *Chin. J. Anim. Nutr.* **2019**, *31*, 5834–5845.
29. Sokol, H.; Seksik, P.; Furet, J.; Firmesse, O.; Mariat, D.; Beaugerie, L.; Dore, J.M. T1238 fecal microbiota alterations in inflammatory bowel disease (IBD): Decrease of Faecali-bacterium Prausnitzii but no quantitative variation of *Escherichia coli*. *Gastroenterology* **2008**, *134*, A-513. [CrossRef]

30. Kobatake, E.; Nakagawa, H.; Seki, T.; Miyazaki, T. Protective effects and functional mechanisms of *Lactobacillus gasseri* SBT2055 against oxidative stress. *PLoS ONE* **2017**, *12*, e0177106. [CrossRef] [PubMed]
31. Lin, X.; Xia, Y.; Wang, G.; Xiong, Z.; Zhang, H.; Lai, F.; Ai, L. Lactobacillus plantarum AR501 alleviates the oxidative stress of D-Galactose-induced aging mice liver by up regulation of Nrf2-Mediated antioxidant enzyme expression. *J. Food Sci.* **2018**, *83*, 1990–1998. [CrossRef]
32. Liu, M. *Study on the Alleviating Effect of Lactobacillus desseri on Inflammatory Bowel Disease of Weaned Piglets*; Hunan Agricultural University: Changsha, China, 2016.
33. Chen, X.L.; Gong, J.P.; Xu, F.L. Effects and mechanisms of glycogen synthase kinase-3 inhibition induced by lipopolysaccharide pretreatment on liver glycogen. *J. South. Med. Univ.* **2014**, *2*, 201–205.
34. Huang, L.H. *Effects of Chitooligosaccharides on Alleviating Hepatic Oxidative Damage in Weaned Piglets*; Sichuan Agricultural University: Chengdu, China, 2016.
35. Li, L.Z.; Li, L.Y. *Key Technologies of Modern Biochemical Drug Production*; Chemical Industry Press: Beijing, China, 2006; pp. 145–150.
36. Song, Y.H.; Feng, T. Clinical application of determination of free L-fucose in urine. *Prog. Biochem. Biophys.* **1999**, *26*, 513–516.
37. Dong, Z.N.; Shen, W.M. Clinical significance of urinary L-fucose determination. *Proc. Acad. Mil. Med. Sci.* **1997**, *4*, 257.
38. Fu, S.E.; Han, W.; Bian, J. Research progress of arachidonic acid. *Yunnan Chem. Ind.* **2004**, *31*, 31–32.
39. Zhong, J. *Effects of Arachidonic Acid (ARA) on the Growth and Hepatoprotective of Monopterus albus*; Jiangxi Agricultural University: Nanchang, China, 2018.
40. Winterbour, N.; Christine, C. Regulation of intracellular glutathione. *Redox Biol.* **2018**, *22*, 101086.

fermentation

Article

Silage Making of Napier Grass and Sugarcane Top at Different Proportions: Evolution of Natural Fermentation Characteristics, Chemical Composition, and Microbiological Profile

Huade Xie, Fanquan Zeng, Xianqing Luo, Zhipei Li, Yuhong Pan, Yanxia Guo, Lijuan Peng, Li Liang, Jingzhen Li, Yuchen Liang and Chengjian Yang *

Guangxi Key Laboratory of Buffalo Genetics, Reproduction and Breeding, Guangxi Buffalo Research Institute, Nanning 530001, China; xiehuade093@163.com (H.X.); 13471010613@163.com (F.Z.); 19101477179@163.com (X.L.); lizhipei78@163.com (Z.L.); 13607810045@163.com (Y.P.); gyxlq0417@163.com (Y.G.); lijuanpeng2000@163.com (L.P.); lilianglily0711@163.com (L.L.); m18877155818@163.com (J.L.); 18269340446@163.com (Y.L.)

* Correspondence: ycj0746@sina.com; Tel.: +86-157-7716-2502

Abstract: The co-ensiling technique is widely used to improve silage quality; however, it remains unclear as to what high-quality silages can be made by co-ensiling Napier grass (NG) with Sugarcane top (ST). The aim of this study was to evaluate the fermentation characteristics, chemical composition, and microbiological profile of silage produced from mixtures of NG and ST in varying ratios. Silage was prepared using a small-scale fermentation system, and treatments were designed as control silage (NG ensiled alone) or with 10%, 20%, 30%, 40%, 50%, 60%, 70%, 80%, 90%, and 100%ST on a fresh matter basis with six replicates, respectively. Increasing ST in the silage reduced the contents of crude protein, ash, acetic acid, butyric acid, ammonia-N, as well as pH, but increased the contents of dry matter, ether extract, neutral detergent fiber, acid detergent fiber, water-soluble carbohydrate, lactic acid, and lactic acid bacteria. *Lactobacillales* and *Enterobacterales* were the dominant orders, with *Lactiplantibacillus* and *Weissella* as the dominant genera. Co-ensiling NG with ST enhanced microbial diversity and richness. ST, as a local by-product, is a viable additive to improve NG silage quality and nutrition. This study suggests that good-quality silages can be produced with NG: ST ratios of 40:60 to 20:80 and that these silages offer an opportunity to optimize the nutrient supply for ruminants.

Keywords: Napier grass; sugarcane top; co-ensiling technique; silage microorganisms

Citation: Xie, H.; Zeng, F.; Luo, X.; Li, Z.; Pan, Y.; Guo, Y.; Peng, L.; Liang, L.; Li, J.; Liang, Y.; et al. Silage Making of Napier Grass and Sugarcane Top at Different Proportions: Evolution of Natural Fermentation Characteristics, Chemical Composition, and Microbiological Profile. *Fermentation* **2024**, *10*, 525. https://doi.org/10.3390/fermentation10100525

Academic Editors: Siran Wang and Wansup Kwak

Received: 16 August 2024
Revised: 14 October 2024
Accepted: 14 October 2024
Published: 15 October 2024

1. Introduction

Sugarcane, belonging to the genus *Saccharum*, is widely cultivated in tropical and subtropical countries. It is the leading driver for the global production of sugar and sweeteners [1–3]. By 2022, the total yield of sugarcane in China reached 103.381 million tons [4]. Sugarcane top (ST), a by-product of the sugar industry, represents 15–25% of the aerial part of sugarcane and is used as the main forage material for milk production and fattening of ruminants in different parts of the world [5–7]. In most Asian countries, including China, livestock rearing is mainly dependent on nutritionally poor crop by-products. The available feed resources in China cannot fully meet the feed requirements of the large livestock population, both quantitatively and qualitatively. To bridge the gap between availability and requirement, by-products from agricultural crops must be utilized more efficiently. ST is an important and integral animal roughage resource in southern China. It has a high water-soluble carbohydrate (WSC) content and is highly nutritive and palatable with good intake preference for livestock; therefore, it should be further developed and utilized [8]. However, owing to the seasonal and concentrated harvesting of sugarcane, as well as the humid and hot climatic factors in cultivated regions, ST is prone to mold, and the utilization rate of ST feed is low. Furthermore, the physical structure of untreated ST is relatively hard, which can affect the feed intake and damage the digestive tract of livestock. However, with strategic supplementation and the adoption of various processing techniques, these feed resources can be efficiently utilized for production purposes. By exploiting sugarcane by-products to their full potential, animal production and productivity can be increased. Silage is less affected by weather conditions, making it an effective storage method for ST as a feed alternative, preventing subsequent molding [9]. Furthermore, the efficient utilization of ST not only promotes the development of the sugarcane industry but also alleviates the shortage of seasonal ruminant animal forage.

Napier grass (NG) is a high-yield warm season forage commonly cultivated in tropical and subtropical regions because of its high biomass yield and growth, rich nutritional value, freshness, juiciness, and palatability [10–13]. It is mainly used as a high-quality green feed for ruminants and can also be used as a raw material for preparing silage feed, thereby addressing the problem of insufficient coarse feed during spring and winter [14,15]. However, NG's low dry matter (DM) and WSC concentrations, along with its high cell wall content, complicate the production of high-quality silage without additives, leading to potential silage spoilage and reduced animal intake [16–18].

Ensiling is a method of preserving high-quality forage initiated by a complex microbial community in anaerobic environments. During silage production, lactic acid bacteria (LAB) rapidly proliferate, establishing subsequent bacterial communities, which metabolize WSC into organic acids, allowing long-term storage while maintaining nutritional quality and enhancing feed palatability [19,20]. Previous studies have reported that co-ensiling improves silage quality and enhances stability during the fermentation process compared with silage alone [21–26]. Moreover, the feeding effect of NG and ST as a single feed source is not ideal, and they are prone to wastage. Thus, mixing them in specific proportions balances the DM content, compensates for nutritional deficiencies in NG silage, increases WSC content, inhibits poor fermentation, and improves overall silage quality, leading to high-quality silage production [27,28]. Furthermore, a detailed comprehension of the microbiome and metabolome during the ensiling process could enable researchers to identify novel strategies for enhancing silage preservation; additionally, this knowledge may facilitate the development of silages enriched with active metabolites that promote animal health and welfare [29]. However, the evolution of the microbiological profile and dynamic changes in fermentation characteristics during ensiling when silage production using NG and ST at different proportions requires specifically further investigation.

Therefore, the objective of the current study was to evaluate the mixed silage quality of NG and ST at different proportions throughout the storage period of 1, 3, 5, 7, 15, 30, and 60 d, considering the fermentation characteristics, chemical composition, and microbiological profile.

2. Materials and Methods

2.1. Ensiling Materials and Silage Preparation

NG, 'Guimu-1' ((*Pennisetum americanum* × *P. purpureum*) × *P. purpureum* cv. Guimu No.1), was harvested at approximately 10 cm above ground level after a three-month regrowth period at the vegetative stage of maturity obtained from the Herbage Base at the Guangxi Zhuang Autonomous Region Buffalo Research Institute. ST, grown for approximately ten months and manually harvested following cane harvest, was sourced from an industrial sugar production region in Nanning, China, in February 2023. The chemical compositions of NG and ST are listed in Table 1. Fresh NG and ST samples were immediately chopped (approximately 2 cm) using a forage cutter. NG was mixed thoroughly without ST (Control, S0) or with 10% (S1), 20% (S2), 30% (S3), 40% (S4), 50% (S5), 60% (S6), 70% (S7), 80% (S8), 90% (S9), and 100% (S10) ST based on fresh matter (FM) to create the silage materials. Approximately 1 kg of the sample from each treatment, with six replicates, was packed into plastic bags (160 mm × 250 mm) and sealed using a vacuum sealer (DZ500; Gzrifu Co., Ltd., Guangzhou, China). Silage samples were preserved in the laboratory at 20–25 °C away from light and sampled for analysis after 1, 3, 5, 7, 15, 30, and 60 d.

Table 1. Chemical composition of Napier grass and sugarcane top before ensiling.

Item	DM	CP	EE	Ash	NDF	ADF	WSC
	(%)	(% DM)					
NG	16.71	12.89	1.32	11.91	69.58	36.15	3.46
ST	26.11	7.33	2.59	6.79	74.15	42.98	11.66

ADF, acid detergent fiber; CP, crude protein; DM, dry matter; EE, ether extract; NDF, neutral detergent fiber; NG, Napier grass; ST, sugarcane top; WSC, water-soluble carbohydrates.

2.2. Chemical Analysis

Samples of NG, ST, and their mixed silages were dried in a forced draft oven (LABO–250; STIK Co., Ltd., Shanghai, China) at 65 °C until a constant weight was obtained. Subsequently, they were ground by passing through a 1 mm screen using a sample mill (FS200; Guangzhou Bomin Mechanical & Electrical Equipment Co., Ltd., Guangzhou, China). The DM, crude protein (CP), ether extract (EE), and ash content were analyzed according to methods 934.01, 976.05, 920.39, and 942.05, respectively, of the Association of Official Analytical Chemists (AOAC, 1990) [30]. Neutral detergent fiber (NDF) and acid detergent fiber (ADF) contents were determined according to the methods described by Van Soest et al. (1991) [31]. The WSC was analyzed according to the methods described by Udén [32].

2.3. Fermentation Analysis

The ensiling fermentation products in the silage were determined using cold-water extracts [33]. Briefly, 20 g of fresh silage was homogenized with 180 mL of sterilized distilled water and stored at 4 °C overnight. The pH was measured using a pH meter (HI 8424; HANNA® instruments, Woonsocket, RI, USA). Organic acid contents, including lactic acid (LA), acetic acid (AA), propionic acid, and butyric acid (BA), were measured using high-performance liquid chromatography (1260 Infinity II; Agilent Technologies, Inc., Waldbronn, Germany) according to the method described by Xie et al. (2023) [34]. The analytical conditions employed were as follows: the column utilized was a Shodex RSpak KC–811 (8.0 mm × 300 mm; Showa Denko K.K., Tokyo, Japan); detection was performed using a DAD detector set at 210 nm; the eluent consisted of 3 mmol/L $HClO_4$ with a flow rate of 1.0 mL/min; the temperature was maintained at 50 °C; and the sample injection volume was 5.0 µL. The ammonia-N (NH_3-N) content was measured using the method described by Broderick and Kang (1980) [35].

2.4. Microbial Populations and Bacterial Community Analyses

Enumeration of LAB, yeasts, and molds in ensiled forage was performed using the plate count method described by Cao et al. (2011) [36]. Briefly, silage samples (10 g) were homogenized in 90 mL of distilled water, and 10^{-1}–10^{-5} were serially diluted in sterilized water. The number of LAB was counted on spread plates using De Man–Rogosa–Sharpe agar (Qingdao Hope Bio-Technology Co., Ltd., Qingdao, China) after incubation in an anaerobic incubator at 30 °C for 48 h. Subsequently, yeasts and molds were counted on potato dextrose agar (Qingdao Hope Bio-Technology Co., Ltd., Qingdao, China) after incubation at 30 °C for 48 h. Yeast were distinguished from mold or bacteria by colony appearance and observation of cell morphology. The number of colonies indicated the number of viable microorganisms (log cfu/g FM).

Silage samples stored for 7, 30, and 60 d were sampled for bacterial community analysis. Microbial DNA was extracted from the silage samples using the Ezup Spin Column Super Plant Genomic DNA Extraction Kit (B518262; Sangon Biotech, Shanghai, China), according to the manufacturer's instructions. The V3–V4 regions of the 16S rRNA gene were subjected to amplification using the primers 338F (ACTCCTACGGGAG-GCAGCAG) and 806R (GGACTACHVGGGTWTCTAAT). Biomarker (BMK) Technologies (Beijing, China) conducted the metagenomic sequencing, which included PCR amplification and DNA extraction, followed by Illumina MiSeq sequencing and final sequencing data processing. Data were analyzed using the free online BMK Cloud Platform (https://international.biocloud.net).

2.5. Statistical Analysis

Data on the chemical composition of silage opened on day 60 were analyzed using one-way ANOVA. Data on microbial populations and fermentation characteristics were analyzed using the generalized linear model procedure of SAS software (Statistical Analysis System, version 9.2), according to the model for a 7 × 11 factorial treatment design:

$$Y_{ij} = \mu + \alpha_i + \beta_j + (\alpha \times \beta)_{ij} + e_{ij} \quad (1)$$

where μ represents the overall mean, α_i represents the ensiling time effect (i = 1, 3, 5, 7, 15, 30, 60 d), β_j represents the mixture ration effect (j = S0, S1, S2, S3, S4, S5, S6, S7, S8, S9, S10), $(\alpha \times \beta)_{ij}$ represents the effect of interaction between ensiling time and mixture ration, and e_{ij} represents random error. Tukey's test was used to identify statistically significant differences ($p < 0.05$) between means. An online platform (https://international.biocloud.net/) was used to analyze the sequencing data of the bacterial community. Correlation heat maps were constructed using the Complot R package.

3. Results

3.1. Chemical Compositions

As shown in Table 1, NG had higher CP and ash contents but lower DM, EE, NDF, ADF, and WSC contents than ST before ensiling. Increasing the proportion of ST in NG silage led to significant ($p < 0.05$) increases in DM, EE, NDF, ADF, and WSC contents, and significant ($p < 0.05$) decreases in CP and ash contents (Table 2).

Table 2. Chemical composition of Napier grass, sugarcane top, and their silage sole or mixed silages after 60 d of storage.

Item	DM	CP	EE	Ash	NDF	ADF	WSC
	(%)	(% DM)					
S0 †	11.02 [j]	9.45 [a]	1.08 [g]	10.65 [a]	66.81 [e]	34.15 [g]	1.51 [f]
S1	12.42 [i]	9.34 [a]	1.20 [g]	10.30 [ab]	67.83 [e]	34.81 [fg]	2.02 [f]
S2	13.21 [h]	8.21 [b]	1.45 [f]	10.21 [ab]	69.67 [d]	35.42 [f]	3.67 [e]
S3	14.74 [g]	7.86 [bc]	1.63 [ef]	9.87 [b]	70.16 [cd]	36.37 [e]	4.46 [de]
S4	16.54 [f]	7.80 [bc]	1.71 [de]	9.20 [c]	70.30 [cd]	37.80 [d]	4.54 [de]
S5	17.34 [e]	7.74 [bc]	1.79 [cde]	8.45 [d]	70.78 [bcd]	38.02 [d]	4.73 [cde]
S6	19.53 [d]	7.64 [bc]	1.88 [bcd]	8.37 [d]	71.12 [abc]	38.39 [dc]	5.38 [cd]
S7	20.90 [c]	7.57 [c]	1.94 [bcd]	7.38 [e]	71.75 [ab]	39.06 [c]	5.98 [bc]
S8	21.85 [b]	7.47 [cd]	1.99 [abc]	6.91 [e]	71.86 [ab]	40.19 [b]	6.71 [b]
S9	22.14 [b]	6.98 [d]	2.09 [ab]	6.78 [ef]	72.17 [a]	40.90 [ab]	7.14 [b]
S10	23.24 [a]	6.27 [e]	2.21 [a]	6.22 [f]	72.31 [a]	41.75 [a]	8.65 [a]
SEM	0.2438	0.1896	0.0771	0.2122	0.4031	0.3064	0.4257
p-Value	<0.0001	<0.0001	<0.0001	<0.0001	<0.0001	<0.0001	<0.0001

† Napier grass was mixed thoroughly without sugarcane top (Control, S0) or with 10% (S1), 20% (S2), 30% (S3), 40% (S4), 50% (S5), 60% (S6), 70% (S7), 80% (S8), 90% (S9), and 100% (S10) sugarcane top based on fresh weight, respectively. ADF, acid detergent fiber; CP, crude protein; DM, dry matter; EE, ether extract; NDF, neutral detergent fiber; SEM, standard error of the mean; WSC, water-soluble carbohydrates. Values with different small letter superscripts mean significant difference ($p < 0.05$), while those with the same or no letter superscripts mean no significant difference ($p > 0.05$).

3.2. Fermentation Quality

Ensiling days (ED), mixture ratio (MR), and their interaction (ED × MR) affected the pH and content of LA, AA, and NH_3-N (Table 3). Compared with S0, the addition of ST increased ($p < 0.05$) LA content and decreased ($p < 0.05$) the pH and contents of AA, BA, and NH_3-N in silage. A comparison of ED showed that the pH of the silage during the later stages of fermentation was lower ($p < 0.05$) than that during the early stages of fermentation. LA, AA, BA, and NH_3-N content in the silages increased ($p < 0.05$) with storage time.

Table 3. Changes in fermentation profile of Napier grass, sugarcane top, and their mixed silages ensiled for 1, 3, 5, 7, 15, 30, and 60 days.

Item	pH	Lactic Acid	Acetic Acid	Propionic Acid	Butyric Acid	NH_3-N
				(g/kg DM)		
Ensiling days mean (ED)						
1	6.23 ^{a}	2.73 ^{f}	0.70 ^{f}	ND	ND	0.16 ^{g}
3	6.01 ^{b}	3.18 ^{f}	0.86 ^{ef}	ND	ND	0.18 ^{f}
5	5.88 ^{c}	7.59 ^{e}	1.28 ^{e}	ND	ND	0.24 ^{e}
7	5.15 ^{d}	8.54 ^{d}	3.70 ^{d}	ND	ND	0.35 ^{d}
15	4.69 ^{e}	12.41 ^{c}	11.50 ^{c}	ND	0.62 ^{b}	0.40 ^{c}
30	4.30 ^{f}	25.69 ^{b}	15.45 ^{b}	ND	0.83 ^{a}	0.48 ^{b}
60	4.19 ^{g}	27.54 ^{a}	18.78 ^{a}	ND	0.85 ^{a}	0.53 ^{a}
Mixture ratios mean (MR)						
S0 †	5.62 ^{a}	7.56 ^{i}	12.25 ^{a}	ND	3.61 ^{a}	0.45 ^{a}
S1	5.56 ^{b}	8.87 ^{h}	11.48 ^{ab}	ND	ND	0.40 ^{b}
S2	5.47 ^{c}	10.08 ^{g}	10.81 ^{b}	ND	ND	0.38 ^{b}
S3	5.38 ^{d}	10.89 ^{fg}	9.41 ^{c}	ND	ND	0.36 ^{c}
S4	5.26 ^{e}	11.35 ^{f}	8.00 ^{d}	ND	ND	0.34 ^{cd}
S5	5.19 ^{f}	12.53 ^{e}	7.04 ^{e}	ND	ND	0.32 ^{de}
S6	5.09 ^{g}	13.78 ^{d}	6.00 ^{f}	ND	ND	0.30 ^{ef}
S7	5.03 ^{h}	14.40 ^{cd}	5.09 ^{g}	ND	ND	0.29 ^{fg}
S8	4.97 ^{i}	15.15 ^{bc}	4.37 ^{gh}	ND	ND	0.28 ^{fg}
S9	4.88 ^{j}	15.95 ^{b}	3.93 ^{h}	ND	ND	0.27 ^{gh}
S10	4.83 ^{k}	17.19 ^{a}	3.73 ^{h}	ND	ND	0.26 ^{h}
SEM	0.0277	0.5742	0.4516	-	0.0362	0.013
Significance of main effects and interactions						
ED	<0.0001	<0.0001	<0.0001	-	<0.0001	<0.0001
MR	<0.0001	<0.0001	<0.0001	-	<0.0001	<0.0001
ED × MR	<0.0001	<0.0001	<0.0001	-	<0.0001	<0.0001

† Napier grass was mixed thoroughly without sugarcane top (Control, S0) or with 10% (S1), 20% (S2), 30% (S3), 40% (S4), 50% (S5), 60% (S6), 70% (S7), 80% (S8), 90% (S9), and 100% (S10) sugarcane top based on fresh weight, respectively. ND, not detected; SEM, standard error of the mean. Values with different small letter superscripts mean significant difference ($p < 0.05$), while those with the same or no letter superscripts mean no significant difference ($p > 0.05$).

3.3. Microbial Populations

ED, MR, and their interaction (ED × MR) influenced the number of LAB (Table 4). However, the number of yeasts was only affected by ED. A comparison among all types of silages revealed that the number of LAB was the highest ($p < 0.05$) in S6, S7, and S8 silages, followed by S9, S4, S5, S3, S10, S2, S1, and S0 silages. The number of yeasts did not differ ($p > 0.05$) among silage types. A comparison of the ED showed that the silage obtained on days 30 and 60 of fermentation had the highest ($p < 0.05$) number of LAB, followed by those obtained on days 15, 7, 5, 3, and 1 of fermentation. However, the number of yeasts was the highest ($p < 0.05$) in silage on day 15 of fermentation, followed by that on days 7, 5, 60, 30, 3, and 1 of fermentation. Molds were not detected in any of the silages on any ED.

Table 4. Changes in counts of viable microorganisms of Napier grass, sugarcane top, and their mixed silages during fermentation.

Item	Lactic Acid Bacteria	Yeasts	Molds
		log cfu/g FM	
Ensiling days mean (ED)			
1	4.41 ^{e}	4.35 ^{e}	ND
3	4.47 ^{d}	4.38 ^{e}	ND
5	4.49 ^{d}	4.56 ^{c}	ND
7	4.61 ^{c}	4.60 ^{b}	ND
15	4.99 ^{b}	4.87 ^{a}	ND
30	6.51 ^{a}	4.43 ^{d}	ND
60	6.52 ^{a}	4.45 ^{d}	ND
Mixture ratios mean (MR)			
S0 †	4.95 ^{e}	4.52	ND
S1	5.08 ^{d}	4.52	ND
S2	5.09 ^{d}	4.52	ND
S3	5.13 ^{bcd}	4.54	ND
S4	5.18 ^{abc}	4.52	ND
S5	5.15 ^{abcd}	4.54	ND
S6	5.22 ^{a}	4.5	ND
S7	5.23 ^{a}	4.5	ND
S8	5.22 ^{a}	4.51	ND
S9	5.21 ^{ab}	4.52	ND
S10	5.12 ^{cd}	4.51	ND
SEM	0.0741	0.0391	-
Significance of main effects and interactions			
ED	<0.0001	<0.0001	-
MR	<0.0001	0.6832	-
ED × MR	<0.0001	0.2111	-

† Napier grass was mixed thoroughly without sugarcane top (Control, S0) or with 10% (S1), 20% (S2), 30% (S3), 40% (S4), 50% (S5), 60% (S6), 70% (S7), 80% (S8), 90% (S9), and 100% (S10) sugarcane top based on fresh weight, respectively. FM, fresh matter; ND, not detected; SEM, standard error of the mean. Values with different small letter superscripts mean significant difference ($p < 0.05$), while those with the same or no letter superscripts mean no significant difference ($p > 0.05$).

3.4. Bacterial Community

The sequencing analysis coverage was above 0.99 (Table 5), ensuring optimal sequencing depth to accurately reflect microbial species and good representativeness in all silages. On day 7, ACE, Chao 1, and Simpson indices were significant in S4 ($p < 0.05$) (Table 4), though most silages showed no significant differences. The highest Shannon index was in S8 ($p < 0.05$), exceeding that of most other silages. On day 30, when NG co-ensiling reached 50% ST, the ACE, Chao 1, Simpson, and Shannon indices were higher ($p < 0.05$) than those of NG silage without ST or with low levels of ST. On day 60, S10 had higher ACE and Chao1 values than the other silages ($p < 0.05$); S0 and S1 had lower ($p < 0.05$) Simpson and Shannon indices than the other silages.

Table 5. Changes in bacterial alpha diversity parameters of Napier grass, sugarcane top, and their mixed silages stored at 7, 30, and 60 d.

Ensiling Days	Mixture Rations	ACE	Chao1	Simpson	Shannon	Coverage
7	S0 †	193.89 ^{ab}	204.58 ^{a}	0.78 ^{abc}	2.95 ^{d}	0.9998
	S1	195.23 ^{ab}	202.68 ^{a}	0.78 ^{abc}	3.30 ^{cd}	0.9998
	S2	205.91 ^{a}	219.40 ^{a}	0.75 ^{bcd}	3.34 ^{bcd}	0.9998
	S3	154.62 ^{ab}	163.00 ^{ab}	0.75 ^{bcd}	3.25 ^{d}	0.9999
	S4	133.30 ^{b}	137.25 ^{b}	0.68 ^{d}	3.09 ^{d}	0.9999
	S5	197.71 ^{ab}	203.94 ^{a}	0.73 ^{cd}	3.38 ^{bcd}	0.9999

Table 5. *Cont.*

Ensiling Days	Mixture Rations	ACE	Chao1	Simpson	Shannon	Coverage
7	S6	171.89 ^{ab}	175.29 ^{ab}	0.72 ^{cd}	3.34 ^{bcd}	0.9998
	S7	188.47 ^{ab}	190.35 ^{ab}	0.80 ^{abc}	3.77 ^{ab}	0.9999
	S8	169.81 ^{ab}	177.46 ^{ab}	0.84 ^{a}	3.86 ^{a}	0.9998
	S9	152.70 ^{ab}	156.42 ^{ab}	0.84 ^{a}	3.75 ^{ab}	0.9999
	S10	187.04 ^{ab}	200.50 ^{a}	0.83 ^{ab}	3.73 ^{abc}	0.9998
p-Value		0.2424	0.1208	0.0034	0.0009	0.1526
30	S0	126.04 ^{c}	133.30 ^{d}	0.67 ^{c}	2.54 ^{d}	0.9998
	S1	129.69 ^{c}	133.56 ^{d}	0.63 ^{c}	2.55 ^{d}	0.9999
	S2	154.05 ^{bc}	158.66 ^{bcd}	0.60 ^{c}	2.67 ^{d}	0.9998
	S3	151.74 ^{bc}	153.76 ^{cd}	0.62 ^{c}	2.76 ^{cd}	0.9999
	S4	171.84 ^{ab}	176.57 ^{abcd}	0.69 ^{bc}	3.23 ^{bc}	0.9998
	S5	191.17 ^{ab}	195.10 ^{abc}	0.77 ^{ab}	3.67 ^{ab}	0.9998
	S6	187.60 ^{ab}	195.98 ^{abc}	0.78 ^{ab}	3.69 ^{ab}	0.9998
	S7	197.78 ^{a}	202.49 ^{ab}	0.85 ^{a}	4.12 ^{a}	0.9997
	S8	205.04 ^{a}	216.09 ^{a}	0.85 ^{a}	4.07 ^{a}	0.9998
	S9	179.04 ^{ab}	189.14 ^{abc}	0.85 ^{a}	4.03 ^{a}	0.9998
	S10	154.35 ^{bc}	157.39 ^{bcd}	0.69 ^{bc}	2.81 ^{cd}	0.9998
p-Value		0.0018	0.0021	<0.0001	<0.0001	0.1633
60	S0	178.94 ^{b}	184.60 ^{b}	0.78 ^{c}	3.32 ^{c}	0.9998
	S1	186.81 ^{b}	196.33 ^{b}	0.75 ^{c}	3.42 ^{c}	0.9998
	S2	204.11 ^{b}	207.90 ^{b}	0.91 ^{ab}	4.73 ^{b}	0.9999
	S3	211.96 ^{b}	215.75 ^{b}	0.93 ^{ab}	4.89 ^{ab}	0.9998
	S4	234.16 ^{b}	234.45 ^{b}	0.95 ^{a}	5.37 ^{a}	0.9999
	S5	178.18 ^{b}	182.56 ^{b}	0.94 ^{a}	5.05 ^{ab}	0.9999
	S6	202.07 ^{b}	202.96 ^{b}	0.93 ^{ab}	5.03 ^{ab}	0.9999
	S7	190.13 ^{b}	194.28 ^{b}	0.93 ^{ab}	4.98 ^{ab}	0.9999
	S8	209.27 ^{b}	212.53 ^{b}	0.94 ^{a}	5.18 ^{ab}	0.9999
	S9	318.48 ^{b}	322.32 ^{b}	0.91 ^{ab}	4.92 ^{ab}	0.9999
	S10	667.09 ^{a}	677.17 ^{a}	0.89 ^{b}	4.91 ^{ab}	0.9999
p-Value		<0.0001	<0.0001	<0.0001	<0.0001	0.3009

† Napier grass was mixed thoroughly without sugarcane top (Control, S0) or with 10% (S1), 20% (S2), 30% (S3), 40% (S4), 50% (S5), 60% (S6), 70% (S7), 80% (S8), 90% (S9), and 100% (S10) sugarcane top based on fresh weight, respectively. Values with different small letter superscripts mean significant difference ($p < 0.05$), while those with the same or no letter superscripts mean no significant difference ($p > 0.05$).

The beta diversity of the bacterial community was determined using principal coordinate analysis between various NG and ST mixed silages on days 7, 30, and 60 (Figure 1). The results showed some differences in the composition of silage microorganisms between the samples collected on different ED and various NG and ST MRs, indicating significant differences in bacterial composition among the samples collected on different fermentation days.

The relative abundances of microorganisms in the various silages under different ensiling times at the order and genus levels are shown in Figures 2 and 3, respectively. At the order level, *Lactobacillales* and *Enterobacterales* were the most dominant in terms of relative abundance in silages stored for 7 d (Figure 2a), 30 d (Figure 2b), and 60 d (Figure 2c). At the genus level, the top two genera in terms of relative abundance in the silages stored at 7 d (Figure 3a), 30 d (Figure 3b), and 60 d (Figure 3c) were *Lactiplantibacillus* and *Weissella*. Furthermore, the relative abundances of *Lactobacillales* and *Lactiplantibacillus* in silages stored for 30 d were higher than in those stored for 7 or 60 d.

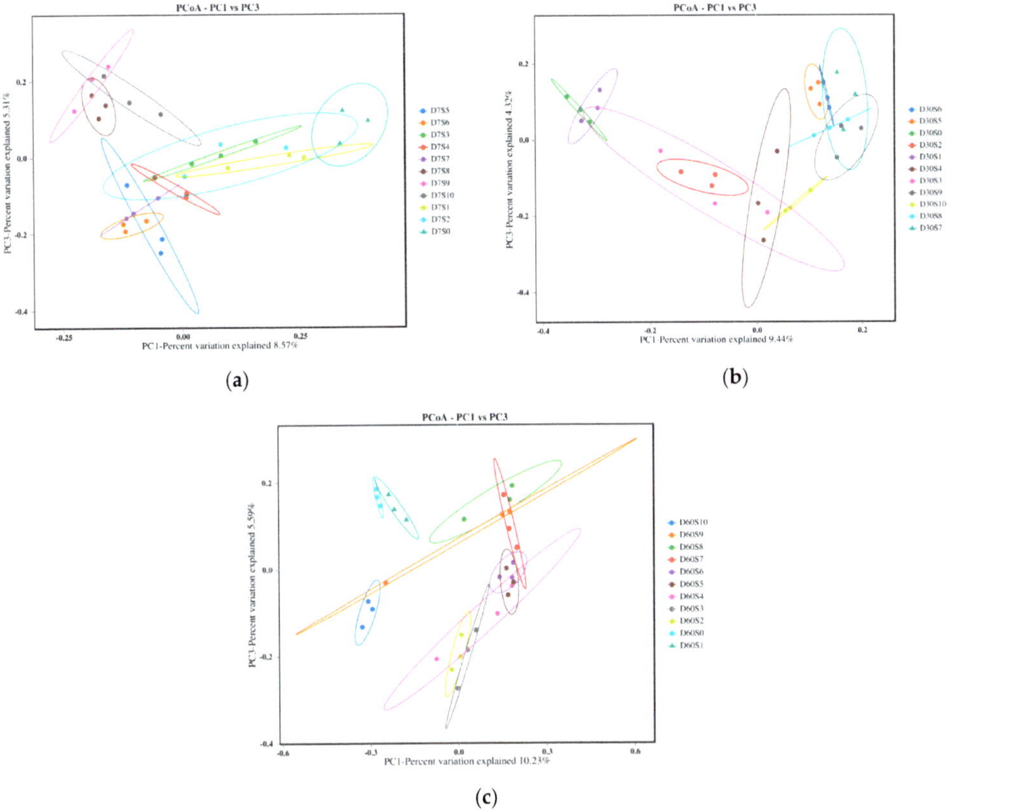

Figure 1. Principal coordinates analysis of the bacterial community in the silage under different ensiling times. (**a**) Silage stored at 7 d; (**b**) silage stored at 30 d; (**c**) silage stored at 60 d.

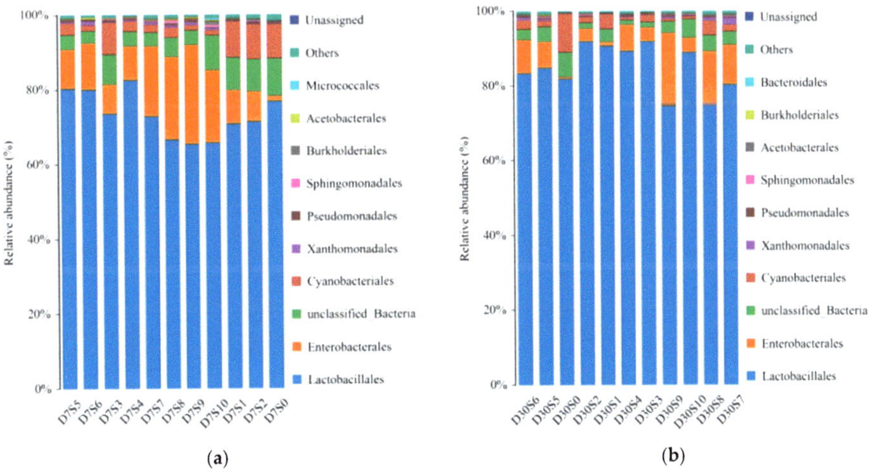

Figure 2. *Cont.*

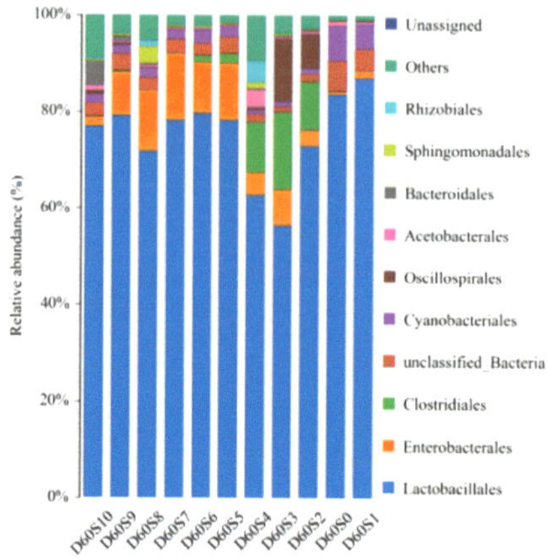

(c)

Figure 2. Bacterial changes at order level of silage under different ensiling times. (**a**) Silage stored at 7 d; (**b**) silage stored at 30 d; (**c**) silage stored at 60 d.

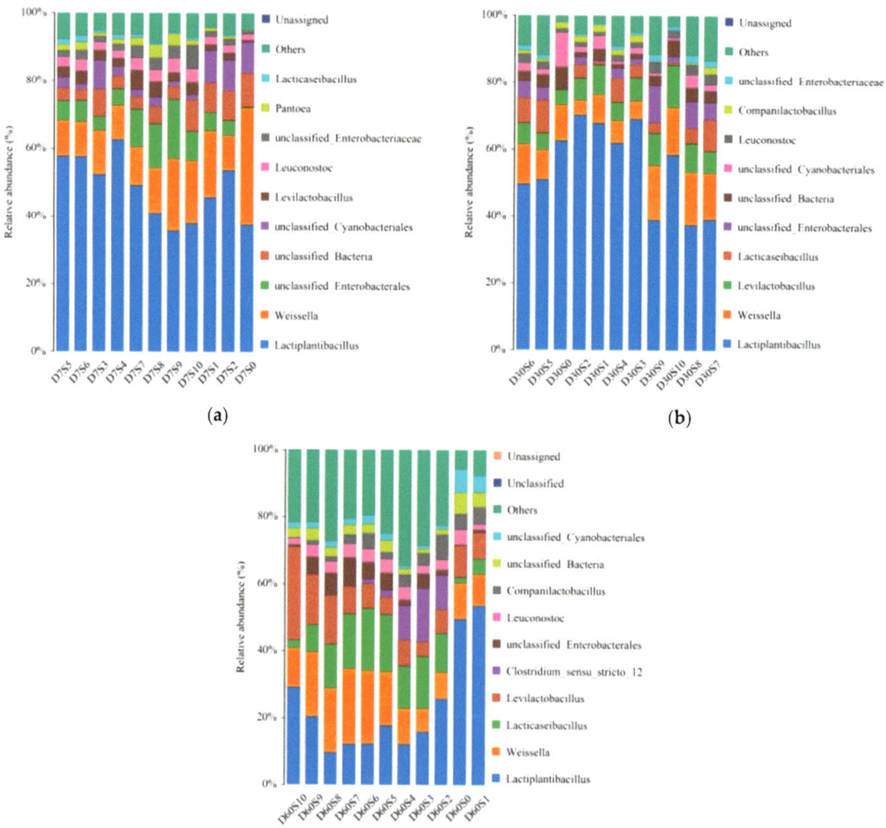

(c)

Figure 3. Bacterial changes at genus level of silage under different ensiling times. (**a**) Silage stored at 7 d; (**b**) silage stored at 30 d; (**c**) silage stored at 60 d.

3.5. Association among Silage Bacteria, Chemical Composition, and Fermentation Parameters

The correlation between the relative abundance of bacterial genera and chemical composition is shown in Figure 4. The CP content was negatively correlated ($p < 0.05$) with *Lentilactobacillus* and *Levilactobacillus* but positively correlated ($p < 0.05$) with *Companilactobacillus, Lactiplantibacillus*, and *unclassified_Cyanobacteriales*. The ash content was negatively correlated ($p < 0.05$) with *Lentilactobacillus, Levilactobacillus*, and *Weissella* but positively correlated ($p < 0.05$) with *Companilactobacillus*. The DM content was positively correlated ($p < 0.05$) with *Lentilactobacillus, Levilactobacillus*, and *Weissella* but negatively correlated ($p < 0.05$) with *Companilactobacillus* and *Lactiplantibacillus*. The WSC content was positively correlated ($p < 0.05$) with *Lentilactobacillus* and *Levilactobacillus* but negatively correlated ($p < 0.05$) with *Companilactobacillus, Lactiplantibacillus*, and *unclassified_Cyanobacteriales*. The ADF content was positively correlated ($p < 0.05$) with *Lentilactobacillus* and *Levilactobacillus* but negatively correlated ($p < 0.05$) with *Companilactobacillus*. Finally, the EE and NDF contents were positively correlated ($p < 0.05$) with *Lentilactobacillus* but negatively correlated ($p < 0.05$) with *Companilactobacillus, Lactiplantibacillus*, and *unclassified_Cyanobacteriales*.

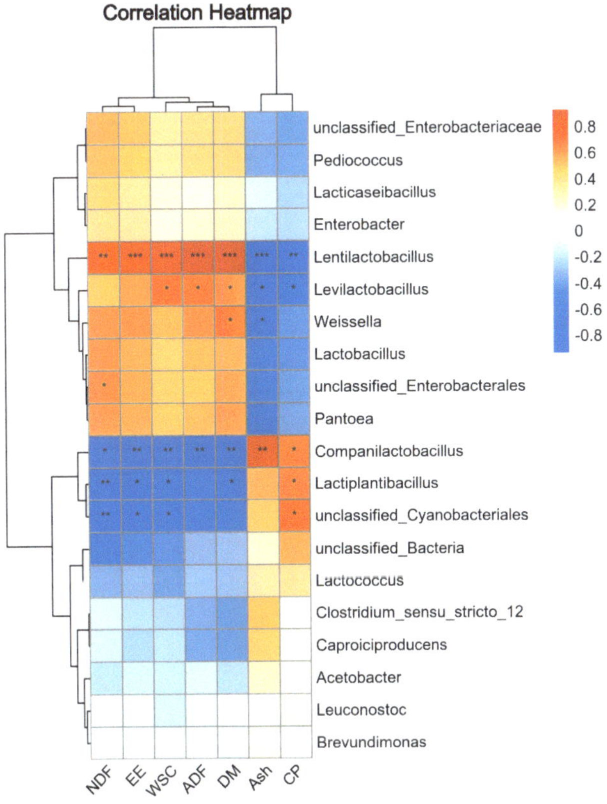

Figure 4. Correlation of bacterial genera with chemical composition of silage stored at 60 d. Note: the change in defined color and its depth indicates the nature and strength of the correlation, respectively. * indicates $0.01 < p \leq 0.05$, ** indicates $0.001 < p \leq 0.01$, *** indicates $p \leq 0.001$.

The correlation between the relative abundance of bacterial genera and fermentation parameters is shown in Figure 5. The pH was negatively correlated ($p < 0.05$) with *Lentilactobacillus* and *Levilactobacillus* but positively correlated ($p < 0.05$) with *Companilactobacillus*. The AA content was negatively correlated ($p < 0.05$) with *Lentilactobacillus, Weissella*, and *Levilactobacillus* but positively correlated ($p < 0.05$) with *Companilactobacillus* and *Lactiplan-*

tibacillus. The NH_3-N content was negatively correlated ($p < 0.05$) with *Lentilactobacillus*, *Weissella*, *unclassified_Enterobacterales*, and *pantoea* but positively correlated ($p < 0.05$) with *Companilactobacillus*, *Lactiplantibacillus*, and *unclassified_Cyanobacteriales*. LA content was positively correlated ($p < 0.05$) with *Lentilactobacillus*, *Weissella*, and *Levilactobacillus* but negatively correlated ($p < 0.05$) with *Companilactobacillus*.

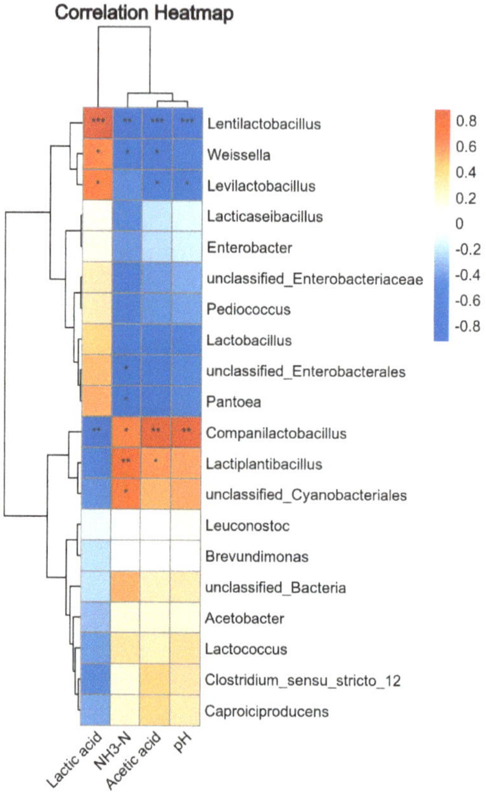

Figure 5. Correlation of bacterial genera with fermentation parameters of silage stored at 60 d. Note: the change in defined color and its depth indicates the nature and strength of the correlation, respectively. * indicates $0.01 < p \leq 0.05$, ** indicates $0.001 < p \leq 0.01$, *** indicates $p \leq 0.001$.

4. Discussion

Essential conditions to produce high-quality silage encompass ensuring that the raw materials have an optimal moisture content and soluble sugar concentration sufficient to meet the requirements for effective substrate fermentation [37,38]. Before ensiling, ST was rich in DM, EE, NDF, ADF, and WSC, indicating that it could be utilized as a feed ingredient for cattle and sheep. In contrast, NG had relatively low DM (16.71%) and WSC (3.46% DM), making high-quality silage production difficult when ensiling NG alone compared to ST. Therefore, inoculants such as LAB, acid agents, molasses, or other beneficial raw materials are commonly used to inhibit the growth of harmful microorganisms and to improve the fermentation quality of ensiled NG [39–42]. ST have high DM and WSC contents (26.11% and 11.66%, respectively), are likely to enhance the dry matter and sugar content, inhibit the proliferation of detrimental microorganisms, and preserve protein integrity in NG silage, ultimately resulting in the production of high-quality silage. We also evaluated the chemical composition of mixed silage and found that DM, EE, NDF, ADF, and WSC contents increased as the proportion of ST in the silage mixture increased, whereas CP and ash contents decreased. This is likely due to the higher DM, EE, NDF, ADF, and WSC and lower CP and ash contents of ST compared to NG, indicating that co-ensiling NG with ST compensates for certain nutritional deficiencies in NG and provides a more balanced diet.

pH is an important parameter for evaluating silage fermentation quality [43]. During the silage process, microorganisms, especially LAB, produce large amounts of organic acids, including LA, from WSC in the substrate. This rapid decrease in pH inhibits poor fermentation and improves silage quality [44]. A steady reduction in pH and stabilization of LA content occurs after storage. Low pH ensures better aerobic stability and prevents forage from further fermentation. In the present study, the pH and NH_3-N content decreased, and the LA content increased as the proportion of ST increased in the mixed silage. The addition of ST likely regulated NG's moisture content and increased sugar and fermentation substrates, resulting in complete fermentation. Therefore, silages with added ST greatly improved LA fermentation. In this study, BA formation was observed in NG silage alone, probably because of AA, ethanol, and CO_2 production or clostridial fermentation [45].

LAB play an important role in silage fermentation, and their relative abundance is used to evaluate their effect on silage quality [46,47]. Generally, silage can be well preserved when the LAB count reaches $\geq 10^5$ cfu/g of FM [48]. In this study, the LAB count reached 10^5 cfu/g of FM in all silages, except for S0, which was NG ensiled alone. Other factors involved in fermentation quality include not only the physiological properties of epiphytic bacteria but also the chemical composition, especially WSC, of the ensiling forage material [49]. Studies have revealed that WSC levels $\geq$ 5% relative to DM are crucial for ensuring acceptable fermentation quality [50,51]. In this study, adding ST at a 60% ratio increased the WSC content to over 5% DM (Table 2), indicating that this proportion sufficiently supplements the substrate for microbial fermentation. Ensiling NG with more than 60% ST ensured better fermentation quality and nutrient preservation. ST, with its high WSC content, may also meet the energy demands of LAB during silage fermentation [52]. Yeast was detected in all silages, probably because raw silage material naturally contains many yeast species. Yeast is acid-resistant and can coexist with LAB in silage.

A detailed understanding of the microbial populations during ensiling could help researchers uncover new ways to improve silage preservation. Therefore, it is important to explore the dynamic changes occurring in epiphytic microbiota during ensiling, as understanding the dynamic changes, interactions, and metabolic pathways of the microbial community during ensiling can provide a theoretical basis for effectively regulating silage fermentation [53]. Furthermore, the silage microbiome may play a key role in the detoxification of plant-derived toxic metabolites [54]. However, to our knowledge, this is the first report of a bacterial community in ensiled NG silage combined with ST. The sequencing coverage for each silage was greater than 0.99, reflecting the true microbial species composition with good representativeness [55]. On day 7, ACE, Chao 1, and Simpson indices were found in S4, and the highest Shannon index was found in S8, possibly due to the presence of too many microbial species in the initial aerobic phase of fermentation, resulting in an unpredictable variation pattern of the alpha diversity parameters [56]. By day 30, when NG was co-ensiled with 50% ST, the ACE, Chao 1, Simpson, and Shannon indices were higher compared to NG silage without or with low-level ST. This may be because ST exhibited higher microbial diversity than NG before ensiling. Principal coordinate analysis results showed some differences in the composition of silage microorganisms in various NG and ST mixed silages, indicating significant differences in bacterial composition among silages stored on different fermentation days. The dominant microorganisms in the silages were Lactobacillales and *Lactiplantibacillus* at the order and genus levels, respectively. The relative abundances of *Lactobacillales* and *Lactiplantibacillus* at 30 d were higher than those at 7 or 60 d. From day 7, under natural fermentation conditions, *Lactobacillus* in Lactobacillales became dominant in all silages, achieving good quality after 30 days of ensiling. This underscores the significance of *Lactobacillus* in NG and ST mixed silage. Future work could focus on isolating and cultivating *Lactobacillus* from naturally fermented silage to further improve silage quality [57,58].

Increasing evidence indicates that changes in bacterial communities and their abundance affect fermentation characteristics during ensiling [59]. In the present study, CP, ash, pH, AA, and NH_3-N were positively correlated with *Lentilactobacillu*. However, DM, EE, ADF, NDF, WSC, and LA were positively correlated with *Lentilactobacillus*. This suggests that chemical composition and fermentation parameters were simultaneously affected by these factors and that enhancing their abundance improved silage quality.

5. Conclusions

Our findings highlight the importance of ST in enhancing the fermentability of NG and the potential of using local by-products as additives for NG silage. Co-ensiling NG with ST increased the diversity and richness of silage microorganisms. Using ST, a local by-product, as an additive in NG silage could be an effective approach to improve the quality and nutritional value of NG silage. This study suggests that good-quality silages can be produced with NG:ST ratios of 40:60 to 20:80 and that these silages offer an opportunity to optimize the nutrient supply for ruminants.

Author Contributions: Conceptualization, C.Y. and H.X.; methodology, H.X. and Y.L.; software, H.X., X.L. and J.L.; validation, C.Y., H.X. and Y.G.; formal analysis, F.Z. and Y.P.; investigation, H.X. and Y.G.; resources, Z.L. and L.P.; data curation, X.L. and L.L.; writing—original draft preparation, H.X.; writing—review and editing, H.X., L.L. and C.Y.; supervision, C.Y.; project administration, H.X.; funding acquisition, H.X. and C.Y. All authors have read and agreed to the published version of the manuscript.

Funding: This research was funded by the Guangxi Natural Science Foundation (2023GXNSFBA026301, 2024GXNSFAA010387), the Guangxi Science and Technology Major Project (GuiKe AA22068099-8), and the National Modern Agricultural Industry Technology System Guangxi Dairy Buffalo Innovation Team Project (nycytxgxcxtd-2021-21-03).

Institutional Review Board Statement: Not applicable.

Informed Consent Statement: Not applicable.

Data Availability Statement: Dataset available on request from the authors.

Conflicts of Interest: The authors declare no conflicts of interest.

References

1. Thorat, B.S.; Pawar, G.R.; Sushir, K.V.; Talekar, S.D.; Repale, J.M.; Kadlag, A.D. Sugarcane: A Climate Resilient Devine Crop. *Int. J. Environ. Clim. Chang.* **2024**, *14*, 555–577. [CrossRef]
2. Guo, X.; Xu, D.; Li, F.; Bai, J.; Su, R. Current approaches on the roles of lactic acid bacteria in crop silage. *Microb. Biotechnol.* **2022**, *16*, 67–87. [CrossRef]
3. Daniel, J.L.P.; Jacovaci, F.A.; Junges, D.; Santos, M.C.; Lima, J.R.; Anjos, I.A.; Landell, M.G.A.; Huhtanen, P.; Nussio, L.G. Fibre digestibility and its relationships with chemical and morphological traits in thirty-two sugarcane varieties. *Grass Forage Sci.* **2016**, *72*, 545–555. [CrossRef]
4. National Bureau of Statistics of China. *NBS: China Statistical Yearbook, 12th, ed.*; China Statistics Press: Beijing, China, 2023.
5. Li, M.; Zi, X.; Yang, H.; Ji, F.; Tang, J.; Lv, R.; Zhou, H. Effects of king grass and sugarcane top in the absence or presence of exogenous enzymes on the growth performance and rumen microbiota diversity of goats. *Trop. Anim. Health Prod.* **2021**, *53*, 106. [CrossRef] [PubMed]
6. Santos, F.; Eichler, P.; Machado, G.; De Mattia, J.; De Souza, G. By-Products of the Sugarcane Industry. In *Sugarcane Biorefinery, Technology and Perspectives*; Elsevier: Amsterdam, The Netherlands, 2020; pp. 21–48.
7. Salinas-Chavira, J.; Almaguer, L.J.; Aguilera-Aceves, C.E.; Zinn, R.A.; Mellado, M.; Ruiz-Barrera, O. Effect of substitution of sorghum stover with sugarcane top silage on ruminal dry matter degradability of diets and growth performance of feedlot hair lambs. *Small Rumin. Res.* **2013**, *112*, 73–77. [CrossRef]
8. Xie, H.; Peng, L.; Li, M.; Guo, Y.; Liang, X.; Peng, K.; Yang, C. Effects of mixed sugarcane tops and napiergrass silages on fermentative quality, nutritional value, and milk yield in water buffaloes. *Anim. Sci. J.* **2023**, *94*, e13824. [CrossRef]
9. Muraro, G.B.; de Almeida Carvalho-Estrada, P.; de Oliveira Pasetti, M.H.; Santos, M.C.; Nussio, L.G. Bacterial dynamics of sugarcane silage in the tropics. *Environ. Microbiol.* **2021**, *23*, 5979–5991. [CrossRef]
10. Islam, M.R.; Garcia, S.C.; Islam, M.A.; Bashar, M.K.; Roy, A.; Roy, B.K.; Sarker, N.R.; Clark, C.E.F. Ruminant Production from Napier Grass (*Pennisetum purpureum* Schum): A Review. *Animals* **2024**, *14*, 467. [CrossRef]

11. Nazar, M.; Wang, S.; Zhao, J.; Dong, Z.; Li, J.; Ali Kaka, N.; Shao, T. Effects of various epiphytic microbiota inoculation on the fermentation quality and microbial community dynamics during the ensiling of sterile Napier grass. *J. Appl. Microbiol.* **2021**, *130*, 1466–1480. [CrossRef]
12. Zhao, G.; Wu, H.; Li, L.; He, J.; Hu, Z.; Yang, X.; Xie, X. Effects of applying cellulase and starch on the fermentation characteristics and microbial communities of Napier grass (*Pennisetum purpureum* Schum.) silage. *J. Anim. Sci. Technol.* **2021**, *63*, 1301–1313. [CrossRef]
13. Li, J.; Tang, X.; Zhao, J.; Chen, S.; Wang, S.; Shao, T. Improvement of fermentation quality and cellulose convertibility of Napier grass silage by inoculation of cellulolytic bacteria from Tibetan yak (*Bos grunniens*). *J. Appl. Microbiol.* **2021**, *130*, 1857–1867. [CrossRef] [PubMed]
14. Gonçalves, T.R.; da Silva Filho, J.R.V.; Pereira, G.A.; Meira, T.M.; de Sá, M.K.N.; Gois, G.C.; de Souza Silva, C.; Campos, F.S.; Yamamoto, S.M.; de Araújo, G.G.L. Potential use of vitiviniculture waste in mixed cactus pear silages with elephant grass in lamb diet. *Anim. Feed. Sci. Technol.* **2024**, *307*, 115824. [CrossRef]
15. Chen, D.; Zheng, M.; Zhou, Y.; Gao, L.; Zhou, W.; Wang, M.; Zhu, Y.; Xu, W. Improving the quality of Napier grass silage with pyroligneous acid: Fermentation, aerobic stability, and microbial communities. *Front. Microbiol.* **2022**, *13*, 1034198. [CrossRef]
16. Sánchez-Guerra, N.A.; Gonzalez-Ronquillo, M.; Anderson, R.C.; Hume, M.E.; Ruiz-Albarrán, M.; Bautista-Martínez, Y.; Zúñiga-Serrano, A.; Nájera-Pedraza, O.G.; Salinas-Chavira, J. Improvements in fermentation and nutritive quality of elephant grass [*Cenchrus purpureus* (Schumach.) Morrone] silages: A review. *Trop. Anim. Health Prod.* **2024**, *56*, 1–16. [CrossRef]
17. Li, H.; Ran, Q.; Jia, Z.; Shuai, Y.; Zhou, Q.; Guan, H. Effect of different dry matter content on fermentation characteristics and nutritional quality of Napier grass silage with novel lactic acid bacteria strains. *Lett. Appl. Microbiol.* **2023**, *76*, ovad018. [CrossRef] [PubMed]
18. Jaipolsaen, N.; Sangsritavong, S.; Uengwetwanit, T.; Angthong, P.; Plengvidhya, V.; Rungrassamee, W.; Yammuenart, S. Comparison of the Effects of Microbial Inoculants on Fermentation Quality and Microbiota in Napier Grass (*Pennisetum purpureum*) and Corn (*Zea mays* L.) Silage. *Front. Microbiol.* **2022**, *12*, 784535. [CrossRef]
19. Mitiku, A.A.; Andeta, A.F.; Borremans, A.; Lievens, B.; Bossaert, S.; Crauwels, S.; Aernouts, B.; Kechero, Y.; Van Campenhout, L. Silage making of maize stover and banana pseudostem under South Ethiopian conditions: Evolution of pH, dry matter and microbiological profile. *Microb. Biotechnol.* **2020**, *13*, 1477–1488. [CrossRef]
20. Bai, J.; Ding, Z.; Ke, W.; Xu, D.; Wang, M.; Huang, W.; Zhang, Y.; Liu, F.; Guo, X. Different lactic acid bacteria and their combinations regulated the fermentation process of ensiled alfalfa: Ensiling characteristics, dynamics of bacterial community and their functional shifts. *Microb. Biotechnol.* **2021**, *14*, 1171–1182. [CrossRef]
21. Zhang, S.J.; Chaudhry, A.S.; Osman, A.; Shi, C.Q.; Edwards, G.R.; Dewhurst, R.J.; Cheng, L. Associative effects of ensiling mixtures of sweet sorghum and alfalfa on nutritive value, fermentation and methane characteristics. *Anim. Feed Sci. Technol.* **2015**, *206*, 29–38. [CrossRef]
22. Ren, H.; Feng, Y.; Liu, T.; Li, J.; Wang, Z.; Fu, S.; Zheng, Y.; Peng, Z. Effects of different simulated seasonal temperatures on the fermentation characteristics and microbial community diversities of the maize straw and cabbage waste co-ensiling system. *Sci. Total Environ.* **2020**, *708*, 135113. [CrossRef]
23. Lee, Y.H.; Ahmadi, F.; Kim, Y.I.; Oh, Y.K.; Kwak, W.S. Co-ensiling garlic stalk with citrus pulp improves the fermentation quality and feed-nutritional value. *Asian-Australas. J. Anim. Sci.* **2020**, *33*, 436–445. [CrossRef] [PubMed]
24. Larsen, S.U.; Hjort-Gregersen, K.; Vazifehkhoran, A.H.; Triolo, J.M. Co-ensiling of straw with sugar beet leaves increases the methane yield from straw. *Bioresour. Technol.* **2017**, *245*, 106–115. [CrossRef] [PubMed]
25. Hillion, M.L.; Moscoviz, R.; Trably, E.; Leblanc, Y.; Bernet, N.; Torrijos, M.; Escudié, R. Co-ensiling as a new technique for long-term storage of agro-industrial waste with low sugar content prior to anaerobic digestion. *Waste Manag.* **2018**, *71*, 147–155. [CrossRef] [PubMed]
26. Du, G.; Zhang, G.; Shi, J.; Zhang, J.; Ma, Z.; Liu, X.; Yuan, C.; Li, X.; Zhang, B. Keystone *Taxa lactiplantibacillus* and *Lacticaseibacillus* Directly Improve the Ensiling Performance and Microflora Profile in Co-Ensiling Cabbage Byproduct and Rice Straw. *Microorganisms* **2021**, *9*, 1099. [CrossRef]
27. Cheng, Q.; Li, P.; Xiao, B.; Yang, F.; Li, D.; Ge, G.; Jia, Y.; Bai, S. Effects of LAB inoculant and cellulase on the fermentation quality and chemical composition of forage soybean silage prepared with corn stover. *Grassl. Sci.* **2021**, *67*, 83–90. [CrossRef]
28. Li, P.; Zhang, Y.; Gou, W.; Cheng, Q.; Bai, S.; Cai, Y. Silage fermentation and bacterial community of bur clover, annual ryegrass and their mixtures prepared with microbial inoculant and chemical additive. *Anim. Feed Sci. Technol.* **2019**, *247*, 285–293. [CrossRef]
29. Xu, D.; Wang, N.; Rinne, M.; Ke, W.; Weinberg, Z.G.; Da, M.; Bai, J.; Zhang, Y.; Li, F.; Guo, X. The bacterial community and metabolome dynamics and their interactions modulate fermentation process of whole crop corn silage prepared with or without inoculants. *Microb. Biotechnol.* **2020**, *14*, 561–576. [CrossRef]
30. AOAC: Official Methods of Analysis. *Association of Official Analytical Chemists*, 15th ed.; AOAC: Arlington, VA, USA, 1990.
31. Van Soest, P.J.; Robertson, J.B.; Lewis, B.A. Methods for Dietary Fiber, Neutral Detergent Fiber, and Nonstarch Polysaccharides in Relation to Animal Nutrition. *J. Dairy Sci.* **1991**, *74*, 3583–3597. [CrossRef]
32. Udén, P. In vitro studies on microbial efficiency from two cuts of ryegrass (Lolium perenne, cv. Aberdart) with different proportions of sugars and protein. *Anim. Feed. Sci. Technol.* **2006**, *126*, 145–156. [CrossRef]

33. Cai, Y. Analysis Method for Silage. In *Japanese Society of Grassland Science*; Field and Laboratory Methods for Grassland Science. Tosho Printing Co., Ltd.: Tokyo, Japan, 2004; pp. 279–282.
34. Xie, H.; Xie, F.; Guo, Y.; Liang, X.; Peng, L.; Li, M.; Tang, Z.; Peng, K.; Yang, C. Fermentation quality, nutritive value and in vitro ruminal digestion of Napier grass, sugarcane top and their mixed silages prepared using lactic acid bacteria and formic acid. *Grassl. Sci.* **2023**, *69*, 23–32. [CrossRef]
35. Broderick, G.A.; Kang, J.H. Automated Simultaneous Determination of Ammonia and Total Amino Acids in Ruminal Fluid and In Vitro Media. *J. Dairy Sci.* **1980**, *63*, 64–75. [CrossRef] [PubMed]
36. Cao, Y.; Cai, Y.; Hirakubo, T.; Fukui, H.; Matsuyama, H. Fermentation characteristics and microorganism composition of total mixed ration silage with local food by-products in different seasons. *Anim. Sci. J.* **2011**, *82*, 259–266. [CrossRef]
37. Borreani, G.I.O.R.G.I.O.; Tabacco, E.R.N.E.S.T.O.; Schmidt, R.J.; Holmes, B.J.; Muck, R.A. Factors affecting dry matter and quality losses in silages. *J. Dairy Sci.* **2018**, *101*, 3952–3979. [CrossRef]
38. Zhang, L.; Li, X.; Wang, S.; Zhao, J.; Dong, Z.; Zhao, Q.; Xu, Y.; Pan, X.; Shao, T. Effect of Sorbic Acid, Ethanol, Molasses, Previously Fermented Juice and Combined Additives on Ensiling Characteristics and Nutritive Value of Napiergrass (*Pennisetum purpureum*) Silage. *Fermentation* **2022**, *8*, 528. [CrossRef]
39. Tao, X.; Chen, S.; Zhao, J.; Wang, S.; Li, J.; Sun, F.; Shao, T. Fermentation and aerobic stability of Napier grass silage treated with different levels of citric acid residue. *Grassl. Sci.* **2021**, *67*, 139–147. [CrossRef]
40. Cai, Y.; Du, Z.; Yamasaki, S.; Nguluve, D.; Tinga, B.; Macome, F.; Oya, T. Influence of microbial additive on microbial populations, ensiling characteristics, and spoilage loss of delayed sealing silage of Napier grass. *Asian-Australas J Anim Sci.* **2020**, *33*, 1103–1112. [CrossRef]
41. Mugabe, W.; Shao, T.; Li, J.; Dong, Z.; Yuan, X. Effect of hexanoic acid, Lactobacillus plantarum and their combination on the aerobic stability of napier grass silage. *J. Appl. Microbiol.* **2020**, *129*, 823–831. [CrossRef]
42. Bureenok, S.; Yuangklang, C.; Vasupen, K.; Schonewille, J.T.; Kawamoto, Y. The effects of additives in napier grass silages on chemical composition, feed intake, nutrient digestibility and rumen fermentation. *Asian-Australas. J. Anim. Sci.* **2012**, *25*, 1248–1254. [CrossRef]
43. Wang, C.; He, L.; Xing, Y.; Zhou, W.; Yang, F.; Chen, X.; Zhang, Q. Effects of mixing Neolamarckia cadamba leaves on fermentation quality, microbial community of high moisture alfalfa and stylo silage. *Microb. Biotechnol.* **2019**, *12*, 869–878. [CrossRef]
44. Zhang, J.; Liu, Y.; Wang, Z.; Bao, J.; Zhao, M.; Si, Q.; Sun, P.; Ge, G.; Jia, Y. Effects of Different Types of LAB on Dynamic Fermentation Quality and Microbial Community of Native Grass Silage during Anaerobic Fermentation and Aerobic Exposure. *Microorganisms* **2023**, *11*, 513. [CrossRef]
45. Tatlı Seven, P.; Yıldırım, E.N.; Seven, İ.; Kaya, C.A.; İflazoğlu Mutlu, S. An evaluation of the effectiveness of sumac and molasses as additives for alfalfa silage: Influence on nutrient composition, in vitro degradability and fermentation quality. *J. Anim. Physiol. Anim. Nutr.* **2024**, *108*, 1096–1106. [CrossRef] [PubMed]
46. Kim, D.H.; Lee, K.D.; Choi, K.C. Role of LAB in silage fermentation: Effect on nutritional quality and organic acid production—An overview. *AIMS Agric. Food* **2021**, *6*, 216–234. [CrossRef]
47. Broberg, A.; Jacobsson, K.; Ström, K.; Schnürer, J. Metabolite profiles of lactic acid bacteria in grass silage. *Appl. Environ. Microbiol.* **2007**, *73*, 5547–5552. [CrossRef]
48. Wang, L.; Guan, L.; Fang, J.; Cai, Y.; Cao, Y. Fermentation characteristics and in vitro ruminal digestion of yacon residue silage with lactic acid bacteria inoculant or beet pulp. *Rev. Bras. Zootec.* **2019**, *48*, e20180152. [CrossRef]
49. Cao, Y.; Cai, Y.; Takahashi, T.; Yoshida, N.; Tohno, M.; Uegaki, R.; Nonaka, K.; Terada, F. Effect of lactic acid bacteria inoculant and beet pulp addition on fermentation characteristics and in vitro ruminal digestion of vegetable residue silage. *J. Dairy Sci.* **2011**, *94*, 3902–3912. [CrossRef]
50. Ni, K.; Zhao, J.; Zhu, B.; Su, R.; Pan, Y.; Ma, J.; Zhou, G.; Tao, Y.; Liu, X.; Zhong, J. Assessing the fermentation quality and microbial community of the mixed silage of forage soybean with crop corn or sorghum. *Bioresour. Technol.* **2018**, *265*, 563–567. [CrossRef]
51. Wu, P.; Li, L.; Jiang, J.; Sun, Y.; Yuan, Z.; Feng, X.; Guo, Y. Effects of fermentative and non-fermentative additives on silage quality and anaerobic digestion performance of Pennisetum purpureum. *Bioresour. Technol.* **2020**, *297*, 122425. [CrossRef]
52. de Moura Zanine, A.; de Sene, O.A.; de Jesus Ferreira, D.; Parente, H.N.; de Oliveira Maia Parente, M.; Pinho, R.M.A.; Santos, E.M.; Nascimento, T.V.C.; de Oliveira Lima, A.G.V.; Perazzo, A.F.; et al. Fermentative profile, losses and chemical composition of silage soybean genotypes amended with sugarcane levels. *Sci. Rep.* **2020**, *10*, 21064. [CrossRef] [PubMed]
53. Tang, X.; Liao, C.; Huang, X.; Chen, C.; Xu, D.; Chen, C. Epiphytic microbiota source stimulates the fermentation profile and bacterial community of alfalfa-corn mixed silage. *Front. Microbiol.* **2024**, *14*, 1247254. [CrossRef]
54. Gharechahi, J.; Kharazian, Z.A.; Sarikhan, S.; Jouzani, G.S.; Aghdasi, M.; Hosseini Salekdeh, G. The dynamics of the bacterial communities developed in maize silage. *Microb. Biotechnol.* **2017**, *10*, 1663–1676. [CrossRef]
55. Keshri, J.; Chen, Y.; Pinto, R.; Kroupitski, Y.; Weinberg, Z.G.; Sela Saldinger, S. Bacterial Dynamics of Wheat Silage. *Front. Microbiol.* **2019**, *10*, 1532. [CrossRef] [PubMed]
56. Mpanza, T.D.E.; Mani, S. Effects of Vachellia mearnsii Tannin Extract as an Additive on Fermentation Quality, Aerobic Stability, and Microbial Modulation of Maize Silage. *Microorganisms* **2023**, *11*, 2767. [CrossRef]
57. Margalho, L.P.; Jorge, G.P.; Noleto, D.A.; Silva, C.E.; Abreu, J.S.; Piran, M.V.; Brocchi, M.; Sant'Ana, A.S. Biopreservation and probiotic potential of a large set of lactic acid bacteria isolated from Brazilian artisanal cheeses: From screening to in product approach. *Microbiol. Res.* **2021**, *242*, 126622. [CrossRef] [PubMed]

58. Gonda, H.; Nikodinoska, I.; Le Cocq, K.; Moran, C.A. Efficacy of six lactic acid bacteria strains as silage inoculants in forages with different dry matter and water-soluble carbohydrate content. *Grass Forage Sci.* **2023**, *78*, 636–647. [CrossRef]
59. Li, D.; Xie, H.; Zeng, F.; Luo, X.; Peng, L.; Sun, X.; Wang, X.; Yang, C. An Assessment on the Fermentation Quality and Bacterial Community of Corn Straw Silagewith Pineapple Residue. *Fermentation* **2024**, *10*, 242. [CrossRef]

fermentation

MDPI

Article

The Production of Marandu Grass (*Urochloa brizantha*) Extracts as a Natural Modifier of Rumen Fermentation Kinetics Using an In Vitro Technique

Rafaela Scalise Xavier de Freitas [1,*,†], Janaina Silveira da Silva [1,*,†], Althieres José Furtado [1], Flavio Perna Junior [2], Alessandra Lopes de Oliveira [3] and Ives Cláudio da Silva Bueno [1]

1 Department of Animal Science, College of Animal Science and Food Engineering, University of São Paulo/FZEA, Pirassununga CEP 13635-900, São Paulo, Brazil; althieresjf@usp.br (A.J.F.); ivesbueno@usp.br (I.C.d.S.B.)

2 Department of Nutrition and Animal Production, College of Veterinary Medicine and Animal Science, University of São Paulo/FMVZ, Pirassununga CEP 13635-900, São Paulo, Brazil; fpernajr@usp.br

3 Department of Food Engineering, College of Animal Science and Food Engineering, University of São Paulo/FZEA, Pirassununga CEP 13635-900, São Paulo, Brazil

* Correspondence: rafascalise@usp.br (R.S.X.d.F.); janaina.silveira.silva@usp.br (J.S.d.S.)

† These authors contributed equally to this work.

Abstract: The ethanolic (EE) and hydroalcoholic (HE) extracts of *Urochloa brizantha* concentrations were developed with the aim of evaluating their effect on rumen fermentation using an in vitro gas production technique. The EE and HE presented 3.62 and 5.38 mg protodioscin/mL, respectively. Ten treatments were evaluated in a completely randomized factorial arrangement ($2 \times 4 + 2$), where the main effects were two extracts (EE and HE) and four levels (50, 100, 150, and 200 mL of extract/kg of DM) plus two controls: one positive (25 ppm of monensin–MON) and one (with no additives–CTL). The extract treatments (EXT, EE, and HE) reduced colonization time by 33.59% compared to the MON. IVDMD ($p < 0.001$) and IVOMD ($p < 0.0001$) were negatively affected by EXT addition when compared to CTL. Additionally, EXT reduced the proportion of propionic acid and increased the proportion of butyric acid in relation to CTL and MON treatments. Both EE and HE extracts of *U. brizantha* were able to alter rumen fermentation kinetic, with HE showing a higher concentration of protodioscin. Further research is needed to optimize extraction methodologies, comprehensively profile secondary compounds, and conduct trials with varying doses to effectively assess the viability of *U. brizantha* extract as an additive.

Keywords: degradability; forage; gas production; plant extract; saponin

Citation: Freitas, R.S.X.d.; da Silva, J.S.; Furtado, A.J.; Perna Junior, F.; de Oliveira, A.L.; Bueno, I.C.d.S. The Production of Marandu Grass (*Urochloa brizantha*) Extracts as a Natural Modifier of Rumen Fermentation Kinetics Using an In Vitro Technique. *Fermentation* **2024**, *10*, 447. https://doi.org/10.3390/fermentation10090447

Academic Editor: Wansup Kwak

Received: 28 July 2024
Revised: 24 August 2024
Accepted: 26 August 2024
Published: 28 August 2024

1. Introduction

The rumen's unique evolutionary advantage, through its symbiosis with microorganisms, allows ruminants to utilize lignocellulosic material and convert non-protein nitrogen into microbial protein [1]. Animal nutritionists frequently employ rumen manipulation to enhance fibrous feed degradability and reduce nitrogen excretion by ruminants, thereby improving their performance [2]. Consequently, various nutritional alternatives have been employed, including ionophore antibiotics, lipids, probiotics, and plant extracts.

Among these, ionophore antibiotics, particularly monensin, have been the most widely used in ruminant production in recent years [3]. However, the international community is increasingly demanding environmental measures, especially concerning the use of additives in ruminant feed. This adds to the complexity of the issue, especially given the restrictions on synthetic drugs [4]. Therefore, plant extracts are particularly promising as an alternative to synthetic drugs due to their secondary compounds—such as saponins, tannins, flavonoids, and essential oils, which exhibit various biological activities [5].

These compounds can modulate ruminal microorganisms, improve animal performance, reduce energy loss, and mitigate GHG emissions [6,7]. Among them, saponins are notable bioactive compounds abundantly present in the plant kingdom [8]. They are classified into steroidal saponins, which are found in monocots, and triterpenoid saponins, which are found in eudicots [9]. However, the use of saponins from various sources as additives has led to inconsistencies in research results [10–13]. These inconsistencies may arise from the significant variability in both the type and concentration of extracts and the nature of the secondary metabolites present [14].

In addition, the most researched extracts have significant human applications, either for use in food or as herbal medicines, which can lead to competition and increased production costs [15]. However, the plant kingdom offers a vast array of species, providing numerous options [16,17], including forage species [18]. Forage is a crucial component of ruminant nutrition, with species such as Urochloa, which is economically important and contains saponins like protodioscin [19].

Although some studies have examined the secondary compounds in the *Urochloa* genus, their primary focus has been on identifying saponins responsible for allelopathic effects [20,21] and inducing photosensitization in ruminants [22,23]. Consequently, several significant gaps remain. These include the development of saponin-rich extracts, the choice of extraction method and solvent, the optimal concentrations of the extract, and the selection of forages with higher nutritional value to maximize the positive impact of the additive on animal nutrition.

Among the *Urochloa* species, *Urochloa brizantha* is the most widely distributed in Brazil due to its excellent digestibility and palatability [24]. We hypothesize that ethanolic and hydroalcoholic extracts of *Urochloa brizantha* containing saponins, when used as natural additives, can modulate the fermentation profile by affecting microbial activity in the rumen. Therefore, our study aimed to investigate the effects of concentrations of extracts from *U. brizantha* cv. Marandu on in vitro fermentation kinetics.

2. Materials and Methods

The study was approved by the ethics committee of the College of Animal Science and Food Engineering, University of São Paulo, Pirassununga, Brazil (Protocol number CEUA: 1344101121).

2.1. Inoculum Donors and Substrate

Ruminal fluid was collected through the rumen cannula from eight male Santa Inês x Dorper sheep (41.11 ± 8.08 kg) fed an experimental diet consisting of 70% corn silage and 30% concentrate, with free access to freshwater and mineral supplements. On the day of incubation, rumen content was collected before the morning feeding. The solid phase was collected manually from the dorsal sac through the cannula, squeezed, placed in plastic bags, and stored in heated boxes at 39 °C. The liquid phase was collected from the ventral sac using a stainless-steel probe (2.5 mm screen) attached to a 500 mL graduated pipette and stored in pre-warmed thermal bottles, which had been previously flushed with CO_2. Both fluids and solids were transported immediately to the laboratory.

Four inocula were prepared, each from two donors. The solid and liquid phases were homogenized in a 1:1 ratio in a blender for 10 s and filtered through two layers of cotton cloth following the method described by Bueno et al. [25]. The inocula was transferred to anaerobic jars, continuously saturated with CO_2, and maintained in a Daisy Incubator (Ankom Technology, Macedon, NY, USA) at 39 °C until use.

The diet was dried and ground using a Willey mill with a 1 mm sieve as a substrate for gas production bioassay. The substrate analysis revealed the following composition: 436.1 g of dry matter (DM; ID 930.15), 55.6 g of mineral matter (ash; ID 942.05), 114.9 g of crude protein (CP; ID 954.01), 29.1 g of ether extract (EE; ID 920.39), 438.6 g of neutral detergent fiber (NDF; ID 973.18), and 245.1 g of acid detergent fiber (ADF; ID 973.18), according to AOAC [26].

2.2. *Urochloa Brizantha* Extract Preparation

U. brizantha cv. Marandu, in a period of growth, was harvested from a pasture established in the experimental area of the Department of Animal Science at FZEA-USP, Pirassununga, São Paulo (21°59′ S latitude, 47°26′ W longitude, and 634 m altitude) The local climate, according to Koppen's classification, is of the Cwa type, characterized by an average annual temperature of over 22 °C, annual rainfall of 1.238 mm, and a relative humidity of 73%. The soil in the experimental area was classified as dystrophic red latosol, gentle to moderately undulating [27].

An initial cut was performed on 10 December 2020, leaving a residue height of 15 cm, followed by fertilization on 18 December to standardize growth between the paddocks. The plots were regularly monitored to assess vegetation growth and control invasive plants. Irrigation was not required due to abundant rainfall during this period. On 15 February 2021, a uniformization cut was performed on all plots to maintain consistent growing conditions. At this time, the plants were 65 days old and fully grown. Each plot was allocated 8 m^2. After the uniformization cut, maintenance fertilization was applied with nitrogen (N, 25 g of NPK/m^2) and phosphorus (200 g of P_2O_5/m^2) to enhance secondary metabolite production [28]. When the plants reached 30 cm in height, they were cut on March 26, 2021, using a tractor mower, leaving a residue height of 15 cm. The collected material was dried in a shaded greenhouse at a controlled temperature (below 40 °C to avoid inactivating secondary metabolites) for 14 days, then ground in a disintegrator (Nogueira, model DPM-4, São João da Boa Vista, SP, Brazil) with 2 mm mesh sieves.

To produce *U. brizantha* extract, the pressurized liquid extraction (PLE) technique on a pilot scale was utilized with 99.5% ethanol (pure) and a hydroalcoholic solution (70% ethanol). A sample of dried and crushed Marandu grass (±400 g) was placed in a 2 L cell. The solvent (ethanolic—99.5% of ethanol and hydroalcoholic—70% ethanol) was pumped into the cell, and the intermittent process began once temperature (80 °C) and pressure (10.35 MPa) were adjusted. Each rinse cycle used 480 mL of solvent. At the end of the process, the purge valve was opened, and nitrogen gas (N_2) was used to transfer all remaining extract from the cell to the collection vessel [29].

The concentrations of saponins in the extracts were determined and quantified as protodioscin by the high-performance liquid chromatographic (HPLC) method described by Lee et al. [30]. The saponin contents in the EE and HE extracts were 3.62 and 5.38 mg protodioscin/mL, respectively.

2.3. Treatments

Ten treatments were tested, involving two extracts: ethanolic (EE) and hydroalcoholic (HE); four levels: 50, 100, 150, and 200 mL of extract/kg of DM; and two controls: one positive (MON, 25 ppm of monensin) and one negative control (CTL, no additives). The four levels used corresponded to dosages of 0.37, 0.75, 1.12, and 1.50 mg of saponin (protodioscin; g/mL of extract).

2.4. In Vitro Gas Production Assay

To evaluate the effect of extracts on fermentation kinetics, the semi-automatic in vitro gas production technique described by Theodorou et al. [31], modified by Mauricio et al. [32], was used. A total of 132 glass bottles, each with a capacity of 160 mL, were prepared for the study, which included four inocula and eleven treatments (ten tested plus a blank), with each treatment tested in triplicates. In the negative control, a hydroalcoholic solution was used to ensure that any observed effects in the experimental treatments could be attributed specifically to the additives rather than the solvent. Additionally, a hydroalcoholic solution was employed to dilute the monensin.

In each bottle we added 1.0 g of substrate (70% silage and 30% concentrate) in filter bags (TNT, 100 g/m^2), 10 mL of rumen inoculum, and 90 mL of nutrient solution. The nutrient solution was composed of micromineral solutions ($CaCl_2 \cdot 2H_2O$; $MnCl_2 \cdot 2H_2O$; $CoCl_2 \cdot 6H_2O$ and $FeCl_3 \cdot 6H_2O$), macrominerals (Na_2HPO_4; KH_2PO_4 and $MgSO_4 \cdot 7H_2O$), buffer (NH_4CO_3 and $NaHCO_3$), reducing agent (cysteine-HCl; 1 M NaOH and $Na_2S \cdot 9H_2O$), and indicator (resazurin), prepared according to the method described by Menke et al. [33]. Blanks were used for each inoculum to measure the fraction of total gas production due to substrate in the inocula; these values were subtracted from the total to obtain net gas production.

The bottles were sealed with butyl rubber stoppers, shaken, and incubated in an oven with forced-air circulation at 39 °C for 96 h. All treatments were incubated simultaneously.

The headspace pressure was measured at 4, 8, 12, 16, 20, 24, 30, 36, 42, 48, 60, 72, and 96 h of incubation using a transducer (Pressure Press Data 800; Piracicaba, SP, Brazil). After each measurement, the accumulated gas was released from all bottles. The total gas volume produced in each bottle was estimated according to the equation V = (4.6788 ∗ P), where V = volume of gases (mL) and P = measured pressure (psi) [34]. This equation is validated for the in vitro experimental conditions for specific fermentation kinetic bioassays at the Ruminal Fermentability Laboratory. Cumulative gas production was determined by summing the gas volumes obtained from each pressure measurement after 96 h of fermentation.

Fermentation kinetics were analyzed using gas production data based on the following model by France et al. [35]:

$$Y = A\left\{1 - \exp\left[-b(t-L) - c\left(\sqrt{t} - \sqrt{L}\right)\right]\right\}$$

$$\mu = \frac{-(b+c)}{\left(2\sqrt{t}\right)}$$

where Y = cumulative production of gases (mL/g DM) in incubation time t (h), A = potential gas production (mL/g DM), L = colonization time (lag time), b and c are mathematical constants of the mode, T1/2 = time to half-asymptote (h), and μ = fractional rate of gas production (/h). The kinetic parameters (A, L, μ, and $T\frac{1}{2}$) were compared in the statistical analysis.

At the end of the 96-h incubation, the bottles were placed in ice containers to stop the fermentation process. Two mL samples of ruminal fluid from each bottle were collected to determine short-chain fatty acid (SCFA) and ammonia nitrogen (NH_3-N).

The filter bags for substrate incubation were washed in running water and dried in an oven with forced-air circulation (105 °C). The in vitro dry matter degradability (IVDMD) was determined by the difference between the weights before and after dry incubation in an oven at 105 °C to determine the residue. The in vitro organic matter degradability (IVOMD) was estimated by the difference in residues (ash) after burning in a muffle furnace at 550 °C for four hours. The partitioning factor (PF), calculated by relating dry matter degradation to total gas production (mL) over 96 h, was used to compare microbial efficiency [36].

2.5. Short-Chain Fatty Acid (SCFA) and Ammonia Nitrogen (NH_3-N) Determination

A short-chain fatty acid (SCFA) profile in ruminal fluid was analyzed using gas chromatography (GC–2014; Shimadzu) as described by Bueno et al. [37].

Ammonia nitrogen (NH_3-N) concentrations in the rumen fluid were determined by colorimetry, according to the methodology of Kulasek [38] and adapted by Foldager et al. [39].

2.6. *Experimental Design and Statistical Analysis*

The experimental design was completely randomized in a factorial arrangement. $2 \times 4 + 2$, with two types of *U. brizantha* extracts (ethanolic—99% and hydroalcoholic—70%), four concentrations (50, 100, 150, and 200 mL/kg of DM), plus two additional treatments, one positive control (monensin, Elanco Brasil, Rumensin®, 25 mg/kg DM), and one negative control (no additive), with three replicates within each experimental unit.

The normality of the residuals was assessed using the Shapiro–Wilk test (PROC UNIVARIATE). Statistical analysis was carried out using the GLIMMIX procedure (SAS 9.4; SAS Institute, Cary, NC, USA) using orthogonal contrasts of the effects of grouped treatments:

(1) Effect of monensin (MON) treatment against control (CTL);
(2) Effect of extracts (EXT, the grouping of EE and HE extracts in all doses) concerning the positive control (MON);
(3) Effect of extracts (EXT, the grouping of EE and HE extracts in all doses) concerning the negative control (CTL);
(4) Effect of type of extract (ethanolic, EE, and hydroalcoholic, HE) at different doses: 50, 100, 150, and 200 mL/kg DM);
(5) Level effect (LVL), comparing each level (50, 100, 150, and 200 mL/kg DM) against the other level of extracts;
(6) EXT vs. LVL interaction effect.

The significance was set at $p \leq 0.05$, and the p values obtained for the contrasts are reported.

3. Results

3.1. *In Vitro Gas Production Assay*

There was no significant difference ($p > 0.05$) between the treatments in maximum gas production potential (A), fractional rate of gas production (μ) at time to half-asymptote ($T\frac{1}{2}$), and $T\frac{1}{2}$ (Table 1). However, the colonization time (L, Lag time) was significant ($p < 0.01$) when comparing the EXT (6.13 h) and MON (9.29 h) treatments, representing a reduction of 33.59%, and there was also a difference ($p = 0.02$) between the EE (5.42 h) and EH (6.84 h) extracts.

IVDMD showed a level effect (NIV) ($p = 0.048$), indicating that increasing the level of inclusion of the extracts in the diet negatively impacted diet degradability. The CTL treatment (79.77%, $p < 0.001$) had the highest IVDMD, followed by MON (77.56%, $p = 0.03$) and EXT (73.94%). Among the extracts, HE (75.73%) had a higher IVDMD ($p < 0.001$) than EE (72.14%). There was also a significant effect ($p < 0.0001$) between EXT (72.69%) and CTL (79.38%) and between ($p < 0.01$) EE (70.82%) and HE (74.57%) extracts.

The partition factor (PF, mg DMD/mL) was significant ($p < 0.01$) in the contrast between MON (2.79 mg DMD/mL) and CTL (2.52 mg DMD/mL). It was also highly significant ($p < 0.001$) when comparing MON (2.79 mg DMD/mL) with EXT (2.40 mg DMD/mL).

The accumulated gas production (mL/g DM) curves after 96 h of fermentation for the treatments, the EE (Figure 1), and HE (Figure 2) extracts showed a similar fermentation pattern.

Table 1. Effect of ethanolic (EE) and hydroalcoholic (HE) extracts of Urochloa brizantha on kinetic parameters of gas production in vitro modeled according to France model, degradability and partitioning factor (FP).

Variables	EE				HE				MON	CTL	SEM	Contrast *p*-Valor					
	50	100	150	200	50	100	150	200				1	2	3	4	5	6
A (mL)	335.7	303.3	294.7	270.3	352.4	325.4	318.9	307.2	302.6	320.6	19.2	0.05	0.63	0.71	0.07	0.35	0.74
L (h)	4.27	6.35	5.32	5.74	7.31	7.49	6.01	6.55	9.29	7.50	1.22	0.14	*	0.10	0.02	0.53	0.94
$\mu_{T\frac{1}{2}}$ (/h)	0.023	0.024	0.022	0.027	0.025	0.032	0.023	0.021	0.027	0.026	0.004	0.87	0.54	0.65	0.53	0.63	0.22
$T\frac{1}{2}$ (h)	38.88	39.95	37.82	36.56	37.60	34.97	37.92	42.71	39.77	40.00	3.99	0.96	0.69	0.59	0.99	0.56	0.32
IVDMD (%)	78.54	71.15	70.12	67.76	80.62	76.70	74.29	71.32	77.56	79.77	1.29	0.27	0.03	**	**	0.048	0.81
IVOMD (%)	75.30	72.37	69.53	66.08	80.20	76.15	71.92	70.00	75.89	79.38	1.75	0.18	0.13	***	*	0.12	0.65
PF (mg DMD/mL)	2.45	2.41	2.44	2.42	2.41	2.39	2.29	2.40	2.79	2.52	0.09	0.01	***	0.11	0.25	0.57	0.38

EE: ethanolic extract; HE: hydroalcoholic extract; MON: 25 ppm monensin; CTL: control (no additive added). SEM: standard error of the mean. * $p < 0.01$; ** $p < 0.001$; *** $p < 0.0001$. France et al. [35]: p = a + b (1 − exp^{-ct}), where p = gas production (mL); in time t; a and b = constants of model; c = production gas rate (h^{-1}); a + b = potential gas production (mL); A: maximum gas production potential; *L*: colonization time (lag time); μ: fractional rate of gas production; $T\frac{1}{2}$: time to half-asymptote; IVDMD in vitro dry matter degradability; IVOMD: in vitro organic matter degradability; PF: partitioning factor. **Contrast:** 1: MON vs. CTL; 2: EXT vs. MON; 3: EXT vs. CTL; 4: EE vs. HE; 5: Level effect; 6: Effect of EXT vs. LVL interaction.

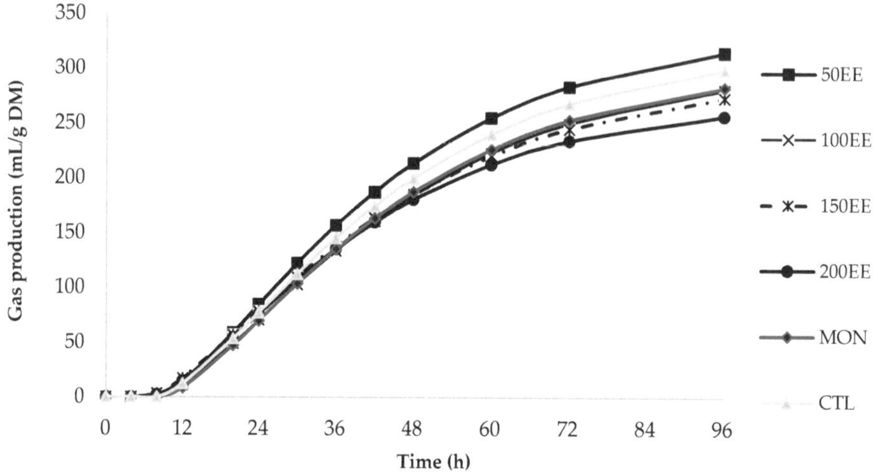

Figure 1. In vitro cumulative gas production curve of ethanolic extract from *Urochloa brizantha* at 96 h according to the France model.

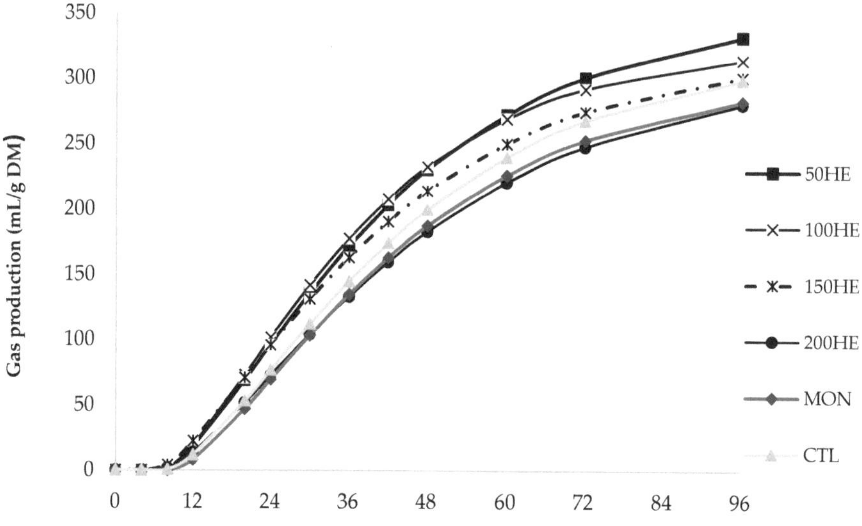

Figure 2. In vitro cumulative gas production curve of hydroalcoholic extract from *Urochloa brizantha* at 96 h according to the France model.

3.2. Short-Chain Fatty Acids and Ammonia Nitrogen

The inclusion of extract ($p > 0.05$) in the diet did not affect the proportion of short-chain fatty acids (SCFAs) (Table 2).

Table 2. Effect of ethanolic (EE) and hydroalcoholic (HE) extracts on the production of short-chain fatty acids (SCFAs) and ammoniacal nitrogen (NH_3-N) at 96 h of incubation in vitro.

Variables	EE				HE				MON	CTL	SEM	Contrast *p*-Valor					
	50	100	150	200	50	100	150	200				1	2	3	4	5	6
SCFA (mM)	39.63	60.75	55.05	60.26	71.00	72.28	49.46	56.34	53.12	47.49	7.56	0.63	0.19	0.59	0.63	0.14	0.06
C2 (molar %)	44.28	42.57	43.17	43.82	46.64	48.31	46.68	48.03	47.57	44.57	1.50	0.16	0.64	0.16	0.57	*	0.90
C3 (molar %)	28.11	25.44	27.09	26.78	26.44	25.05	25.45	25.36	33.48	30.69	2.25	0.09	**	**	0.23	0.17	0.97
C4 (molar %)	21.96	24.93	24.49	24.17	19.81	20.74	22.27	20.95	13.64	18.41	1.62	0.02	0.01	**	0.13	0.02	0.62
IC4 (molar %)	0.82	0.70	0.65	0.73	0.86	0.71	0.78	0.72	0.63	0.82	0.08	0.06	0.21	0.16	0.12	0.26	0.56
IC5 (molar %)	1.55	1.73	1.55	1.62	1.83	1.75	1.61	1.50	1.48	1.61	0.19	0.42	0.77	0.20	0.64	0.38	0.17
C5 (molar %)	3.36	3.54	3.15	2.88	4.42	3.67	3.33	3.44	3.25	3.63	0.83	0.56	0.72	0.66	0.73	0.38	0.22
C2:C3	1.60	1.96	1.61	1.69	1.81	2.02	1.89	1.98	1.49	1.47	0.23	0.94	0.03	0.07	0.32	0.07	0.80
NH_3-N (mg/100 mL)	39.77	40.92	41.48	43.19	41.94	43.93	43.69	45.31	44.77	41.51	1.26	0.04	0.07	0.33	*	0.10	0.96

EE: ethanolic extract; HE hydroalcoholic extract; MON: 25 ppm monensin; CTL: control (no additive added). SEM: standard error of the mean. * $p < 0.01$; ** $p < 0.001$. C2: acetic acid; C3: propionic acid; C4: butyric acid; IC4: isobutyric acid; IC5: isovaleric acid; C5: valeric acid. **Contrast:** 1: MON vs. CTL; 2: EXT vs. MON; 3: EXT vs. CTL; 4: EE vs. HE; 5: Level effect; 6: Effect of EXT vs. LVL interaction.

The inclusion of extracts ($p > 0.05$) in the diet in vitro did not affect the proportion of acetic acid (C2) compared to the control treatment (CTL) and monensin (MON) (Table 2). The addition of the EXT levels had a significant effect ($p < 0.01$) on concentrations, with higher levels at 50 mL/kg for EE (44.28) and at 100 mL/kg for EH (48.31).

The MON treatment (33.48%) provided a higher proportion of propionic acid (C3) than the CTL treatment (30.69%, $p < 0.001$) and EXT (23.34%, $p < 0.001$). The reduction of C3 with the addition of extracts resulted in an 18.13% increase in the C2:C3 ratio compared to the MON treatment.

Concerning butyric acid (C4), the MON (13.64%) treatment significantly reduced the proportion of C4 compared to the CTL treatments ($p = 0.02$) and EXT treatments ($p = 0.01$). Additionally, CTL (18.41%) was significantly lower ($p < 0.01$) than EXT (22.42%). Furthermore, the addition of EXT had a significant effect ($p = 0.02$), with concentrations reaching 24.93% at 100 mL/kg for EE and 22.27% at 150 mL/kg for EH.

The treatments had no significant effect on the proportions of isobutyric (IC4), valeric (C5), and isovaleric (IC5) acids.

The ammoniacal nitrogen (NH_3-N) had a significant effect ($p = 0.04$) when comparing the MON treatment (44.77 mg/mL) with CTL (41.51 mg/mL). The concentration of NH_3-N was different ($p < 0.01$) between EE (41.34 mg/mL) and HE (43.72 mg/mL).

4. Discussion

4.1. In Vitro Gas Production Assay

While some studies have evaluated the effects of plant extracts containing saponins on in vitro rumen fermentation kinetics, none have specifically assessed the effect of *Urochloa brizantha* extracts on degradability, fermentability, and microbial efficiency.

According to Seo et al. [40], the degradation of fiber by rumen microorganisms is a time-dependent process, with the colonization time corresponding to the time between adhesion and colonization at the start of incubation and the degradation of the substrate [41].

In this study, the inclusion of EXT reduced colonization time by 33.59% compared to the MON. This finding is consistent with the results of Freitas et al. [42], who observed that adding 75, 150, and 250 g/L m of ethanolic extracts of *Urochloa humidicola* decreased colonization time.

Similarly, Jiménez-Peralta et al. [43], observed a reduction in colonization time with Leucaena (*Leucaena leucocephala*) and Weeping Willow (*Salix babylonica*) extracts when using 1.8 mL/g DM. Additionally, they noted an improvement in the cumulative gas profile after 72 h. In contrast, Abarghuei, Rouzbeha, and Salem [44] found no significant effect of various pomegranate peel extracts (PEH: water; PEM: methanol, ethanol, and water) on cumulative gas production after 120 h. However, they did observe a decrease in the fractional rate of gas production with increasing doses of extracts.

These findings suggest that the extract may have reduced the colonization time, thereby improving the fermentation efficiency of the substrate. According to Zhou et al. [45], the addition of *Neolamarckia cadamba* leaves, which contain saponins, enhanced the digestibility of the diet, suggesting a similar potential for improving fermentation efficiency. This enhancement may be attributed to the extraction process, which removed the soluble fraction from the intracellular content, making it readily available to the microorganisms.

In our study, we observed a decrease in IVDMD and IVOMD with extract addition. This finding corroborates with the results of Hu et al. [46], who observed a reduction in IVOMD by adding 0.8 mg/mL of *Tea saponin*. In contrast, Abarghuei, Rouzbeha, and Salem [44] observed that the addition of PEH and PEM does not affect the IVOMD. However, Jiménez-Peralta et al. [43] observed that the two highest doses of extracts of Leucaena and Weeping Willow (1.2 and 1.8 mL/g DM) increased IVOMD values. Therefore, these variations can be attributed to the chemical nature, activities, and concentrations of the active compounds present in the extracts [47].

Furthermore, the type of solvent influences the composition of the secondary compounds. In this study, it was possible observe a greater reduction in degradability with the addition of EE than HE, which may have been due to the type of secondary compound extracted. Only the saponin protodioscin was quantified in our study, but different secondary compounds may be present, as the combination of ethanol and water enhances extraction efficiency [48].

Patra et al. [2] tested plant extracts using different solvents (methanol, ethanol, and water) and found that all extracts reduced feed degradability by 6 to 7%. The authors attributed this reduction to the impact of certain secondary compounds on the rumen microbiota. There may be some concern that the alcohol in alcoholic extracts could negatively affect the degradability of the substrate and mask the possible effects of the secondary compounds. However, Raun and Kristensen [49] suggest that microorganisms in the rumen can metabolize ethanol.

According to Leal et al. [19], protodioscin present in *U. brizantha* negatively correlates with the degradability of grasses by microorganisms, thereby adversely affecting IVDMD and IVOMD and prolonging bacterial colonization time. However, in our study, we did not observe an extended colonization time despite the decreased degradability following the addition of the extracts. This discrepancy may be because we used *U. brizantha* extract rather than the whole plant, as the whole plant retains its fibrous carbohydrate structure and has a more diluted concentration of secondary compounds compared to the extract.

4.2. Short-Chain Fatty Acids and Ammonia Nitrogen

SCFA production in the rumen is highly dependent on the degree and rate of fermentation [50], and saponins' effects on the production of SCFA are directly associated with the diet provided [5]. This effect tends to be more pronounced in forage-based diets, as the mechanism of action of saponins is like that of ionophores, which have a greater impact on gram-positive bacteria than on gram-negative bacteria [51].

This effect of saponin on SCFA production is further supported by a meta-analysis which showed reductions in acetate and increases in propionate proportions [52]. Corroborating the findings of Kang et al. [53], they tested *Momordica charantia* saponins using the in vitro gas production technique and observed, after 48 h of incubation, an increase in the molar proportion of propionate and a reduction in the proportion of acetate, resulting in a lower C2:C3 ratio. According to Wina et al. [54], saponin inclusion reduces butyric acid (C4) because they are the main products of protozoan fermentation, while propionic acid (C3) competes with methanogenic archaea for the use of H_2. However, despite expectations, in the present study, the addition of extracts resulted in an increase in C4 and a reduction in C3. Similarly, Patra and Yu [55] observed that butyrate increased linearly with increasing doses of vanillin. Therefore, some extracts may inhibit microorganisms that compete with butyrate-producing bacteria, thereby reducing the production of other short-chain fatty acids, such as propionic acid, and redirecting more substrates towards butyrate production.

The observed changes in SCFA production can also be influenced by the dietary characteristics of feeds, which influence rumen pH, rumen microorganisms, and, consequently, the concentration of ammonia nitrogen and the proportions of SCFA [56]. According to Hino and Russell [57], the deamination and decarboxylation of amino acids in the rumen produce branched-chain fatty acids. Branched-chain amino acids are crucial elements in protein synthesis [58]. However, when they are deaminated into branched-chain fatty acids, they are used by cellulolytic microorganisms to increase the activity of fiber-degrading enzymes, thereby improving the degradation of dry matter, neutral detergent fiber, and acid detergent fiber in bulk in vitro experiments [59]. This demonstrates that the MON treatment did not promote the deamination of amino acids, which allows other bacteria to utilize them, leading to an increased protein concentration in the rumen fluid [60].

In addition to these effects on fermentation, protodioscin has been studied for its various biological activities, including anti-inflammatory, neuroprotective, and antitumor effects [61–63]. Furthermore, the diverse methods for extracting and determining saponins, as well as a wide range of saponin structures, can have different effects on rumen microorganisms, complicating the comparison of studies [64]. However, to date, no work has produced the extract in pressurized liquid to test its effect on rumen fermentation, so we cannot infer whether it harms or improves ruminant production.

Therefore, developing a new additive from *U. brizantha* is a complex challenge involving the selection of the plant, the compound of interest, the solvent, and the extraction technique used. However, gaps remain that warrant future research, mainly optimizing the extraction process of protodioscin from *U. brizantha*. New extraction approaches should be explored, especially concerning the timing of plant harvest and the extraction conditions used. Additionally, it is crucial to analyze the composition, quantify the extract's secondary metabolites, and concentrate on the desired active ingredient to understand the effectiveness of *U. brizantha* extract as an additive.

5. Conclusions

The addition of 100, 150, and 200 mL/kg DM of *Urochloa brizantha* extracts in vitro decreased IVDMD, IVOMD, and propionate production while increasing butyric acid production. Both ethanolic and hydroalcoholic extracts were able to modulate rumen fermentation. Further research is required to optimize the extraction methodologies, comprehensively profile the secondary compounds, and conduct additional trials with different doses to effectively assess the viability of *U. brizantha* extracts as an additive.

Author Contributions: Conceptualization, I.C.d.S.B., J.S.d.S. and R.S.X.d.F.; methodology, R.S.X.d.F.; software, J.S.d.S. and R.S.X.d.F.; validation, I.C.d.S.B. and R.S.X.d.F.; formal analysis, J.S.d.S. and R.S.X.d.F.; investigation, R.S.X.d.F. and J.S.d.S.; resources, A.L.d.O., F.P.J. and A.J.F.; data curation, R.S.X.d.F. and J.S.d.S.; writing—original draft preparation, R.S.X.d.F. and J.S.d.S.; writing—review and editing, J.S.d.S., R.S.X.d.F., F.P.J. and A.J.F.; visualization, J.S.d.S. and R.S.X.d.F.; supervision, J.S.d.S. and A.L.d.O.; project administration, J.S.d.S. and I.C.d.S.B.; funding acquisition, I.C.d.S.B. and J.S.d.S. All authors have read and agreed to the published version of the manuscript.

Funding: The authors are grateful to the Fundação de Amparo à Pesquisa do Estado de São Paulo–FAPESP (Process nº 2021/14739-4) for their financial support of this research and the Conselho Nacional de Desenvolvimento Científico e Tecnológic–CNPq (Grant nº 152217/2022-3).

Institutional Review Board Statement: The animal study protocol was approved by Ethics Committee of the Department of Animal Science, College of Animal Science and Food Engineering, University of São Paulo (Protocol number CEUA: 1344101121 approved in January of 2022).

Data Availability Statement: The data presented in this study are available upon request from the corresponding author.

Acknowledgments: We would like to thank Bernardo Dias Ribeiro, from the Chemistry Department at UFRJ, for his availability and assistance in his laboratory.

Conflicts of Interest: The authors declare no conflicts of interest.

References

1. Newbold, C.J.; Ramos-Morales, E. Review: Ruminal Microbiome and Microbial Metabolome: Effects of Diet and Ruminant Host. *Animal* **2020**, *14*, S78–S86. [CrossRef]
2. Patra, A.K.; Kamra, D.N.; Agarwal, N. Effect of Plant Extracts on in Vitro Methanogenesis, Enzyme Activities and Fermentation of Feed in Rumen Liquor of Buffalo. *Anim. Feed Sci. Technol.* **2006**, *128*, 276–291. [CrossRef]
3. e Silva, S.N.S.; Chabrillat, T.; Kerros, S.; Guillaume, S.; Gandra, J.R.; de Carvalho, G.G.P.; Silva, F.F.d.; Mesquita, L.G.; Gordiano, L.A.; Camargo, G.M.F.; et al. Effects of Plant Extract Supplementations or Monensin on Nutrient Intake, Digestibility, Ruminal Fermentation and Metabolism in Dairy Cows. *Anim. Feed Sci. Technol.* **2021**, *275*, 114886. [CrossRef]
4. da Silva, C.S.; de Souza, E.J.O.; Pereira, G.F.C.; Cavalcante, E.O.; de Lima, E.I.M.; Torres, T.R.; da Silva, J.R.C.; da Silva, D.C. Plant Extracts as Phytogenic Additives Considering Intake, Digestibility, and Feeding Behavior of Sheep. *Trop. Anim. Health Prod.* **2017**, *49*, 353–359. [CrossRef] [PubMed]
5. Jayanegara, A.; Yogianto, Y.; Wina, E.; Sudarman, A.; Kondo, M.; Obitsu, T.; Kreuzer, M. Combination Effects of Plant Extracts Rich in Tannins and Saponins as Feed Additives for Mitigating in Vitro Ruminal Methane and Ammonia Formation. *Animals* **2020**, *10*, 1531. [CrossRef]
6. Dhanasekaran, D.K.; Dias-Silva, T.P.; Abdalla Filho, A.L.; Sakita, G.Z.; Abdalla, A.L.; Louvandini, H.; Elghandour, M.M.M.Y. Plants Extract and Bioactive Compounds on Rumen Methanogenesis. *Agrofor. Syst.* **2020**, *94*, 1541–1553. [CrossRef]

7. Hassan, A.; Abu Hafsa, S.H.; Elghandour, M.M.Y.; Kanth Reddy, P.R.; Salem, M.Z.M.; Anele, U.Y.; Ranga Reddy, P.P.; Salem, A.Z.M. Influence of Corymbia Citriodora Leaf Extract on Growth Performance, Ruminal Fermentation, Nutrient Digestibility, Plasma Antioxidant Activity and Faecal Bacteria in Young Calves. *Anim. Feed Sci. Technol.* **2020**, *261*, 114394. [CrossRef]
8. Ashour, A.S.; El Aziz, M.M.A.; Gomha Melad, A.S. A Review on Saponins from Medicinal Plants: Chemistry, Isolation, and Determination. *J. Nanomed. Res.* **2019**, *7*, 282–288. [CrossRef]
9. Chiocchio, I.; Mandrone, M.; Tomasi, P.; Marincich, L.; Poli, F. Plant Secondary Metabolites: An Opportunity for Circular Economy. *Molecules* **2021**, *26*, 495. [CrossRef]
10. Śliwiński, B.J.; Kreuzer, M.; Wettstein, H.R.; Machmüller, A. Rumen Fermentation and Nitrogen Balance of Lambs Fed Diets Containing Plant Extracts Rich in Tannins and Saponins, and Associated Emissions of Nitrogen and Methane. *Arch. Anim. Nutr./Arch. Fur Tierernahr.* **2002**, *56*, 379–392. [CrossRef]
11. Zhang, F.; Li, B.; Ban, Z.; Liang, H.; Li, L.; Zhao, W.; Yan, X. Evaluation of Origanum Oil, Hydrolysable Tannins and Tea Saponin in Mitigating Ruminant Methane: In Vitro and In Vivo Methods. *journal Anim. Physiol. Anim. Nutr.* **2021**, *105*, 630–638. [CrossRef] [PubMed]
12. Trotta, R.J.; Kreikemeier, K.K.; Foote, S.; McLeod, K.R.; Harmon, D.L. Influence of Anti-Coccidial Compounds and Phytogenic Saponin Extracts on In Vitro and In Vivo Ruminal Fermentation and Methane Production of Cattle. *Animals* **2023**, *13*, 2308. [CrossRef] [PubMed]
13. Niu, H.; Xu, Z.; Yang, H.E.; McAllister, T.A.; Acharya, S.; Wang, Y. In Vitro Ruminal Fermentation of Fenugreek (Trigonella Foenum-Graecum, L.) Produced Less Methane than That of Alfalfa (*Medicago sativa*). *Anim. Biosci.* **2021**, *34*, 584–593. [CrossRef] [PubMed]
14. Preez, D.A.D.; Akanmu, A.M.; Adejoro, F.A.; Hassen, A. The Effect of Monensin vs. Neem, and Moringa Extracts on Nutrient Digestibility, Growth Performance, Methane, and Blood Profile of Merino Lambs. *Animals* **2023**, *13*, 3514. [CrossRef]
15. Freitas, R.S.X.; Nepomuceno, D.d.D.; Modesto, E.C.; de Oliveira, D.R.; Pereira, T.P.; de Morais, L.F.; de Carvalho Almeida, J.C.; de Carvalho, M.G. Caracterização Do Extrato Metanólico de Urochloa Humidicola Como Aditivo Fitogênico Para Ruminantes. *Biosci. J.* **2017**, *33*, 1586–1591. [CrossRef]
16. Gupta, A.; Naraniwal, M.; Kothari, V. Modern Extraction Methods for Preparation of Bioactive Plant Extracts. *Int. J. Appl. Nat. Sci.* **2012**, *1*, 8–26.
17. Erb, M.; Kliebenstein, D.J. Plant Secondary Metabolites as Defenses, Regulators, and Primary Metabolites: The Blurred Functional Trichotomy. *Plant Physiol.* **2020**, *184*, 39–52. [CrossRef] [PubMed]
18. Niderkorn, V.; Jayanegara, A. Opportunities Offered by Plant Bioactive Compounds to Improve Silage Quality, Animal Health and Product Quality for Sustainable Ruminant Production: A Review. *Agronomy* **2021**, *11*, 86. [CrossRef]
19. Leal, E.S.; Ítavo, L.C.V.; Do Valle, C.B.; Ítavo, C.C.B.F.; Dias, A.M.; Dos Santos Difante, G.; Barbosa-Ferreira, M.; Nonato, L.M.; De Melo, G.K.A.; Gurgel, A.L.C. Influence of Protodioscin Content on Digestibility and in Vitro Degradation Kinetics in Urochloa Brizantha Cultivars. *Crop Pasture Sci.* **2020**, *71*, 278–284. [CrossRef]
20. Feitoza, R.B.B.; Varela, R.M.; Torres, A.; Molinillo, J.M.G.; Lima, H.R.P.; Moraes, L.F.D.; Da Cunha, M.; Macías, F.A. Evaluation of the Phytotoxicity of Urochloa Humidicola Roots by Bioassays and Microscopic Analysis. Characterization of New Compounds. *J. Agric. Food Chem.* **2020**, *68*, 4851–4864. [CrossRef]
21. de Oliveira, D.R.; Nepomuceno, D.D.; Castro, R.N.; Braz Filho, R.; Carvalho, M.G.d. Special Metabolites Isolated from Urochloa Humidicola (Poaceae). *An. Da Acad. Bras. De Ciências* **2017**, *89*, 789–797. [CrossRef] [PubMed]
22. Brum, K.B.; Haraguchi, M.; Lemos, R.A.A.; Riet-Correa, F.; Fioravanti, M.C.S. Crystal-Ystal-Associated Cholangiopathy in Sheep Grazing Brachiaria Decumbens Containing the Saponin Protodioscin Protodioscin. *Pesqui. Vet. Bras.* **2007**, *27*, 39–42. [CrossRef]
23. da Costa, M.C.M.; Ítavo, L.C.V.; Ítavo, C.C.B.F.; Dias, A.M.; dos Santos Difante, G.; Buschinelli de Goes, R.H. de T.; de Souza Leal, E.; Nonato, L.M.; Kozerski, N.D.; de Moraes, G.J.; et al. Natural Intoxication Caused by Protodioscin in Lambs Kept in Brachiaria Pastures. *Trop. Anim. Health Prod.* **2021**, *53*, 1–9. [CrossRef]
24. Oliveira, M.W.; Lopes Goretti, A.; Lana, R.d.P.; Camacho Rodrigues, T. DRY MATTER AND PROTEIN ACCUMULATION AS A FUNCTION OF NITROGEN FERTILIZATION IN Brachiaria Brizantha CV. MARANDU (Urochloa Brizantha). *Rev. Bras. De Agropecuária Sustentável* **2022**, *12*, 10–18. [CrossRef]
25. Bueno, I.C.S.; Cabral Filho, S.L.S.; Gobbo, S.P.; Louvandini, H.; Vitti, D.M.S.S.; Abdalla, A.L. Influence of Inoculum Source in a Gas Production Method. *Anim. Feed Sci. Technol.* **2005**, *123–124 Pt 1*, 95–105. [CrossRef]
26. AOAC. *Association of Official Analytical Chemists (AOAC)*, 18th ed.; AOAC International, Ed.; AOAC: Gaithersburg, MD, USA, 2011.
27. dos Santos, H.G.; Jacomine, P.K.T.; dos Anjos, L.H.C.; de Oliveira, V.Á.; Lumbreras, J.F.; Coelho, M.R.; de Almeida, J.A.; de Araújo Filho, J.C.; de Oliveira, J.B.; Cunha, T.J.F. *Sistema Brasileiro de Classificação de Solos*, 5th ed.; Embrapa, Ed.; Embrapa: Brasília, Brasil, 2018; ISBN 978-85-7035-800-4.
28. Omer, E.A.; Hussein, E.A.A.; Hendawy, S.F.; El-din, A.A.E.; El-Gendy, A.G. Effect of Nitrogen and Potassium Fertilizers on Growth, Yield, Essential Oil and Artemisinin of Artemisia Annua L Plant. *Int. Res. J. Hortic.* **2014**, *2*, 11. [CrossRef]
29. Rodrigues, L.d.C.; Bodini, R.B.; Caneppele, F. de L.; Dacanal, G.C.; Crevelin, E.J.; de Moraes, L.A.B.; de Oliveira, A.L. Pressurized Liquid Extraction (PLE) in an Intermittent Process as an Alternative for Obtaining Passion Fruit (*Passiflora edulis*) Leaf Hydroalcoholic Extract (Tincture). *Processes* **2023**, *11*, 2308. [CrossRef]
30. Lee, S.T.; Mitchell, R.B.; Wang, Z.; Heiss, C.; Gardner, D.R.; Azadi, P. Isolation, Characterization, and Quantification of Steroidal Saponins in Switchgrass (*Panicum virgatum*, L.). *J. Agric. Food Chem.* **2009**, *57*, 2599–2604. [CrossRef]

31. Theodorou, M.K.; Williams, B.A.; Dhanoa, M.S.; McAllan, A.B.; France, J. A Simple Gas Production Method Using a Pressure Transducer to Determine the Fermentation Kinetics of Ruminant Feeds. *Anim. Feed Sci. Technol.* **1994**, *48*, 185–197. [CrossRef]
32. Mauricio, R.M.; Mould, F.L.; Dhanoa, M.S.; Owen, E.; Channa, K.S.; Theodorou, M.K. A Semi-Automated In Vitro Gas Production Technique for Ruminant Feedstuff Evaluation. *Anim. Feed Sci. Technol.* **1999**, *79*, 321–330. [CrossRef]
33. Menke, K.H.; Raab, L.; Salewski, A.; Steingass, H.; Fritz, D.; Schneider, W. The Estimation of the Digestibility and Metabolizable Energy Content of Ruminant Feedingstuffs from the Gas Production When They Are Incubated with Rumen Liquor In Vitro. *J. Agric. Sci.* **1979**, *93*, 217–222. [CrossRef]
34. Fagundes, G.M.; Benetel, G.; Santos, K.C.; Welter, K.C.; Melo, F.A.; Muir, J.P.; Bueno, I.C.S. Tannin-Rich Plants as Natural Manipulators of Rumen Fermentation in the Livestock Industry. *Molecules* **2020**, *25*, 2943. [CrossRef]
35. France, J.; Dhanoa, M.S.; Theodorou, M.K.; Lister, S.J.; Davies, D.R.; Isac, D. A Model to Interpret Gas Accumulation Profile Associated with In Vitro Degradation of Ruminant Feeds. *J. Theor. Biol.* **1993**, *163*, 99–111. [CrossRef]
36. Blummel, B.Y.M.; Becker, K. The Degradability Characteristics of Fifty-Four Roughages and Roughage Neutral-Detergent Fibres as Described by In Vitro Gas Production and Thies Relationship to Voluntary Feed Intake. *Br. J. Nutr.* **1997**, *77*, 757–768. [CrossRef]
37. Bueno, I.C.S.; Brandi, R.A.; Fagundes, G.M.; Benetel, G.; Muir, J.P. The Role of Condensed Tannins in the In Vitro Rumen Fermentation Kinetics in Ruminant Species: Feeding Type Involved? *Animals* **2020**, *10*, 635. [CrossRef] [PubMed]
38. Kulasek, G. A Micromethod for Determining Urea in Blood Plasma, Whole Blood and Blood Corpuscles with the Use of Urease and Phenol Reagent. *Pol. Arch. Weter.* **1972**, *15*, 801–810.
39. Foldager, J. Protein Requirement and Non-Protein Nitrogen for High Producing Cows in Early Lactation. Ph.D. Thesis, Michigan State University, East Lansing, MI, USA, 1977.
40. Seo, S.; Lee, S.C.; Lee, S.Y.; Seo, J.G.; Ha, J.K. Degradation Kinetics of Carbohydrate Fractions of Ruminant Feeds Using Automated Gas Production Technique. *Asian-Australas. J. Anim. Sci.* **2009**, *22*, 356–364. [CrossRef]
41. de Oliveira, J.S.; Zanine, A.D.M.; Santos, E.M. Diversidade Microbiana No Ecossitema Ruminal. *Rev. Electrónica De Vet.* **2007**, *8*, 1–12.
42. de Freitas, R.S.X.; Nepomuceno, D.D.D.; Modesto, E.C.; Pereira, T.P.; Almeida, J.C.d.C.; de Morais, L.F.; Pereira, L.G.R.; Machado, F.S.; Campos, M.M.; Tomich, T.R. Methanolic Extract of Urochloa Humidicola on In Vitro Rumen Fermentation. *Pesqui. Agropecuária Bras.* **2018**, *53*, 504–513. [CrossRef]
43. Jiménez-Peralta, F.S.; Salem, A.Z.M.; Mejia-Hernández, P.; González-Ronquillo, M.; Albarrán-Portillo, B.; Rojo-Rubio, R.; Tinoco-Jaramillo, J.L. Influence of Individual and Mixed Extracts of Two Tree Species on In Vitro Gas Production Kinetics of a High Concentrate Diet Fed to Growing Lambs. *Livest. Sci.* **2011**, *136*, 192–200. [CrossRef]
44. Abarghuei, M.J.; Rouzbehan, Y.; Salem, A.F. The Influence of Pomegranate-Peel Extracts on In Vitro Gas Production Kinetics of Rumen Inoculum of Sheep. *Turk. J. Vet. Anim. Sci.* **2014**, *38*, 212–219. [CrossRef]
45. Zhou, W.; Pian, R.; Yang, F.; Chen, X.; Zhang, Q. The Sustainable Mitigation of Ruminal Methane and Carbon Dioxide Emissions by Co-Ensiling Corn Stalk with Neolamarckia Cadamba Leaves for Cleaner Livestock Production. *J. Clean. Prod.* **2021**, *311*, 127680. [CrossRef]
46. Hu, W.; Liu, J.; Wu, Y.; Guo, Y.; Ye, J. Effects of Tea Saponins on In Vitro Ruminal Fermentation and Growth Performance in Growing Boer Goat. *Arch. Anim. Nutr.* **2006**, *60*, 89–97. [CrossRef]
47. Hart, K.J.; Yáñez-Ruiz, D.R.; Duval, S.M.; McEwan, N.R.; Newbold, C.J. Plant Extracts to Manipulate Rumen Fermentation. *Anim. Feed Sci. Technol.* **2008**, *147*, 8–35. [CrossRef]
48. Yusoff, I.M.; Mat Taher, Z.; Rahmat, Z.; Chua, L.S. A Review of Ultrasound-Assisted Extraction for Plant Bioactive Compounds: Phenolics, Flavonoids, Thymols, Saponins and Proteins. *Food Res. Int.* **2022**, *157*, 111268. [CrossRef] [PubMed]
49. Raun, B.M.L.; Kristensen, N.B. Metabolic Effects of Feeding Ethanol or Propanol to Postpartum Transition Holstein Cows. *J. Dairy Sci.* **2011**, *94*, 2566–2580. [CrossRef]
50. Cone, J.W.; Becker, P.M. Fermentation Kinetics and Production of Volatile Fatty Acids and Microbial Protein by Starchy Feedstuffs. *Anim. Feed Sci. Technol.* **2012**, *172*, 34–41. [CrossRef]
51. Patra, A.K.; Saxena, J. The Effect and Mode of Action of Saponins on the Microbial Populations and Fermentation in the Rumen and Ruminant Production. *Nutr. Res. Rev.* **2009**, *22*, 204–219. [CrossRef]
52. Jayanegara, A.; Wina, E.; Takahashi, J. Meta-Analysis on Methane Mitigating Properties of Saponin-Rich Sources in the Rumen: Influence of Addition Levels and Plant Sources. *Asian-Australas. J. Anim. Sci.* **2014**, *27*, 1426–1435. [CrossRef]
53. Kang, J.; Zeng, B.; Tang, S.; Wang, M.; Han, X.; Zhou, C.; Yan, Q.; Liu, J.; Tan, Z. Effects of Momordica Charantia Polysaccharide on In Vitro Ruminal Fermentation and Cellulolytic Bacteria. *Ital. J. Anim. Sci.* **2017**, *16*, 226–233. [CrossRef]
54. Wina, E.; Muetzel, S.; Becker, K. The Impact of Saponins or Saponin-Containing Plant Materials on Ruminant Production—A Review. *J. Agric. Food Chem.* **2005**, *53*, 8093–8105. [CrossRef] [PubMed]
55. Patra, A.K.; Yu, Z. Effects of Vanillin, Quillaja Saponin, and Essential Oils on In Vitro Fermentation and Protein-Degrading Microorganisms of the Rumen. *Appl. Microbiol. Biotechnol.* **2014**, *98*, 897–905. [CrossRef]
56. Faniyi, T.O.; Adegbeye, M.J.; Elghandour, M.M.M.Y.; Pilego, A.B.; Salem, A.Z.M.; Olaniyi, T.A.; Adediran, O.; Adewumi, M.K. Role of Diverse Fermentative Factors towards Microbial Community Shift in Ruminants. *J. Appl. Microbiol.* **2019**, *127*, 2–11. [CrossRef]
57. Hino, T.; Russell, J.B. Effect of Reducing-Equivalent Disposal and NADH/NAD on Deamination of Amino Acids by Intact Rumen Microorganisms and Their Cell Extracts. *Appl. Environ. Microbiol.* **1985**, *50*, 1368–1374. [CrossRef]

58. An, J.; He, H.; Lan, X.; Liu, L.; Wang, Z.; Ge, Y.; Shen, W.; Cheng, A.; Wan, F. Branched-Chain Amino Acids in Ruminant Nutrition: Function Effects and Summary of Recent Advances. *Anim. Feed Sci. Technol.* **2024**, *312*, 1–15. [CrossRef]
59. Roman-Garcia, Y.; Denton, B.L.; Mitchell, K.E.; Lee, C.; Socha, M.T.; Firkins, J.L. Conditions Stimulating Neutral Detergent Fiber Degradation by Dosing Branched-Chain Volatile Fatty Acids. I: Comparison with Branched-Chain Amino Acids and Forage Source in Ruminal Batch Cultures. *J. Dairy Sci.* **2021**, *104*, 6739–6755. [CrossRef] [PubMed]
60. Yang, C.M.; Russell, J.B. The Effect of Monensin Supplementation on Ruminal Ammonia Accumulation In Vivo and the Numbers of Amino Acid-Fermenting Bacteria. *J. Anim. Sci.* **1993**, *71*, 3470–3476. [CrossRef]
61. Liu, J.Y.; Hou, Y.L.; Cao, R.; Qiu, H.X.; Cheng, G.H.; Tu, R.; Wang, L.; Zhang, J.L.; Liu, D. Protodioscin Ameliorates Oxidative Stress, Inflammation and Histology Outcome in Complete Freund's Adjuvant Induced Arthritis Rats. *Apoptosis* **2017**, *22*, 1454–1460. [CrossRef] [PubMed]
62. Yang, G.; Liu, P.; Shi, H.; Fan, W.; Feng, X.; Chen, J.; Jing, S.; Wang, L.; Zheng, Y.; Zhang, D.; et al. Identification of Anti-Inflammatory Components in Dioscorea Nipponica Makino Based on HPLC-MS/MS, Quantitative Analysis of Multiple Components by Single Marker and Chemometric Methods. *J. Chromatogr. B Anal. Technol. Biomed. Life Sci.* **2022**, *1213*, 123531. [CrossRef]
63. Huang, C.F.; Hsieh, Y.H.; Yang, S.F.; Kuo, C.H.; Wang, P.H.; Liu, C.J.; Lin, R.C. Mitophagy Effects of Protodioscin on Human Osteosarcoma Cells by Inhibition of P38MAPK Targeting NIX/LC3 Axis. *Cells* **2023**, *12*, 395. [CrossRef]
64. Belanche, A.; Pinloche, E.; Preskett, D.; Newbold, C.J. Effects and Mode of Action of Chitosan and Ivy Fruit Saponins on the Microbiome, Fermentation and Methanogenesis in the Rumen Simulation Technique. *FEMS Microbiol. Ecol.* **2016**, *92*, fiv160. [CrossRef]

MDPI AG
Grosspeteranlage 5
4052 Basel
Switzerland
Tel.: +41 61 683 77 34

Fermentation Editorial Office
E-mail: fermentation@mdpi.com
www.mdpi.com/journal/fermentation

www.ingramcontent.com/pod-product-compliance
Lightning Source LLC
Chambersburg PA
CBHW041346210926
51850CB00013B/3003